FISHERIES MANAGEMENT

Progress Towards Sustainability

FISHERIES MANAGEMENT

Progress Towards Sustainability

Edited by

Tim R. McClanahan and Juan Carlos Castilla

Blackwell
Publishing

Blackwell Publishing editorial offices:
Blackwell Publishing Ltd, 9600 Garsington Road, Oxford OX4 2DQ, UK
Tel: +44 (0) 1865 776868
Blackwell Publishing Professional, 2121 State Avenue, Ames, Iowa 50014-8300, USA
Tel: +1 515 292 0140
Blackwell Publishing Asia Pty Ltd, 550 Swanston Street, Carlton, Victoria 3053, Australia
Tel: +61 (0)3 8359 1011

First published 2007 by Blackwell Publishing Ltd

ISBN-13: 978-1-4051-3932-8

Library of Congress Cataloging-in-Publication Data

Fisheries management : progress towards sustainability / edited by T.R. McClanahan,
J.C. Castilla.
p. cm.
Includes bibliographical references and index.
ISBN-13: 978-1-4051-3932-8 (hardback : alk. paper)
ISBN-10: 1-4051-3932-3 (hardback : alk. paper) 1. Fishery management. I. McClanahan, T. R.
II. Castilla, J. C.
SH328.F57 2007
338.3'727 – dc22
2006012264

A catalogue record for this title is available from the British Library

Set in 11/13pt Times
by TechBooks, New Delhi, India
Printed and bound in Singapore
by COS Printers Pte Ltd

The publisher's policy is to use permanent paper from mills that operate a sustainable forestry policy, and which has been manufactured from pulp processed using acid-free and elementary chlorine-free practices. Furthermore, the publisher ensures that the text paper and cover board used have met acceptable environmental accreditation standards.

For further information on Blackwell Publishing, visit our website: www.blackwellpublishing.com

Contents

Part II Small-scale and Artisanal Fisheries

List of Contributors

Angel C. Alcala, Director, Silliman University – Angelo King Center for Research and Environmental Management (SUAKCREM), Silliman University Marine Laboratory, Bantayan, Dumaguete City, Philippines.
E-mail: suakcrem@yahoo.com

Jim Anderson, PO Box 78736, Dar es Salaam, Tanzania.
E-mail: anderson@africaonline.co.tz

Nancy Barahona, Instituto de Fomento Pesquero (IFOP), Blanco 839, Casilla 8-V, Valparaíso, Chile. E-mail: nbarahon@ifop.cl

George M. Branch, Marine Biology Research Institute, Zoology Department, University of Cape Town, Private Bag X3, Rondebosch 7701, South Africa.
E-mail: gmbranch@egs.uct.ac.za

David T. Brewer, CSIRO Marine Research, PO Box 120, Cleveland, Queensland, 4163, Australia. E-mail: david.brewer@csiro.au

Arthur Bull, Bay of Fundy Marine Resource Centre, PO Box 273, Cornwallis Park, Nova Scotia, B0S 1H0, Canada. E-mail: arthbull@tartannet.ns.ca

Nick Caputi, Fisheries Research Division, Department of Fisheries Western Australia, Western Australian Fisheries and Marine Research Laboratories, PO Box 20, North Beach, WA 6020, Australia. E-mail: ncaputi@fish.wa.gov.au

Juan Carlos Castilla, Centro de Estudios Avanzados en Ecología y Biodiversidad, Facultad de Ciencias Biológicas, Pontificia Universidad Católica de Chile, Casilla 114-D, Santiago, Chile. E-mail: jcastill@bio.puc.cl

Anthony Charles, Management Science/Environmental Studies, Saint Mary's University, Halifax, Nova Scotia, B3H 3C3, Canada.
E-mail: tony.charles@smu.ca

Ana Cinti, Grupo de Ecología y Manejo de Recursos, Departamento de Biología Marina, Universidad Católica del Norte, Casilla 117, Coquimbo, Chile.
E-mail: anacinti@unc.cl

Néstor F. Ciocco, Centro Nacional Patagónico, 9120 Puerto Madryn, Argentina.
E-mail: ciocco@cenpat.edu.ar

Barry M. Clark, Anchor Consultancy, Department of Zoology, University of Cape Town, Private Bag X3, Rondebosch 7701, South Africa.
E-mail: bclark@botzoo.uct.ac.za

Omar Defeo, Facultad de Ciencias, Montevideo, Uruguay.

Catherine M. Dichmont, CSIRO Marine Research, PO Box 120, Cleveland, Queensland, 4163, Australia. E-mail: cathy.dichmont@csiro.au

Stefan Gelcich, Centro de Estudios Avanzados en Ecología y Biodiversidad, Facultad de Ciencias Biológicas, Pontificia Universidad Católica de Chile, Casilla 114-D, Santiago, Chile. E-mail: sgelcich@bio.puc.cl

Edgardo Gomez, Marine Science Institute, University of the Philippines, Diliman, Philippines. E-mail: edgomezph@yahoo.com

Jean M. Harris, Ezemvelo KwaZulu-Natal Wildlife, PO Box 13053, Cascades, 3202, KwaZulu-Natal, South Africa. E-mail: jmharris@iafrica.com

John Kearney, John F. Kearney and Associates, 5064 Doctor's Brook, RR# 3, Antigonish, Nova Scotia, Canada, B2G 2L1.
E-mail: john.kearney@ns.sympatico.ca

Hassan Kalombo, Tanga Coastal Zone Conservation and Development Programme, PO Box 5036, Tanga, Tanzania. E-mail: kalombo@kaributanga.com

Neil R. Loneragan, Centre for Fish and Fisheries Research, Murdoch University, South St, Murdoch, Western Australia, 6150.
E-mail: n.loneragan@murdoch.edu.au

Mary Livingston, Ministry of Fisheries, PO Box 1020, Wellington, New Zealand. E-mail: m.livingston@niwa.co.nz

Helene Marsh, School of Tropical Environment Studies and Geography, James Cook University, Douglas, Queensland, Australia, 4810 and CRC Reef Research, PO Box 772, Townsville, Queensland, Australia, 4810.
E-mail: helene.marsh@jcu.edu.au

Roy Melville-Smith, Fisheries Research Division, Department of Fisheries Western Australia, Western Australian Fisheries and Marine Research Laboratories, PO Box 20, North Beach, WA 6020, Australia.
E-mail: rmsmith@fish.wa.gov.au

Tim R. McClanahan, Wildlife Conservation Society, Marine Programs, Bronx, NY. E-mail: tmcclanahan@wcs.org

Chris Milley, Nexus Coastal Resource Management, 21 Jerico Road, Hubley, Nova Scotia, B3Z 1B8, Canada. E-mail: cmilley@mcpei.ca

Carlos Molinet, Instituto de Ecología y Evolución, Universidad Austral de Chile, Casilla 567, Valdivia, Chile. E-mail: cmolinet@uach.cl

Solomon Makoloweka, Tanga Coastal Zone Conservation and Development Programme, PO Box 5036, Tanga, Tanzania.
E-mail: makoloweka@kaributanga.com

Carlos A. Moreno, Instituto de Ecología y Evolución, Universidad Austral de Chile, Casilla 567, Valdivia, Chile. E-mail: cmoreno@uach.cl

J.M. (Lobo) Orensanz, Centro Nacional Patagónico, 9120 Puerto Madryn, Argentina. E-mail: lobo@cenpat.edu.ar

Ana M. Parma, Centro Nacional Patagónico, 9120 Puerto Madryn, Argentina.
E-mail: lobo@cenpat.edu.ar

Bruce F. Phillips, Department of Environmental Biology, Curtin University of Technology, GPO Box U1987, Perth, WA 6845, Australia.
E-mail: b.phillips@curtin.edu.au

Ian R. Poiner, AIMS, PMB No 3, Townsville MC, 4810, Queensland, Australia.
E-mail: i.poiner@aims.gov.au

Garry Russ, School of Marine Biology and Aquaculture, James Cook University, Townsville, Australia, 4810. E-mail: garry.russ@jcu.edu.au

Melita Samoilys, IUCN Eastern African Regional Office, PO Box 68200, Nairobi 00200, Kenya. E-mail: MAS@iucnearo.org

Sandile C. Sibiya, Marine and Coastal Management, Department of Environment Affairs and Tourism, Foreshore, Cape Town, South Africa.
E-mail: ssibiya@deat.gov.za

Kevin Sullivan, Ministry of Fisheries, PO Box 1020, Wellington, New Zealand.
E-mail: m.livingston@niwa.co.nz

Sue Wells, 56 Oxford Road, Cambridge, CB4 3PW, United Kingdom.
E-mail: suewells100@tiscali.co.uk

Alan T. White, Fisheries Improved for Sustainable Harvests Project, Tetra Tech EM Inc. and Coastal Conservation and Education Foundation, Cebu City, Philippines. E-mail: awhite@mozcom.com

Alejandro Zuleta, CEPES, Universidad Austral de Chile, Casilla 567, Valdivia, Chile. E-mail: alzuleta@entelchile.net

Acknowledgements

Members of the Pew Fellows Program in Marine Conservation organized this book through their annual meetings and a discussion group on community-based fisheries management. Discussion group members asked knowledgeable co-authors to assist with the development of their chapters, and other authors and chapters were invited to participate based on measures of management progress reported in the larger literature. We are grateful for the support of the Pew Institute for Ocean Science (Director: Ellen Pikitch) through the Pew Charitable Trust for organizing these meetings. Other important donors and collaborators that supported the authors include the Andrew Mellon Foundation, Asociación de Industriales Ericeros, Aycon Consultores Ltd, Asociación de Pescadores Artesanales de Puerto Madryn, Australian Agency for International Development, Australian Research Council, Canadian Maritimes, CENPAT, Chilean MIDEPLAN doctoral scholarship program, Chinquihue Foundation, Coastal Resource Management Project in the Philippines, CoMPeB, The David and Lucille Packard Foundation, Fisheries and Natural Resources Officers of Muheza and Pangani Districts and Tanga City, Tanzania, FONDAP 1501-0001 (CASEB), Great Barrier Reef Marine Park Authority, The Green Trust, Irish Aid, International Union for Conservation of Nature, Living Marine Resource Fund of the Marine and Coastal Management, Department of Environment Affairs and Tourism, South Africa, National Research Foundation of South Africa, Natural Sciences and Engineering Research Council of Canada, Philippine Department of Environment and Natural Resources, Social Science and Humanities Research Council of Canada, Silliman University, SubPesca, United Board for Christian Education in Asia, Wildlife Conservation Society, World Fish Center, World Wide Fund for Nature, and the United States Agency for International Development. We also thank a number of individuals that helped in many ways, these include José Alonso, José Ascorti, Mebrahtu Ateweberhan, Rodrigo Azócar, Carlos (El Loco) Beloso, Juan Berón, Pedro Brunetti, Dave Carter, Luis De Francesco, Hormiga Díaz, Inés Elías, José Fondacaro, Patricio Galvez, María Eva Góngora, Raúl Gili, Malcolm Haddon, Burke Hill, Chris Horrill, Luigin Isola, Annie Jarrett, Gabriel Jerez, Richard Kenchington, Mike O'Brien, Dr Santiago R. Olivier, Mr Makota, Joseph Maina, Gustavo San Martín, Vivienne Mawson, Malevo Medina, Hernán Miranda, Moses Mwambogo, Pedro Oroquieta, Vivian Pezo, Dr. Mariano Pizarro,

Santiago and Zeito Picallo, Marta Piñeiro, Corinna Ravilious, Andrea Rodríguez, Peter Rothlisberg, Bocha Rúa, Nercio Saldivia, Kitty Sarsa, Leonardo Sasso, El Lobo Suárez, Gonzalo Covarrubias Darío Rivas, Claudio Romero, Guillermo Roa, Pedro Rubilar, J. Valencia, Dave Vance, Eric Verheij, and Claudio Vicencio. We especially thank the many fishers who shared their experience and participated in interviews, questionnaires, and discussions. We hope that this book will help to produce the needed management to avert the crises.

Preface

It is now common knowledge that marine fisheries are in trouble and this trouble extends from near to offshore areas and affects most regions of the Earth. The problems have arisen in the context of a well-developed fisheries science that has largely focused on the resource and the biophysical aspects that control the resource, but has paid less attention to the social aspects of the management that lead to success. This focus has changed considerably in the last decade and a number of innovative means to manage fisheries that are socially acceptable have developed and led to success in human organization, increasing fisheries catches or, at minimum, reversing declines. In order to increase the chances for success in other fisheries, it would help practitioners if they understand the history of the successes and some of the factors that led to reversing failures. This book will provide historical case studies of a number of fisheries that have improved management. Consequently, it is expected that this book will form an important supplementary text for fisheries management and marine conservation courses that require a contemporary survey of recent advances.

The book begins with a brief chapter describing the status of the world's fisheries and the factors that have led to the global decline. This chapter is followed by a series of case studies that focus on the history of management progress. The book concludes with a synthesis of these and other published case studies organized around general principles of success. Each author has participated in the fisheries they describe and presents a personal and historical view of their case study, such that the chapters are narratives describing the biophysical, cultural, political, economic, and management context and how this environment created challenges for management and influenced decisions, reorganization, and progress. Aspects of the trial and error process and the conflicts and resolutions are emphasized. This book will offer a readable account of contemporary and progressive movements in fisheries management that will be useful for participants and observers of the current fisheries crises.

Tim R. McClanahan and Juan Carlos Castilla
Mombasa, Kenya and Santiago, Chile

Part I
Introduction

Chapter 1
A Brief History of Fisheries Management with Emphasis on Societal Participatory Roles

Omar Defeo, Tim R. McClanahan, and Juan Carlos Castilla

Fishery science and management

Fisheries management requires the integration of the biology and ecology of fish resources with the socio-economic, resource user, and management institutional factors that affect the behavior of fishers and policy-makers (Seijo et al. 1998). The purpose of this multidisciplinary field is to aid decision-making to achieve a sustainable development of the activity. Fishery sustainability has, however, been far more difficult to achieve than is commonly thought: fish populations are becoming increasingly limited, world catch has begun to drop and 75% of the individual fish stocks around the world are fully to heavily exploited, overexploited, or depleted (FAO 2004). In this chapter we present a brief history of fisheries science, focusing on the changes in thought and the problems encountered in the establishment of effective management procedures. We identify some of the pitfalls and shortcomings and also discuss some of the biological, legal, social, and economic factors that could help or hinder the success of fisheries management. The review forms the basis for understanding the chapters that follow and frames the individual case studies within the broader historical context of fisheries and its science and particularly within co-management.

The theory: assessing and managing fisheries systems

The fishery system can be broken down into three interacting subsystems: (1) resource; (2) resource users (fishers); (3) resource management (Fig. 1.1). These subsystems have idiosyncrasies that change from fishery to fishery and from place to place (even within a single fishery). The resource subsystem includes: (1) the life cycle of a species (including reproductive biology, recruitment, growth and mortality); (2) environmental factors affecting abundance and spatio-temporal distribution of the species; (3) ecological interdependencies. The resource users' subsystem

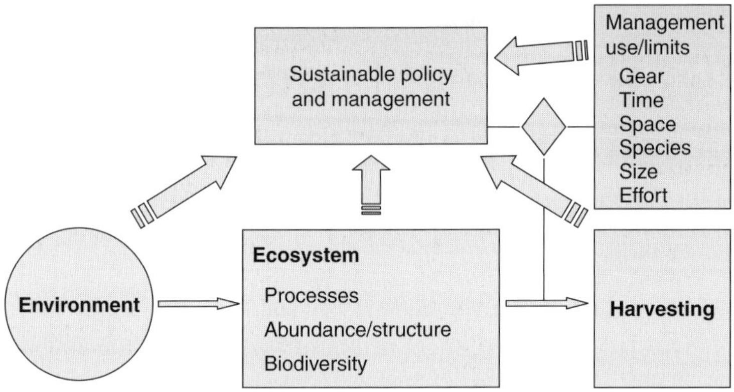

Figure 1.1 Model of the various subcomponents of the fishery and the management decision options. Dashed arrows represent information flowing from system to management while solid arrows represent physical flows of resources.

includes the fleets operating in the fishery, their spatial dynamics, the composition of the catch (including different species or population components and selectivity curves), and economic functions. Resource management captures the whole complex dynamics of the first two subsystems plus external forces such as markets, politics, lobbies, and societal interests. It also considers ways of intervention, institutional development, selection criteria of management strategies, enforcement mechanisms, and the way of contending with multiple criteria in the selection of management instruments. The intrinsic nature of the fishery system varies according to different life histories, harvesting practices, and management options. Thus the interaction of the three subsystems produces the unique behavior of the overall system (Castilla and Defeo 2001).

During most of the last century, until the 1980s, most stock-assessment techniques have been indistinctly applied to a wide array of species, ignoring their life history traits (Caddy 1975). Moreover, the same stock-assessment and management tools, strongly biased towards industrial fisheries, were also used, often inappropriately, in small-scale artisanal fisheries which have been erroneously lumped into the industrial component. In the following sections we provide a historical overview of traditional stock-assessment techniques and management options and how they evolved through time.

Single stocks

Single-species stock assessments have been performed since the early 1930s. Surplus production theory arose after the development of Graham's reformulation of the logistic equation and the former Baranov's first unified theory of fishing (Caddy 1999). The classical view of sustainability, prevalent during the first eight decades of the last century, was developed from deterministic and static production models in which equilibrium abundance or biomass is derived as a function of

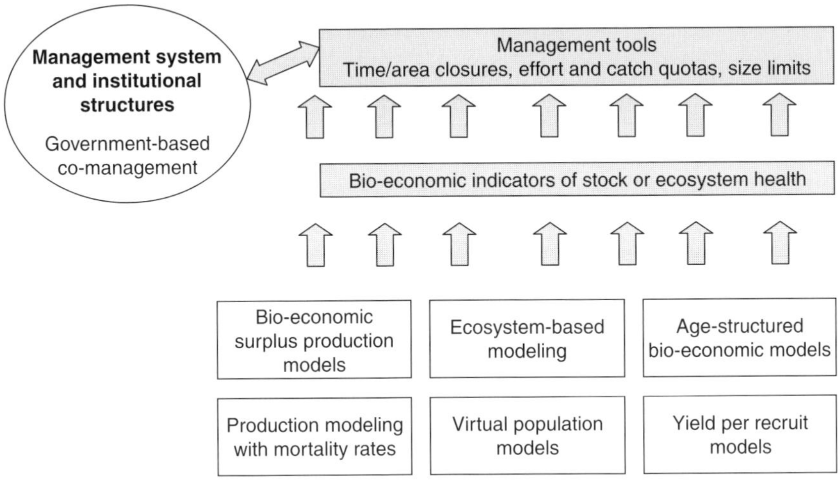

Figure 1.2 Simplified flow diagram showing some traditional fish stock assessment methodologies, together with the most commonly used management tools and different criteria of state intervention (government-based management and co-management). The models highlighted here have gained in complexity, ranging from the former static, deterministic and single-species to the dynamic, stochastic, and multispecies versions applied nowadays.

fishing mortality and leads to the notion of optimum fishing mortality and the concept of maximum sustainable yield (MSY: reviewed in Quinn and Collie 2005). The development of surplus production models (Fig. 1.2) was formalized by Schaefer (1954) on the basis of the Verhulst logistic model of population growth, and found its first management context in fisheries at the time that catch and effort data were increasingly available and age estimates of fish were very difficult to pursue. These models depend heavily on the relation between catch per unit effort (CPUE) and stock size, with catch rates being an index of population abundance (Ricker 1975). Adding relevant economic/social factors affecting the fishery system complemented surplus production models, as biological factors are not the only ones governing fisheries management (Gordon 1953; Anderson 1986; Charles 1989).

Production modeling, using fishery yield but total mortality rate in place of fishing effort (yield–mortality models) was also used to estimate MSY conditions and to model changes in yield against mortality over time (Defeo and Caddy 2001). The advantages of these models are evident when only annual catch and overall mortality estimated from regularly sampled size frequency data are available but no data on fishing effort. Surplus production models produce biologically-meaningful bio-economic reference points (BRPs) for managing fisheries including the classical concepts of MSY, yield at maximum biological production (Y_{MBP}), maximum economic yield (MEY), and the associated mortality levels. Virtual population analysis (VPA) and 'cohort analysis', either using age- or size-based population structure, are widely used models for the analysis of fished populations. These models evaluate the effect that a fishery has had on particular year classes of a stock (Sparre and Venema 1998).

VPA techniques look at the past, using historical catch data together with estimates of natural mortality and fishing mortality for the oldest age or length classes. These models vary in the way they use data and fit the model rather than in the form of the model itself (Lassen and Medley 2000). As an alternative to VPA models, yield per recruit (Y/R) and Thompson and Bell models constitute 'predictive models' used to predict future yields and biomass as a function of different fishing effort levels (Sparre and Venema 1998). Y/R models compute the yield that produces one recruit at different levels of fishing mortality. The Thompson and Bell model (1934) is widely used in regions where VPA models are applied, because Y/R models are usually based on the findings provided by VPA. The relationship between the spawning stock and recruitment (stock-recruitment relationship (SRR): Ricker 1975; Shepherd 1982) also relates to these models. Although SRR is not a proper assessment method, it has been widely used to understand the long-term behavior of the population and the identification of the factors driving the recruitment process and population variability through time. In this setting, recruitment variability seems to be produced by the intertwined forces of density-dependent factors such as intraspecific competition, density-independent factors such as oceanographic conditions, winds, temperature, also harvesting as a perturbation source, operating together (Fogarty et al. 1991; Lima et al. 2000). Availability of computing power enabling the calculations of the size-based counterparts of the age-based models detailed above, meant the method became very popular because of the inherent simplicity in the calculations of the growth and mortality parameters (Gayanilo and Pauly 1997).

These indirect approaches to stock assessment are based on fishery-dependent data, such as catch, age or length, and fishing effort statistics. Conversely, direct methods are based on fishery-independent data mainly from stock surveys. The direct methods attempt to avoid the biases derived from commercial sampling and have been traditionally used to estimate abundance, demography, age structure, reproductive biology, food items, and other factors. The models above were designed to represent the dynamics of finfish fisheries. Although criticized (Pauly et al. 2002), such models provided the basis for forecasts of yield stocks under stable birth and death processes or dynamic conditions with different levels of exploitation.

Analytical and surplus production models assume the almost universal application of the 'dynamic pool assumption' (Caddy 1996). This has simplified standard calculations on unit stocks (whether VPA, Y/R calculations or production modeling) by allowing the treatment of a unit stock as perfectly mixed age groups with no spatial components within the stock distribution area. Therefore, they assume perfect mixing of the survivors between successive intervals when fishing effort is applied. Hence the harvesting activity should reflect the current population characteristics such as mean abundance, growth, and mortality rates. Central to dynamic pool calculations are (Caddy 1996): (1) that growth and natural mortality rates are unrelated to stock density and are independent of age; (2) that recruitment is independent of stock size and assumed constant for all values of stock size; (3) that fishing mortality is distributed over the entire exploitable stock and the rate of exploitation is unchanged throughout life; (4) that processes governing population dynamics can be integrated over a series of annual cycles to provide a cumulative evaluation of the potential yield of the cohorts; (5) that changes in species interactions such as competition and predation are not included for different levels of harvesting. These assumptions are not realistic when assessing the dynamics of the stocks and fishing process (Caddy 1975, 2004; McClanahan 1992).

Early in modern fisheries science, models were static and often focused on a single species in isolation, and fisheries management was guided by calculations that assume steady state or equilibrium conditions. During the last three decades however, there were concerns about the uncritical use of traditional equilibrium assumptions. Nowadays there are several dynamic models, among which biomass-dynamic models constitute the most generalized approach in assessing the dynamics of an exploited population through time (Hilborn and Walters 1992). These models usually keep track of the age and sex structure of the population and are widely used to model catch rates, allowing the estimation of the different BRPs outlined above, plus stock status estimates of biomass at MSY, also current biomass and its relation to virgin biomass among others (Punt and Hilborn 1996).

These models were formerly applied in a deterministic way, assuming there was no variation in the parameters. Moreover, many parameters used as inputs in assessment models, even in methods that purport to analyze yearly data as VPA, are assumed to be time-invariant, and variability around mean estimates are neither reported nor used to perform sensitivity analyses. This tends to underestimate uncertainty in the parameters and, consequently, in the outcomes provided by the models. Hence, the success of the models as predictors depends on the extent to which their assumptions are met and whether the parameters are constant. Further results and theoretical development demonstrated that the models are commonly not appropriate in their deterministic forms, notably taking into account the several sources of uncertainty in all the three fishery subsystems (Hilborn and Peterman 1996; Charles 2002). The effects of uncertainty have been formerly evaluated through sensitivity analysis, using marginal changes of $\pm$ 5% in model parameters and rerunning the stock assessment. However, it has been difficult to summarize the management implications of sensitivity tests that simultaneously exhibit considerable variation in several parameters. Nowadays, the Bayesian approach to stock assessment appears to be one of the most powerful techniques accounting for uncertainties related to models and input parameter values, incorporating prior information about biological parameters and processes (Punt and Hilborn 1997, 2002; Maunder 2003).

Multispecies

The theory and analytical bio-economic framework for fisheries management was developed during the initial approach to fisheries management that assumed the exploitation of a single stock by a homogeneous fishing fleet. Classically, these single-species assessments lead to single-species quotas as a management tool. Nevertheless, multispecies fishing gears are the norm and single-species assessments and management promote the misreporting and discarding of non-target species. Furthermore, the catch can include species from a filled quota and therefore generate an 'unexpected' overfishing syndrome even in non-marketable species (Hall 1996; Pauly et al. 1998). Consequently, sharp changes in the abundance of targeted and non-targeted species, as well as the composition of exploited species, have been detected worldwide as a result of a growing fishing intensity (Hollingworth 2000). This is true for both industrial fisheries (Chuenpagdee et al. 2003) and coastal artisanal fisheries (Blaber et al. 2000; Rueda and Defeo 2003).

Overfishing has been observed with a progressive decrease in catches of the most valuable species commonly corresponding to the highest levels of the food web (Pauly et al. 2002; Myers and Worm 2003). This has led to the recognition that fishing affects many species regardless of whether they are the target, incidental, or unharvested resource (Alverson et al. 1994; Hall and Mainprize 2005). The marked decline in commercial catches of top carnivorous species and the general increase in catches of species low in the food chain suggest potential trophic effects of fishing caused by the concurrent increase in fishing technology and effort. This agrees with the 'fishing down the food web' hypothesis of Pauly and colleagues (1998), and a number of examples provided strong support for the sequential depletion hypothesis, including overexploitation of target species first and incidental ones later on (Orensanz et al. 1998; Baum et al. 2003; Milessi et al. 2005). This suggests that a multispecies or ecosystem framework should be used for assessment. Among the several alternatives available (Rose et al. 1996), some approaches for modeling multispecies are expansions of the stock-assessment methods that take into account technological and ecological interdependencies through dynamic simulation modeling (Seijo et al. 1998), multispecies VPA (MSVPA: Sparre 1991; Magnusson 1995) and ecological modeling based on mass balance and a food webs approach (McClanahan 1995; Pauly et al. 2000). Despite these trends, the utility of multispecies models in the evaluation of the performance of fisheries is difficult. Multispecies models have large data requirements, there are difficulties regarding the correct identification of relationships among species, the appropriate selection of the most relevant species, influences of exogenous environmental variables, and adequate descriptors of technological impacts between fleets in technologically interdependent fisheries (Seijo et al. 1998; Sainsbury et al. 2000).

The fishery research agenda: edge issues for the twenty-first century

Spatial considerations: closed areas and habitat protection

The dynamic pool assumption (DPA) is not universally applicable and it depends heavily on the life history characteristics of exploited species and the dynamics of the fishing process. A potentially useful way to deal with departures from the

DPA, which includes the aggregated distribution of fishing effort in the densest patches and strong gradients in growth, mortality, and recruitment that follows density-dependent and environmental variations, is to divide the whole analyzed area into a number of sub-areas of sizes small enough that uniform conditions exist. The absence of movement is a fundamental assumption in population models, becoming highly relevant when the non-uniform distribution of fishing effort is taken into account. Here, DPA is violated and thus CPUE does not provide a reliable measure of abundance, because fishing effort is not applied homogeneously over the entire stock distribution; rather, local beds are fished down in a mining strategy, to economic extinction (Caddy 1975; Prince 2003).

A fishery management regime is expected to go through a series of stages including a transition from open access to various restrictions (Caddy and Defeo 2003). In this context, permanent and temporary closed areas, times and restrictions on gear, sizes, and species extracted are management tools well suited to deal with the heterogeneous distribution of resources and effort (Fig. 1.1; Orensanz and Jamieson 1998; Caddy 2004). The use of habitat closures to improve yields can be subsumed under the establishment of marine parks or marine protected areas (MPAs) for coastal species, or zoning of oceanic areas directed to enhance depleted or overexploited fish stocks (Murray et al. 1999; Castilla, 2000; Russ et al. 2004; McClanahan and Graham 2005). Optimum placement and the estimate of the appropriate size of marine reserves is a high priority need for fishery management (Roberts et al. 2003). Short-term area closures and reproductive refuges, as opposed to permanently established MPAs, may be used to protect areas that historically have utmost importance as recruitment or spawning sources.

Ecosystem-based approaches

The decreasing trend in global catch rates as a result of overfishing and changes in habitat has damaged the productivity of stocks and the associated ecosystems (Blaber et al. 2000). Industrial bottom-trawling fisheries have been shown to be wasteful and environmentally harmful, not only by the removal of biomass and diversity but also by the effects on the habitat and the high amount of discards (Alverson et al. 1994; Hall et al. 2000). Thus, during the last three decades, fishery scientists have taken a more holistic approach to management including aspects of the ecosystem, evaluating the ecological interactions and technological development effects on stocks (Pauly et al. 2002). Ecosystem-based fishery management (EBFM) has recently been proposed as a tool for solving fishery problems (Pauly et al. 2003; Browman and Stergiou 2004; Pikitch et al. 2004). EBFM is conceived as a new direction for fishery management, oriented towards reversing the order of management priorities starting with the ecosystem rather than the target species. EBFM implies several objectives for protecting ecosystem attributes from ecosystem-level effects of fishing: (1) to maintain predator–prey relationships, energy flow and balance, and diversity (Livingston et al. 2005); and (2) to balance diverse societal objectives, by taking into account the knowledge and uncertainties about biotic, abiotic, and human components of ecosystems and their interactions within ecologically meaningful boundaries (FAO 2003; Garcia and Cochrane 2005). Indicators of ecosystem status, such as the mean trophic level of the catch are being developed and included in models to predict future trends (Pauly et al. 2000).

Even though EBFM is a potentially powerful approach to managing fisheries, several shortcomings have been highlighted. For example, the response of some indicators may depend more on environmental conditions and historic fishing regimes than on current fishing conditions. Therefore, ecosystem-based indicators may often be unsuitable to assess the effects of short-term and small-scale management actions (Piet and Jennings 2005). Environmental effects and long-term climate regime shifts could drive recruitment levels and could be particularly important at low spawning stock biomass, significantly affecting catch levels in the following years (Brander 2005; Jenkins 2005). The importance of these cycles is in most cases unknown and thus it becomes very difficult to discern fishing from environmental effects, particularly when interannual cycles influence production (Caddy 1996). Moreover, rapid implementation of EBFM is difficult in artisanal fisheries in poor countries where ecosystem information is scarce (Castilla and Defeo 2005; Defeo and Castilla, 2005). Future implementation challenges include the improvement of predictive models and the identification of meaningful ecosystem indicators so as to provide simple and effective indices of stock and ecosystem status (Livingston et al. 2005). EBFM should nevertheless be accompanied by the outcomes resulting from single-species approaches regarding effort or catch limits. Judgements as to which indicators to choose, and how their abundance should correspond to a limit reference point (see below), needs to be discussed between scientists, managers, and stakeholders.

Reference points, fishery indicators, and the precautionary approach

Investigators have identified different phases of fisheries before management becomes sustainable (Caddy and Gulland 1983; Perry et al. 1999; Castilla and Defeo 2001). These are: (1) a 'developing' phase; (2) a 'crisis' or overexploitation phase, leading to a call for implementation of license limitation; (3) a 'sustainable' phase, often leading to closures in time and space and, more frequently in recent years, to the granting of quotas or other limits on access (Caddy 2004). During at least the first two phases, the classical data-intensive approaches are unlikely to be available, and thus the development of indicator series based on BRPs is useful to measure key aspects of the resource and fishery (Caddy and Mahon 1995). BRPs take the form of target or threshold fishing mortality rates or target or threshold spawning stock biomass levels (Rose and Cowan 2003). BRPs and control rules are presently a common basis for marine fisheries management (Smith and Sainte Marie 2004). BRPs are defined by measures of biomass, fishing mortality, fleet performance, and economic yield; each considered as fishery indicators. Measuring one or more indicators on an annual basis can provide an index of stock and fishery status (Caddy and Defeo 2003). Comprehensive fisheries data consisting of catch and effort and stock surveys are required. Alternatively, a precautionary management framework intended for use in data-poor situations may be set (Mahon 1997; Caddy 1999).

Control rules use values such as standing stock or mortality rates as BRPs to determine the status of the stock and the fishery in order to avoid the main risks (Francis 1992; Mace 1994). The two main risks are: (1) that yields fall below the MSY or MEY; (2) that the stock falls below some minimum and viable population size (Caddy and Mahon 1995). As the precautionary approach is based not only on BRPs but also on the risk of exceeding these risks, a problem arises when essential estimates of risk and uncertainty cannot be routinely obtained in the context of data-poor situations such as exist in most developing countries (Mahon 1997). In this context, Caddy (2002) developed a traffic light (TL) or threshold framework based on reliable but simple indicators and with minimal use of modeling to define BRPs. Each indicator value is then assigned by the classic three-colour system 'green–yellow–red', to identify cut-off values. The management response would then be graduated in intensity depending on the proportion of indicators shown as red. When a fishery reaches the red level then all fishing must stop until the BRPs are above the selected threshold.

Adaptive and experimental management

Controlled field experimentation has unfortunately played a minor role in developing fisheries management (Castilla and Defeo 2001) and most management manipulations have not been originally designed to distinguish rigorously between alternative hypotheses (McAllister and Peterman 1992). A step forward on the matter is recognized in the adaptive management approach (Holling 1978; Walters 1986, 1997). Adaptive management can be defined as a structured dynamic process of 'learning by doing' involving a modeling step and the design of management experiments (Walters and Holling 1990). An adaptive process regularly evolves from the recommendation of additional management experiments, leading to the interchangeable use of the concepts of 'adaptive management' and 'experimental management' (Walters 1997). Even though adaptive management is a key issue to be considered in future fishery management, it has been difficult to design experimental management programs through an adaptive process.

Few large-scale fishery experiments play an important role in the evaluation of alternative management policies (Sainsbury et al. 1997, 2000), especially when they explicitly involve the participation of fishers in field experimentation (Castilla et al. 1998). For example, in Chile the exclusion of humans from rocky shores allowed testing effects of fishing on shellfish abundance and community elasticity (Castilla 1999, 2000). Long-term results demonstrated that humans generate ecological cascading effects that affect the structure and functioning of communities (Castilla 1999). Nonetheless, this system has yet to determine the long-term effects of various levels of exploitation. Varying extraction intensity on species of different trophic levels may translate into different community structures, which emphasizes the importance of linkages and strengths of ecological interactions (Tegner and Dayton, 2000; Micheli et al. 2005). Thus, at least in theory, multispecific experimental approaches may provide a more reliable means of guiding community structure to a desired state than by basing management decisions solely on the predictions of a group of separately viewed single-species models (Sainsbury 1988). However, experimental and manipulative approaches to ecosystem management have not had a high level of success so far. The complex dynamics in multispecies assemblages generally precludes a firm forecast of the ecological outcome of a single manipulation of species abundance (Sainsbury et al. 1997; Castilla 2000). Thus, fishery experiments seem worth pursuing further where the ecosystem is sufficiently simple to allow some predictions to be made (examples in Sainsbury et al. 2000 and Caddy and Defeo 2003). They may benefit from parallel efforts at ecosystem modelling, including lessons learned from the experimental fishing (McClanahan 1995).

The limits to science: when management fails

Fisheries sustainability has been difficult to achieve. An increasing number of fisheries around the world are becoming increasingly limited, catch has begun to drop, and stocks are fully to heavily exploited, overexploited, or depleted (Botsford et al. 1997; FAO 2004), without any evident signs of recovery (but see Caddy and Agnew 2004). Among the reasons explaining this fact are the inherent variable and complex characteristics of the resource, fisher behavior, and market forces (Castilla and Defeo 2001). Most fishery systems still remain poorly understood regarding the linkages between the structure of the stock and the dynamics and bio-economic features of the fishing process (Orensanz and Jamieson 1998).

Western technocratic models: abuse and misuse

Current fishery models are more sophisticated than their original versions. Despite this, assessment models have failed in cases involving rapid stock declines and, in particular, the use of catch–effort data has underestimated the severity of the decline (Pauly et al. 2002). This is particularly true when CPUE decreases more slowly than stock abundance because of the efficient search of localized patches (Hilborn and Walters 1992). Moreover, poorly measured factors such as climate, prey or predators, and effects of various gears on habitat can all amplify single-species assessment errors, particularly when predicting future recruitment rates. This is aggravated when some stocks have shown recruitment failure after severe decline of the parental stock (Caddy 1999).

This failure in achieving sustainability has intensified the scrutiny and criticism both of the data used to configure models and of the models themselves, particularly during the last ten years (Rose and Cowan 2003). Criticisms of the single-species approaches also include the fact that they do not account for the indirect effects of fishing, including bottom fishing on habitat quality (Quinn and Collie 2005). Despite the above criticisms, the advances made under the single-species approach should not be abandoned, but rather enhanced and combined with new approaches that acknowledge EBFM issues and economic realms (Mace 2004; Quinn and Collie 2005). In many cases the problems are not with the data or models, but with the lack of adherence among managers and fishers to the recommendations derived from them. Moreover, it is difficult to develop effective short-term regulatory systems to achieve sustainability on the basis of the outcomes provided by stock-assessment models (Pauly et al. 2002).

In sessile/sedentary invertebrate fisheries, marked discrepancies between reality and model assumptions – usually those based on the DPA – have led to poor long-term predictions and management failures (Prince 2003; Caddy 2004). The dynamic pool assumption that local effects of fishing effort applied anywhere in a sedentary population are 'mixed' over time is misleading. Here, spatially-discriminated estimations of population parameters, fishing effort, and catches by sub-areas of the whole population range must be considered and can be integrated afterwards to give an overall picture of the stock and fishery.

The marked decline in commercial catches of top predator species and the general increases in species low in the food chain suggest trophic effects of fishing caused by the concurrent increase in fishing technology and effort (Pauly et al. 1998). Major but often hidden changes in fishing power, coupled with uncertainties in stock estimates and risk-prone management attitudes, have led to collapses of stocks, resulting in the fishing fleet becoming overcapitalized. Unselective gears accelerated the overexploitation syndrome, making industrial fisheries wasteful and environmentally harmful, not only by the removal of biomass and diversity, but also by the potential impact on the habitat and the high amount of discards (Hall et al. 2000; Tegner and Dayton 2000; Chuenpagdee et al. 2003). This has caused serious flaws when trying to predict the long-term behavior of the fishing process (Caddy 1999).

Most of the world's fishery science is devoted and applied to large stocks. It is seldom feasible to use the techniques on small stocks, artisanally exploited or for self-subsistence (Mahon 1997; Castilla and Defeo 2005). Fishing intensity increased even under diminishing catch rates, because of low operating costs. The easy access to nearshore resources makes regulatory efforts expensive and ineffective and this has not been accompanied by the concomitant increase of scientific information (Johannes 1998; Castilla and Defeo 2001; Sale et al. 2005). If sophisticated stock-assessment models are to be applied and management strategies, such as seasonal/area closures, are to be used as management tools, then the existence of significant gaps in scientific knowledge must be filled. However, filling them will require a significant investment in research and monitoring compliance that is beyond the finances of most management agencies in developing countries. These facts preclude the application of sophisticated assessment models in these fisheries and call for a more simplistic, short-term approach, based on easy to use fishery indicators in a precautionary management context, which would require strong collaborations among the scientific, management, and fishing communities (Mahon 1997).

The forgotten social side of management and attitudes

Overfishing has been the result of the interacting economic, technological, legal, political, socio-cultural, and scientific drivers. In particular, increasing harvest rates in response to incessant socio-political pressure for greater harvests that are coupled with the high uncertainties inherent in fisheries systems. This has promoted overexploitation and collapse of fish stocks (Botsford et al. 1997). The human component of ecosystems tends to come under the umbrella of the political and cultural aspects of the problem and it is recognized that: '[human] institutions are not configured to manage [fisheries] at the ecosystem scale' (US Commission on Ocean Policy 2004). This causes mismatch and disconnection between human and 'natural' aspects of ecosystems (Defeo and Castilla 2006). Managers strive to find a fishing strategy that either achieves both the fishing mortality and biomass levels defined by the target BRPs or strikes a workable compromise between the two (Mace 1994). Assessment results that implied limitation on fishing mortality levels have often been ignored.

Political pressure from the fishing industry delayed or even ceased the implementation of management measures. Thus scientific advice about catch quotas or limitation of the number of vessels is relegated to second place in favor of political compromises and socio-political resistance to both the management actions and the quality and accuracy of scientific data and assessment estimates (Rosenberg 2003; Mace 2004). The politics of fishery management favor continued exploitation and the political incentive is to maintain the status quo, on the excuse that the overexploitation evidence is not precise enough, and so there is no need to impose an economically-painful restriction on fishing (Charles 2002; Pauly et al. 2002; Rosenberg 2003). The 'burden of proof' problem has become a critical issue in fisheries where the human element within the management subsystem is a critical factor.

Institutional structures, societal participatory roles, and co management

The management subsystem (Fig. 1.2) is a critical component of the fishery system. To achieve the objectives and goals proposed in a management plan, a consideration of possible ways of intervention and of contending with multiple criteria in the selection of management strategies is needed (Seijo et al. 1998). In this context, in the last century, the top-down approach based on scientific information gathered in a centralized management control has been the most common institutional arrangement followed by fishery management bodies. This approach has been questioned because predictions derived from fisheries stock assessment have immediate societal and economic effects and so must deal with the scrutiny of stakeholders as one of the most important components of the fisheries system (Hanna 1994; Castilla and Defeo 2005).

Co-management constitutes a proactive institutional arrangement where fishers, scientists and managers interact to improve the quality of the regulatory process, serving to sustain fisheries over time (Berkes et al. 2001; Gelcich et al. 2005). The main factors supporting fisheries co-management are: (1) a comparatively reduced scale of fishing operations and well-defined boundaries for a management unit; (2) the allocation of institutionalized co-ownership authority to fishers; (3) the voluntary participation of the fishers in enforcing regulations and the legitimization of this participation in the management process is seen as the only way to promote compliance with regulations; (4) the improvement of scientific information to consolidate the management schemes; (5) the incorporation of community traditions and idiosyncrasies; (6) the allocation of territorial use rights for fisheries under a collaborative/voluntary community framework (Castilla and Defeo 2001). The development of co-management has been slowed by institutional constraints, notably via the establishment of legal frameworks, formalizing community responsibility in the management process. Thus, once this strategic institutional arrangement is in place, additional, risk-averse, precautionary management schemes could be gradually introduced (Castilla and Defeo 2001).

Scientists and policy-makers must learn from the various forms of community-based management followed for centuries by traditional fisher communities and not assume that traditional approaches must be discarded, as opposed to updated (Johannes 1998, 2002; Caddy and Defeo 2003). The fruitful interaction between fishers, scientists, policy-makers, extension workers and politicians should provide a comprehensive course of action in scope, and should promote the development of adaptive co-management systems (Olsson et al. 2004). Here, self-organized local groups learn and actively adapt to new courses of management actions derived from dynamic acquisition of scientific knowledge (Castilla and Defeo 2001; Wiber et al. 2004; Hughes et al. 2005).

Summary

The status of the world's fisheries is worrying and the factors that have led to the global decline are biological, social, political, and cultural in nature. Marine fisheries are in trouble and this trouble extends from nearshore to offshore areas and across most of the regions of the Earth. The trouble has occurred in the context of a well-developed fisheries science that has largely focused on the resource and the biophysical aspects that control them but with less focus on the societal aspects of resource management. This has changed considerably in the last decade via innovative and proactive ways to manage fisheries that are socially acceptable and have led to successes: increasing fisheries catches or reversing declines. Since fisheries are a natural service for humans, we hereby sustain that a more comprehensive, realistic, and truly holistic nature–human approach urgently needs to be developed. This will not be easily achieved. This book will provide historical case studies for a number of fisheries that have resulted in improved management. To increase the chances for success in fisheries it would help practitioners to be aware of failures and successes in fisheries histories in different parts of the world.

References

Alverson, D.L., Freeberg, M.H., Murawski, S.A. and Pope, J.G. (1994) *A Global Assessment of Fisheries Bycatch and Discards*. FAO Fisheries Technical Paper, 339.

Anderson, L.G. (1986) *The Economics of Fisheries Management* (revised and enlarged edition). The John Hopkins University Press, Baltimore.

Baum, J.K., Myers, R.A., Kehler, D.G., Worm, B., Harley, S.J. and Doherty, P.A. (2003) Collapse and conservation of shark populations in the northwest Atlantic. *Science*, **299**: 389–92.

Berkes, F., Mahon, R., McConney, P., Pollnac, R. and Pomeroy, R. (2001) *Managing Small-scale Fisheries. Alternative Directions and Methods.* International Development Research Centre, Ottawa.

Blaber, S.J.M., Cyrus, D.P., Albaret, J.J., Ching, C.V., Day, J.W., Elliott, M., Fonseca, M.S., Hoss, D.E., Orensanz, J., Potter, I.C. and Silvert, W. (2000) Effects of fishing on the structure and functioning of estuarine and nearshore ecosystems. *ICES Journal of Marine Science*, **57**: 590–602.

Botsford, L.W., Castilla, J.C. and Peterson, C.H. (1997) The management of fisheries and marine ecosystems. *Science*, **277**: 509–15.

Brander, K.M. (2005) Cod recruitment is strongly affected by climate when stock biomass is low. *ICES Journal of Marine Science*, **62**: 339–43.

Browman, H.I. and Stergiou, K.I. (Eds) (2004) Perspectives on ecosystem-based approaches to the management of marine resources. *Marine Ecology Progress Series*, **274**: 269–303.

Caddy, J.F. (1975) Spatial model for an exploited shellfish population, and its application to the Georges Bank scallop fishery. *Journal of the Fisheries Research Board of Canada*, **32**: 1305–28.

Caddy, J.F. (1996) Regime shifts and paradigm changes: is there a place for equilibrium thinking? *Fisheries Research*, **25**: 219–30.

Caddy, J.F. (1999) Fisheries management in the twenty-first century: will new paradigms apply? *Reviews in Fish Biology and Fisheries*, **9**: 1–43.

Caddy, J.F. (2002) Viewpoint: limit reference points, traffic lights, and holistic approaches to fisheries management with minimal stock assessment input. *Fisheries Research*, **56**: 133–7.

Caddy, J.F. (2004) Current usage of fisheries indicators and reference points, and their potential application to management of fisheries for marine invertebrates. *Canadian Journal of Fisheries and Aquatic Sciences*, **61**: 1307–24.

Caddy, J.F. and Agnew, D.J. (2004) An overview of recent global experience with recovery plans for depleted marine resources and suggested guidelines for recovery planning. *Reviews in Fish Biology and Fisheries*, **14**: 43–112.

Caddy, J.F. and Defeo, O. (2003) *Enhancing or Restoring the Productivity of Natural*

Populations of Shellfish and other Marine Invertebrate Resources. FAO Fisheries Technical Paper, 448.

Caddy, J.F. and Gulland, J.A. (1983) Historical patterns of fish stocks. *Marine Policy*, **7**: 267–78.

Caddy, J.F. and Mahon, R. (1995) *Reference Points to Fisheries Management.* FAO Fisheries Technical Paper, 347.

Castilla, J.C. (1999) Coastal marine communities: trends and perspectives from human-exclusion experiments. *Trends in Ecology and Evolution*, **14**: 280–3.

Castilla, J.C. (2000) Roles of experimental marine ecology in coastal management and conservation. *Journal of Experimental Marine Biology and Ecology*, **250**: 3–21.

Castilla, J.C. and Defeo, O. (2001) Latin-American benthic shellfisheries: emphasis on co-management and experimental practices. *Reviews in Fish Biology and Fisheries*, **11**: 1–30.

Castilla, J.C. and Defeo, O. (2005) Paradigm shift needed for world fisheries. *Science*, **309**: 1324–5.

Castilla, J.C., Manríquez, P., Alvarado, J., Rosson, A., Pino, C., Espoz, C., Soto, R., Oliva, D. and Defeo, O. (1998) The artisanal caletas as unit of production and basis for community-based management of benthic invertebrates in Chile. *Canadian Special Publication of Fisheries and Aquatic Sciences*, **125**: 407–13.

Charles, A.T. (1989) Bio-socio-economic fishery models: labour dynamics and multi-objective management. *Canadian Journal of Fisheries and Aquatic Sciences*, **46**: 1313–22.

Charles, A.T. (2002) The precautionary approach and 'burden of proof' challenges in fishery management. *Bulletin of Marine Science*, **70**: 683–94.

Chuenpagdee, R., Morgan, L.E., Maxwell, S.M., Norse, E.A. and Pauly, D. (2003) Shifting gears: assessing collateral impacts of fishing methods in US waters. *Frontiers in Ecology and the Environment*, **1**: 517–24.

Defeo, O. and Caddy, J.F. (2001) Evaluating a dynamic approach to yield-mortality models. *ICES Journal of Marine Science*, **58**: 1253–60.

Defeo, O. and Castilla, J.C. (2005) More than one bag for the world fishery crisis and keys for co-management successes in selected artisanal Latin American shellfisheries. *Reviews in Fish Biology and Fisheries*, **15**: 265–83.

FAO (2003) *Fisheries Management. The Ecosystem Approach to Fisheries.* FAO Technical Guidelines for Responsible Fisheries, 4 (Supplement 2).

FAO (2004) *The State of World Fisheries and Aquaculture*. FAO, Rome.

Fogarty. M.J., Sissenwine, M.P. and Cohen, E.W. (1991) Recruitment variability and the dynamics of exploited marine populations. *Trends in Ecology and Evolution*, **6**: 241–6.

Francis, R.I.C.C. (1992) Use of risk analysis to assess fishery management strategies: a case study using orange roughy (*Hoplostethus atlanticus*) on the Chatham Rise, New Zealand. *Canadian Journal of Fisheries and Aquatic Sciences*, **49**: 922–30.

Garcia, S.M. and Cochrane, K.L. (2005) Ecosystem approach to fisheries: a review of implementation guidelines. *ICES Journal of Marine Science*, **62**: 311–18.

Gayanilo, F.C., Jr and Pauly. D. (1997) *FAO–ICLARM Stock Assessment Tools (FISAT).* FAO Computerized Information Series (Fisheries), 8.

Gelcich, S., Edwards-Jones, G. and Kaiser, M.J. (2005) Importance of attitudinal differences among artisanal fishers toward co-management and conservation of marine resources. *Conservation Biology*, **19**: 865–75.

Gordon, H.S. (1953) An economic approach to the optimum utilization of fishery resources. *Journal of the Fisheries Research Board of Canada*, **10**: 442–7.

Hall, M.A. (1996) On bycatches. *Reviews in Fish Biology and Fisheries*, **6**: 319–52.

Hall, S.J. and Mainprize, B.M. (2005) Managing by-catch and discards: how much progress are we making and how can we do better? *Fish and Fisheries*, **6**: 134–55.

Hall, M.A., Alverson, D.L. and Metuzals, K.I. (2000) By-catch: problems and solutions. *Fisheries Research*, **41**: 1–6.

Hanna, S.H. (1994) Co-management. In: K.L. Gimbel (Ed.) *Limiting Access to Marine Fisheries: Keeping the Focus on Conservation*. Center for Marine Conservation and World Wildlife Fund, Washington.

Hilborn, R. and Peterman, R.M. (1996) The development of scientific advice with incomplete information in the context of the precautionary approach. *Precautionary Approach to Fisheries. Part 2: Scientific Papers*. FAO Fisheries Technical Paper (350/2).

Hilborn, R. and Walters, C.J. (1992) *Quantitative Fisheries Stock Assessment. Choice, Dynamics and Uncertainty*. Chapman and Hall, New York.

Holling, C.S. (Ed.) (1978) *Adaptive Environmental Assessment and Management*. John Wiley, New York.

Hollingworth, C.E. (Ed.) (2000) Ecosystem effects of fishing. Proceedings of an ICES/SCOR Symposium. *ICES Journal of Marine Science*, **57**: 465–792.

Hughes, T.P., Bellwood, D.R., Folke, C., Steneck, R.S. and Wilson, J. (2005) New paradigms for supporting the resilience of marine ecosystems. *Trends in Ecology and Evolution*, **20**: 380–6.

Jenkins, G.P. (2005) Influence of climate on the fishery recruitment of a temperate, seagrass-associated fish, the King George whiting *Sillaginodes punctata*. *Marine Ecology Progress Series*, **288**: 263–71.

Johannes, R.E. (1998) The case for data-less marine resource management: examples from tropical nearshore finfisheries. *Trends in Ecology and Evolution*, **13**: 243–6.

Johannes, R.E. (2002) The renaissance of community-based resource management in Oceania. *Annual Review of Ecology and Systematics*, **33**: 317–40.

Lassen, H. and Medley, P. (2000) *Virtual Population Analysis. A Practical Manual for Stock Assessment*. FAO Fisheries Technical Paper, 400.

Lima, M., Brazeiro, A. and Defeo, O. (2000) Population dynamics of the yellow clam *Mesodesma mactroides*: recruitment variability, density-dependence and stochastic processes. *Marine Ecology Progress Series*, **207**: 97–108.

Livingston, P.A., Aydin, K., Boldt, J., Ianelli, J. and Jurado-Molina, J. (2005) A framework for ecosystem impacts assessment using an indicator approach. *ICES Journal of Marine Science*, **62**: 592–7.

Mace, P.M. (1994) Relationships between common biological reference points used as thresholds and targets of fisheries management strategies. *Canadian Journal of Fisheries and Aquatic Sciences*, **51**: 110–22.

Mace, P. (2004) In defence of fisheries scientists, single-species models and other scapegoats: confronting the real problems. In: H.I. Browman and K.I. Stergiou (Eds) *Perspectives on Ecosystem-based Approaches to the Management of Marine Resources. Marine Ecology Progress Series*, **274**.

Magnusson, K. (1995) An overview of the multispecies VPA – theory and applications. *Reviews in Fish Biology and Fisheries*, **5**: 195–212.

Mahon, R. (1997) Does fisheries science serve the needs of managers of small stocks in developing countries? *Canadian Journal of Fisheries and Aquatic Sciences*, **54**: 2207–13.

Maunder, M.N. (2003) Paradigm shifts in fisheries stock assessment: from integrated analysis to Bayesian analysis and back again. *Natural Resources Modelling*, **16**: 465–75.

McAllister, M.K. and Peterman, R.M. (1992) Experimental design in the management of fisheries: a review. *North American Journal of Fisheries Management*, **12**: 1–18.

McClanahan, T.R. (1992) Resource utilization, competition and predation: a model and example from coral reef grazers. *Ecological Modelling*, **61**: 195–215.

McClanahan, T.R. (1995) A coral-reef ecosystem-fisheries model – impacts of fishing intensity and catch selection on reef structure and processes. *Ecological Modelling*, **80**: 1–19.

McClanahan, T.R. and Graham, N.A.J. (2005) Recovery trajectories of coral reef fish assemblages within Kenyan marine protected areas. *Marine Ecology Progress Series*, **294**: 241–8.

Micheli, F., Benedetti-Cecchi, L., Gambaccini, S., Bertocci, I., Borsini, C., Osio, G.C. and Roman, F. (2005) Cascading human impacts, marine protected areas, and the structure of Mediterranean reef assemblages. *Ecological Monographs*, **75**: 81–102.

Milessi, A., Arancibia, H., Neira, S. and Defeo, O. (2005) The mean trophic level of Uruguayan landings during the period 1990–2001. *Fisheries Research*, **74**: 223–31.

Murray, S.N., Ambrose, R.F., Bohnsack, J.A., Botsford, L.W., Carr, M.H., Davis, G.E., Dayton, P.K., Gotshall, D., Gunderson, D.R., Hixon, M.A., Lubchenco, J., Mangel, M., MacCall, A., McArdle, D.A., Ogden, J.C., Roughgarden, J., Starr, R.M., Tegner, M.J. and Yoklavich, M.M. (1999) No-take reserve networks: sustaining fishery populations and marine ecosystems. *Fisheries*, **24**: 11–25.

Myers, R.A. and Worm, B. (2003) Rapid worldwide depletion of predatory fish communities. *Nature*, **423**: 280–3.

Olsson, P., Folke, C. and Berkes, F. (2004) Adaptive comanagement for building resilience in social–ecological systems. *Environmental Management*, **34**: 75–90.

Orensanz, J.M. and Jamieson, J. (1998) The assessment and management of spatially structured stocks. *Canadian Special Publication of Fisheries and Aquatic Sciences*, **125**: 441–59.

Orensanz, J.M., Armstrong, J., Armstrong, D. and Hilborn, R. (1998) Crustacean resources are vulnerable to serial depletion – the multifaceted decline of crab and shrimp fisheries in the greater Gulf of Alaska. *Reviews in Fish Biology and Fisheries*, **8**: 117–76.

Pauly, D., Christensen, V., Dalsgaard, J., Froese, R. and Torres, F. (1998) Fishing down marine food webs. *Science*, **279**: 860–3.

Pauly, D., Christensen, V. and Walters, C. (2000) Ecopath, Ecosim and Ecospace as tools for evaluating ecosystem impact of fisheries. *ICES Journal of Marine Science*, **57**: 697–706.

Pauly, D., Christensen, V., Guenette, S., Pitcher, T.J., Sumaila, U.R., Walters, C.J., Watson, R. and Zeller, D. (2002) Towards sustainability in world fisheries. *Nature*, **418**: 689–95.

Pauly, D., Alder, J., Bennett, E., Christensen, V., Tyedmers, P. and Watson, R. (2003) The future for fisheries. *Science*, **302**: 1359–61.

Perry, R.I., Walters, C.J. and Boutillier, J.A. (1999) A framework for providing scientific advice for the management of new and developing invertebrate fisheries. *Reviews in Fish Biology and Fisheries*, **9**: 125–50.

Piet, G.J. and Jennings, S. (2005) Response of potential fish community indicators to fishing. *ICES Journal of Marine Science*, **62**: 214–25.

Pikitch, E.K., Santora, C., Babcock, E.A., Bakun, A., Bonfil, R., Conover, D.O., Dayton, P., Doukakis, P., Fluharty, D., Heneman, B., Houde, E.D., Link, J., Livingston, P.A., Mangel, M., McAllister, M.K., Pope, J. and Sainsbury, K.J. (2004) Ecosystem-based fishery management. *Science*, **305**: 346–7.

Prince, J.D. (2003) The barefoot ecologist goes fishing. *Fish and Fisheries*, **4**: 359–71.

Punt, A.E. and Hilborn, R. (1996) *Biomass Dynamic Models. User's Manual.* FAO Computerized Information Series (Fisheries), 10.

Punt, A.E. and Hilborn, R. (1997) Fisheries stock assessment and decision analysis: the Bayesian approach. *Reviews in Fish Biology and Fisheries*, **7**: 35–63.

Punt, A.E. and Hilborn, R. (2002) *BAYES-SA – Bayesian Stock Assessment Methods in Fisheries – User's Manual.* FAO Computerized Information Series (Fisheries), 12.

Quinn, T.J., II and Collie, J.S. (2005) Sustainability in single-species population models. *Philosophical Transactions of the Royal Society B*, **360**: 147–62.

Ricker, W.E. (1975) Computation and interpretation of biological statistics of fish populations. *Bulletin of the Fisheries Research Board of Canada*, **191**.

Roberts, C.M., Andelman, S., Branch, G., Bustamante, R.H., Castilla, J.C., Dugan, J., Halpern, B.S., Lafferty, K.D., Leslie, H., Lubchenco, J., McArdle, D., Possingham, H.P., Ruckelshaus, M. and Warner, R.R. (2003) Ecological criteria for evaluating candidate sites for marine reserves. *Ecological Applications*, **13**: S199–S214.

Rose, K.A. and Cowan, J.H., Jr (2003) Data, models, and decisions in U.S. marine fisheries management: lessons for ecologists. *Annual Reviews in Ecology and Systematics*, **34**: 127–51.

Rose, K.A., Tyler, J.A., SinghDermot, D. and Rutherford, E.S. (1996) Multispecies modeling of fish populations. In: B.A. Megrey and E. Moksness (Eds) *Computers in Fisheries Research*. Chapman and Hall, London.

Rosenberg, A.A. (2003) Managing to the margins: the overexploitation of fisheries. *Frontiers in Ecology and the Environment*, **1**: 102–6.

Rueda, M. and Defeo, O. (2003) Linking fishery management and conservation in a tropical estuarine lagoon: biological and physical effects of an artisanal fishing gear. *Estuarine, Coastal and Shelf Science*, **56**: 935–42.

Russ, G.R., Alcala, A.C., Maypa, A.P., Calumpong, H.P. and White, A.T. (2004) Marine reserve benefits local fisheries. *Ecological Applications*, **14**: 597–606.

Sainsbury, K.J. (1988) The ecological basis of multispecies fisheries and management of a demersal fishery in Tropical Australia. In: J. Gulland (Ed.) *Fish Population Dynamics*. John Wiley and Sons, New York.

Sainsbury, K.J., Campbell, R.A., Lindholm, R. and Whitelaw, A.W. (1997) Experimental management of an Australian multispecies fishery: examining the possibility of trawl-induced habitat modification. In: E.L. Pikitch, D.D. Huppert and M.P. Sissenwine (Eds) *Global Trends: Fisheries Management*. American Fisheries Society Symposium, 20, Bethesda, Maryland.

Sainsbury, K.J., Punt, A.E. and Smith, A.D.M. (2000) Design of operational management strategies for achieving fishery ecosystem objectives. *ICES Journal of Marine Science*, **57**: 731–41.

Sale, P.F., Cowen, R.K., Danilowicz, B.S., Jones, G.P., Kritzer, J.P., Lindeman, K.C., Planes, S., Polunin, N.V.C., Russ, G.R., Sadovy, Y.J., Steneck, R.S. (2005) Critical science gaps impede use of no-take fishery reserves. *Trends in Ecology and Evolution*, **20**: 74–80.

Schaefer, M.B. (1954) Some aspects of the dynamics of populations important to the management of commercial marine fisheries. *Bulletin of the Inter-American Tropical Tuna Commission*, **1**: 27–56.

Seijo, J.C., Defeo, O. and Salas, S. (1998) *Fisheries Bioeconomics. Theory, Modelling and Management*. FAO Fisheries Technical Paper, 368.

Shepherd, J.G. (1982) A versatile new stock-recruitment relationship for fisheries, and the construction of sustainable yield curves. *Journal du Conseil International pour l'Exploration de la Mer*, **40**: 67–75.

Smith, S.J. and Sainte-Marie, B. (2004) Biological reference points for invertebrate fisheries: introduction. *Canadian Journal of Fisheries and Aquatic Sciences*, **61**: 1303–6.

Sparre, P. (1991) Introduction to multispecies virtual population analysis. *ICES Marine Science Symposium*, **193**: 12–21.

Sparre, P. and Venema, S.C. (1998) *Introduction to Tropical Fish Stock Assessment, Part 1: Manual*. FAO Fisheries Technical Paper 306/1 (Rev. 2).

Tegner, M.J. and Dayton, P.K. (2000) Ecosystem effects of fishing in kelp forest communities. *ICES Journal of Marine Science*, **57**: 579–89.

Thompson, W.F. and Bell, F.H. (1934) *Biological Statistics of the Pacific Halibut Fishery. 2. Effect of Changes in Intensity upon Total Yield and Yield Per Unit of Gear*. International Pacific Halibut Commission, Report 8.

US Commission on Ocean Policy (2004) *Preliminary Report on the US Commission on Ocean Policy, Governor's Draft*. Washington DC.

Walters, C.J. (1986) *Adaptive Management of Renewable Resources*. MacMillan Press, New York.

Walters, C.J. (1997) Challenges in adaptive management of riparian and coastal ecosystems. *Conservation Ecology*, **1** (2): 1. http://www.consecol.org/vol1/iss2/art1

Walters, C.J. and Holling, C.S. (1990) Large-scale management experiments and learning by doing. *Ecology*, **71**: 2060–8.

Wiber, M., Berkes, F., Charles, A. and Kearney, J. (2004) Participatory research supporting community-based fishery management. *Marine Policy*, **28**: 459–68.

Part II

Small-scale and Artisanal Fisheries

Chapter 2

Successes, Lessons, and Projections from Experience in Marine Benthic Invertebrate Artisanal Fisheries in Chile

Juan Carlos Castilla, Stefan Gelcich, and Omar Defeo

Introduction

There is a crisis in the world's oceans. Marine fisheries are in trouble (Botsford et al. 1997; Pauly et al. 2003). This is occurring in the context of what many fishery scientists and national, regional and international organizations thought was a well-developed and tuned 'fisheries science–management scenario'. The crisis refers not only to the depletion of oceanic and coastal resources, but has also rippled into modifications of natural ecosystems (Pauly et al. 1998; Myers and Worm 2003). There has been a series of reviews focusing on the cause and solutions for the crisis. For instance, Pauly and colleagues (2002) indicated that fisheries have rarely been sustainable and that serial depletions by fisheries have long been masked by improved technology, geographic expansion, and exploitation of previously spurned species low in the food web. Furthermore, the solution lies in reducing fishing capacity to appropriate levels and the reduction of subsidies. In addition, zoning the oceans into unfished marine reserves (Roberts and Polunin 1993; Castilla 1999; Roberts et al. 2001, 2002; Palumbi 2003) and areas with limited or controlled levels of fishing effort would allow sustainable fisheries for the future (Castilla and Defeo 2001).

We believe the fishery crisis is more multifaceted than has been portrayed and that key elements such as the role of social sciences, specifically the role of fishers and common property rights, have not been duly highlighted in the proposed solutions. For instance, small-scale coastal inshore fisheries, mainly based on artisanal fishers, including invertebrate and algae divers, small-scale cooperatives catching finfish, and subsistence fishers, have their own challenges and cannot be accounted for in the same way as the highly industrialized fisheries, trans-boundary highly migratory fisheries, or bottom-trawling fisheries. These complexities of the fishery crisis and its multifaceted nature need to be recognized if specific scientific management orientated advice on resource sustainability are to be given (Defeo and Castilla 2005).

In this review we focus on the small-scale dive-artisanal fishery for benthic resources in Chile and describe open access (pre-policy) scenarios, new legislation introduced in the early 1990s, the co-management strategies developed and the results following the legalization and granting of territorial user rights for fisheries (TURFs) implemented in the country. We emphasize the lessons we can extract from this fishery policy process regarding collaboration between fishers, fishers' technical consultants, scientists, and state institutions. Finally, we discuss the need to implement experimentation ('learning by doing') management approaches and the need for spatially-explicit networks where management and conservation practices can be merged.

Chilean benthic artisanal fisheries

Artisanal fisheries in Chile supply a significant fraction of high-valued finfish, benthic invertebrate, and algal resources, much of which is exported. For instance, in 2000, 110 000 metric tons (MT) of shellfish were landed, accounting for approximately US$50 million in export revenues (Sernapesca 2004). This activity is also important from a social and employment perspective as there has been an explosive increase in the artisanal fisheries' workforce over recent years from approximately 17 000 registered fishers in 1975 to over 48 000 in 2000 (San Martin 2001). Out of these, 22 600 fishers are registered as divers or coastal intertidal and shallow subtidal food-gatherers, which mainly exploit benthic shellfish as part of their livelihood.

Benthic artisanal fishers extract most species of shellfish through manual collection during low tides, through skin diving and using semi-autonomous or 'hooka' diving gears (Bustamante and Castilla 1987; Castilla and Defeo 2001). Hooka gear (known in Chile as 'material'), includes: a 5–9 m wooden or glass fiber boat , a 10–45 hp outboard motor, air compressor and hoses and a crew of three or four, including a boatman, one assistant, and one or two divers (Fig. 2.1). Diving trips are normally run during the day, usually less than 25 km from the base port and diving occurs no deeper than 25–30 m (Castilla and Defeo 2001).

At least 60 benthic species of invertebrates are exploited in Chile, including crustaceans, mollusks, sea urchins, and tunicates (Bustamante and Castilla 1987). Economically, the most important benthic Chilean artisanal resources are the murcid snail *loco* (*Concholepas concholepas*), the *erizo* (red sea urchin, *Loxechinus albus*), and *lapas* or keyhole limpets (several species of the genus *Fissurella*). The *loco* is currently the most important economic shellfish in Chile,[1] while *erizos* and *lapas* are relatively important in terms of landings in southern and northern regions respectively (Sernapesca 2004).

[1] King crab (*Lithodes antarticus*) and crayfish (*Jasus frontalis*) are more expensive but are restricted to very local geographical areas and therefore do not have such a big influence on a national scale.

Figure 2.1 Diver with hooka equipment preparing to extract keyhole limpets (*lapas*) at El Quisco, Central Chile.

As *loco* is the single most economically important shellfish in Chile we will refer to its fishery, throughout this chapter, as a basic case study illustrating the implementation of changes in Chilean fishery legislation. We divide this fishery into two main stages: a pre-policy and policy process stage (Fig. 2.2). The pre-policy stage includes an initial period of landings exclusively for domestic consumption, followed by an increase in landings in the mid-1970s due to the fact that *loco* became a major export product, mainly to Asian countries (Castilla and Defeo 2001; Leiva and Castilla 2002). Due to the rise in fishing effort, the *loco* experienced overexploitation and strong socio-economic derived conflicts in the late 1980s. These conflicts have been the driving factors for the changes in Chilean fishery legislation with reference to benthic resources. The changes were institutionalized in the 1991 Fisheries and Aquaculture Law (FAL), which we review in the policy process section of the chapter.

Pre-policy stage

During the pre-policy stage, the *loco* fishery showed three phases. The first (1960–1975) was characterized by small landings of around 2000–6000 MT, used mainly

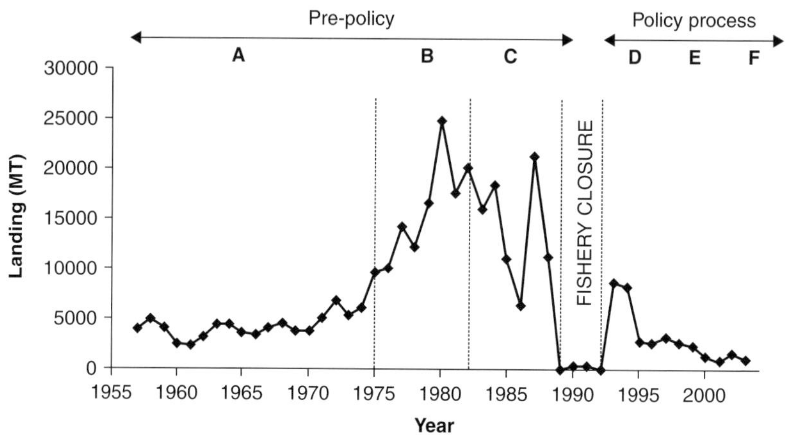

Figure 2.2 *Loco* landings for the period 1957–2003. Three pre-policy and three policy process phases of the fishery are identified. These are: (a) domestic consumption phase, (b) the export phase, (c) regulation problem phase, (d) the pre-development policy phase, (e) the development policy phase, and (f) the maturation phase.

for domestic consumption (Fig. 2.2A). Then, Chile adopted a neo-liberal policy framework; this, together with the implementation of an aggressive exchange rate policy in 1974/75, substantially improved fishing export earnings and produced the necessary incentives for Chile to become the region's leading fish and shellfish exporter (Thorpe et al. 1999). For instance, between 1976 and 1981 *loco* landings abruptly increased, reaching a peak of 24 800 MT in 1980 (Fig. 2.2B). According to Vial (1991), fisheries were the fastest growing economic sector in Chile during the 1980s.

Demand for shellfish, mainly *loco* and sea urchins, was constantly increasing from Asian markets. Local credit programs created by the government meant favorable investment opportunities for new boats, diving gears, and processing plants, thereby stimulating product demand even further (Schurman 1996). At that time, most fisheries in Chile operated under an open access policy and small-scale artisanal fishers, although based at specific artisanal coves or *caletas* (Castilla et al. 1998), used to migrate along the country. As the new export markets grew, fishers intensified their migrations to take advantage of the new opportunities. Buyers began recruiting groups of divers from *caletas* and transporting them to distant fishing grounds targeting high valued species such as the *loco* (Meltzoff et al. 2002). Thousands of divers moved around Chile, mainly to the southern regions, sparking fights between locals and outsiders in what was named at the time the '*loco war*' or '*loco fever*' (Reyes 1988; Meltzoff et al. 2002). From 1982 to 1988, *loco* landings decreased (Fig. 2.2C), probably due to overexploitation, resulting in a complex series of management steps (Castilla and Fernández 1998). The governmental fisheries service could not prevent clandestine catches and smuggling. Reyes (1991, cited in

Meltzoff et al. 2002) stated that smuggling efforts even included labeling *loco* as frozen strawberries for export purposes.

According to a fisheries department official, the open-access state of benthic resource fishing in Chile and the newly opened export markets, were enough to lead to a 'tragedy of the commons' (interview of San Martin in Gelcich 2005). Consequently the *loco* fishery was completely closed between 1989 and 1992 (Fig. 2.2) until the government developed a new regulatory framework: the Fisheries and Aquaculture Law (FAL) No 18.892 (Decreto 430, approved in September 1991) (Ley de Pesca y Acuicultura 1991).

Policy process

The 1991 FAL included the allocation of TURFs to artisanal fishers in what are known as management and exploitation areas for benthic resources (MEABRs). This process was not straightforward. In the following sections we describe the policy process that we divide into three main phases: pre-development, development, and maturation (Fig. 2.2D, E, F).

Pre-development phase

This early phase in the policy process began in 1988 with the implementation of experimental no-take zones by marine ecologists and fisher syndicates, christened as 'natural shellfish restocking via rotational exploited areas' (Castilla 1988). At this time there was no formal related policy and fishers themselves established regulations for the management of these informal management areas. The basic ecological and fishery concepts pertaining to shellfish restocking came from the experience of a single and small 5 ha intertidal and subtidal system at the University No-take Coastal Reserve (Estación Costera de Investigaciones Marinas (ECIM), Las Cruces, 33° 33' S, 71° 36' W) in central Chile. Research at this station between 1982–1988 showed that if shellfish extraction was prevented, then benthic resources such as the *loco*, sea urchin, keyhole limpets, and kelp, may be restored via 'natural seeding', over a period of about 3–5 years (Castilla 1988, 1989, 1990). Inside the reserve the commercial resources reached greater abundances (up to 10 times) and larger sizes compared with nearby exploited fishery grounds (Castilla and Durán 1985; Durán and Castilla 1989; Castilla 1989, 1990, 1999).

It is important to highlight that Castilla et al. (1998) showed that the number of *loco* caught per hour of diving, or catch per unit effort (CPUE), was significantly higher in the 1993 winter season harvest within the Caleta El Quisco experimental natural restocking area, compared to historical fishing grounds in the nearby *caletas* of Algarrobo and Las Cruces (Table 2.1). The mean size of the individuals was also higher in El Quisco, and therefore prices were higher. Searching and traveling time

Table 2.1 Catch per unit effort (CPUE), size range, and prices of *loco* from an experimental no-take zone (Caleta El Quisco) and historical fishery grounds (Caleta Algarrobo and Caleta Las Cruces) in central Chile during the winter of 1993.

Locality (Caleta)	Mean CPUE (*locos*/hour)	Size (cm)	Unit price (US$)
El Quisco (experimental restocking zone)	148.3 ± 40.70	10.7–11.8	1.39–1.94
Algarrobo	64.89 ± 25.83	10.3–10.8	0.91–1.17
Las Cruces	30.93 ± 10.17	10.3 – 10.8	0.86–1.84

for diving were significantly reduced within the restocking area as compared to historical fishing grounds (Castilla et al. 1998).

These successful pilot natural restocking experiments were used as models for the implementation of TURFs in the 1991 FAL. Nevertheless, details as to how marine tenure was going to be given to fishers was lacking until 1997, when the decrees for legalizing TURFs in the form of MEABRs were approved. Between the publication of the FAL in 1991 and the development of enabling regulations in 1997, fisher syndicates who wished to engage with MEABR policy could only do so on an informal basis. During these years of informality, any fishery syndicate or cooperative wishing to conserve and manage their resources used to be helped and supported by teams of university-linked biologists and marine technicians.

Development phase

During the development phase, the policy gained support from the Fisheries Undersecretary and was officially implemented. TURFs for benthic fisheries were finally legalized and given to artisanal fisher syndicates in the form of MEABRs. This stage began in 1997 with the government's approval of the regulations, which established the proceedings and technical criteria for managing MEABRs. It continued up to the year 2000 when MEABRs were already being legally harvested and more than 230 MEABR decrees were being established (Subpesca 2004). The co-management arrangements behind the MEABR policy, in place since 1997, establish that in order to be awarded a MEABR, a fisher syndicate union or association must co-finance a baseline study for their MEABR from which resource catch quotas and management plans are created. The arrangement also establishes that fishers must pay for yearly follow-up assessments to certified consultants in order to determine changes in the total allowable catches and the evolution of the management extractive plan. In addition to this, a yearly management area permit must be paid once the area has been harvested for four years (Subpesca 2002).

During the development phase, the state committed to promote, popularize, and co-finance the implementation of MEABRs, pushing to formalize them for every fishery syndicate in the country. Up until the year 2000, approximately US$1 million had been spent on co-financing MEABR studies (Subpesca 2002). As a government official stated: 'MEABRs were to be the basis for managing benthic fisheries with special emphasis on the lucrative *loco*'. The government's MEABR approach during this stage aimed at generating an increased sense of exclusive use and ownership among fishers (Meltzoff et al. 2002). This created great expectation among fishers as their perception was that they were receiving the equivalent of a 'land grant', which in this case had the form of a highly productive subtidal area (Bernal et al. 1999). Additional strategies that favored the success in implementation of government-supported co-management included publicizing the government's desire to collaborate, concentrating efforts on cohesive localities (Jentoft 1989) and a decision to declare a ban on *loco* extraction, apart from MEABRs with approved management plans.

During these 'development' years, 206 MEABR decrees were processed, 93 of them during 1997, which accounted mainly for fisher communities working with universities and experts on managing areas during the pre-development phase, and who were eager to get formal user rights. The National Fishing Service and private consultants supported the remaining ones. Species that were included in the MEABR plans varied between fishery communities although *loco*, sea urchin, and keyhole limpets remain as the most important (Fig. 2.3).

It is important to highlight that the Fisheries Undersecretary sees these years as a great success for benthic resources, particularly in northern and central Chile. Fisher communities were self-organizing in syndicates and applying for MEABRs, creating partnerships with the government, universities, and consultants. In this way, artisanal fishing coves were being consolidated responding to government

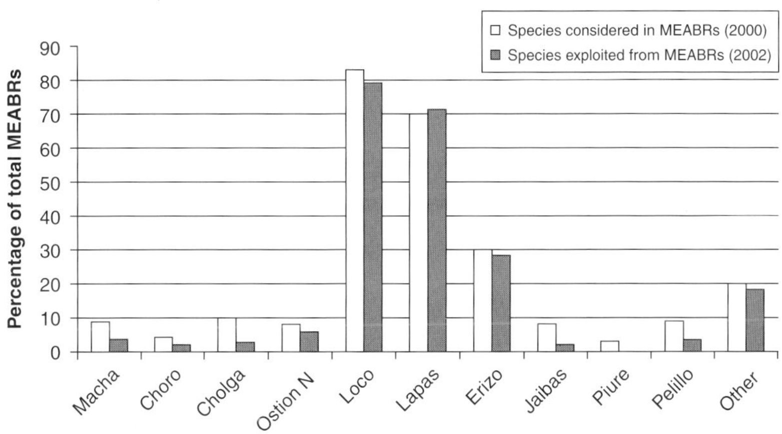

Figure 2.3 Percentage of MEABR (out of 206 areas) that included different benthic species as part of their management plan in the year 2000. The black columns represent the percentage of areas (out of 107) that harvested different species out of their MEABRs until 2002. *Macha* (*Mesodesma donacium*), *choro* (*Choromytilus chorus*), *cholga* (*Aulacomya ater*), *ostión del norte* (*Argopecten purpuratus*), *loco* (*Concholepas concholepas*), *lapas* (*Fissurella* spp.), *erizo* (*Loxechinus albus*), *jaibas* (*Homalaspis plana; Cancer* spp.), *piure* (*Pyura chilensis*), and *pelillo* (*Gracilaria* sp).

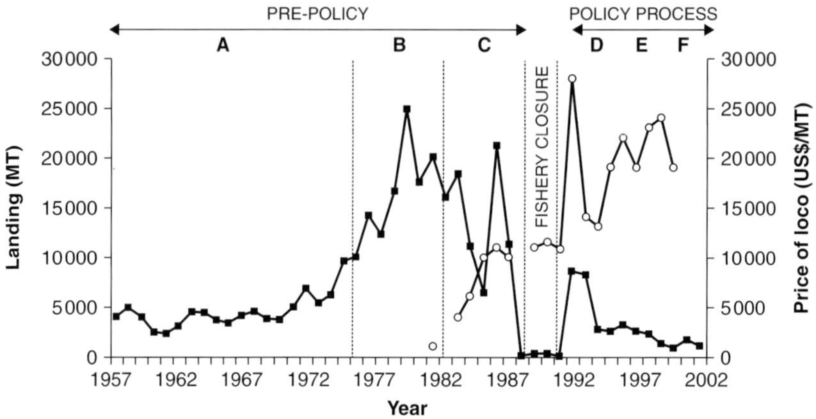

Figure 2.4 Landing (MT) and price (US$/MT) of *loco*. The squares represent landings and the circles the price paid for *loco*.

incentives and common-property theory. There was a reinforcement of fisher associations and a strengthening of leadership that led to the implementation, by fishers themselves, of surveillance procedures to stop poaching within MEABRs and to establish participatory and regulatory rules within the communities. A sense of ownership, responsibility, pride, and hope for sustainability arose among fishers engaging with the policy (Castilla and Defeo 2001). Biological and economic success of MEABRs was proclaimed through government documents (although socio-economic problems have appeared in southern Chile). These studies showed a significant increase in abundance and individual size of resources increased within MEABRs in comparison with open-access sites (Subpesca 2000). The price of *loco* was also seen to increase after the implementation of MEABRs following an inverse relationship with respect to landings (Fig. 2.4). Thus, global markets showed a willingness to pay higher prices for *loco*.

During this stage, fishers were effectively encouraged to become non-migrating, small businessmen who could earn a living through the sale of self-managed resources, and co-management was promoted as the innovative change that helped this process by sharing power between the government and fishing communities. During this development phase two main issues were resolved: (1) the consolidation of the 'pre-development' stage through the formalization of property rights, thus changing *de facto* into *de jure* measures based on biological assessments; and (2) the definition and expansion of the property rights policy concept (MEABRs) with the support and commitment of the Fisheries Undersecretary and the positive response of artisanal fisher associations.

Maturation phase

From the year 2001 to the present, fisher associations have gradually been adapting to their new lifestyle as non-migrating businessmen. By the end of 2005, MEABRs were consolidated throughout many regions of Chile: 1031 applications have been

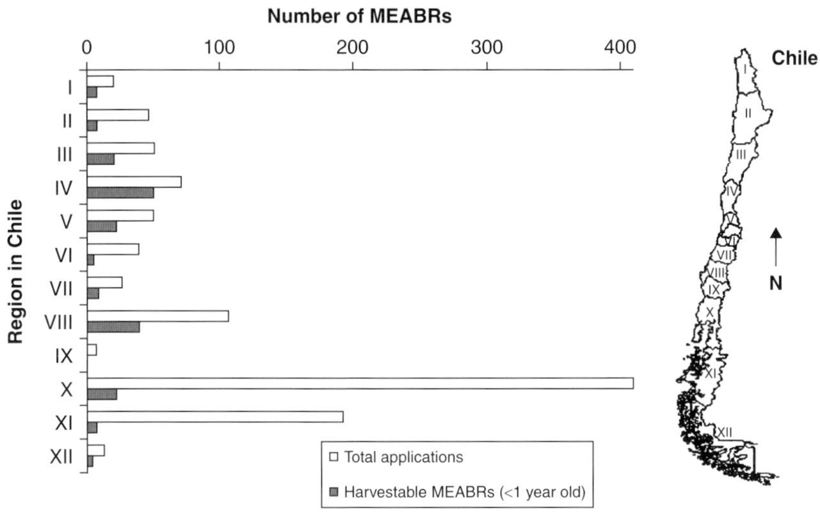

Figure 2.5 Number of MEABR applications in Chile until November 2003 separated by region. The shaded columns represent the number of areas that have been officially decreed for more than a year.

processed and 547 MEABR approved decrees issued, 301 management plans have been implemented (Fig. 2.5), five areas are entering their fifth follow-up assessment, and these are therefore due to pay a management area permit for the first time (Subpesca 2004). Differences between the policy uptake speeds from different regions were observed. Regions IV and V in central Chile proved to be quite advanced with most of their MEABRs implemented and in full extraction. However, many associations of Region X in southern Chile were still in the application processes stage (Fig. 2.5).

During the maturation phase, some fisher association leaders and small-scale fishers have started to view the MEABRs as more than a marine tenure. Now they see them as a way to organize and facilitate fisheries and non-fisheries related business activities, such as tourism and seafood restaurants. An important driving force for this was the fact that MEABR collective resource extractive catches were given to fisher unions or associations (communities of place or interest) which internally decided individual or group extraction quotas. These collective harvesting catches, as opposed to individual catches, gave the right incentives for cooperation.

Fishers have also attached other important non-economic values to the existence and ownership of MEABRs, such as pride and accountability. As part of MEABR consolidation, innovative strategies that account for fishers' entrepreneurship included an attempt to sell management area resources collectively between associations, for instance in the form of a 'selling cooperative', as in the 'Pacificoop' in Region V (Gelcich, personal observation). This cooperative was formed by fishers engaging 15 organizations from central Chile that represent 1170 artisanal fishers. In 2003, the associated unions have offered about 50% of their MEABR production to the cooperative, resulting in approximately 650 000 *loco* and 50 000 kg of *lapa* for sale. According to the Pacificoop Cooperative Director: 'The ultimate objective behind this association is to consolidate MEABRs and form strategic alliances with shellfish exporters in order to get a fair price for our products' (Gelcich 2005).

It is important to highlight the fact that decisions within the fishers' cooperative are taken at three levels: (1) the assembly, which is formed by members of all the unions and has the power to resolve and sanction decisions; (2) the directorates, who are trusted by the assembly to take administrative decisions; and (3) the commissioners, who take operational decisions concerning discipline, commercialization, and are accountable to the directorates and the assembly. This initiative, although so far unique in the country, shows how the MEABR policy has opened new ways for the fishers' long-term engagement as resource stewards and how it has encouraged self-empowerment to solve fishery problems. In this particular case, the avoidance of the competition that associations were experiencing between themselves to sell resources, produced price reductions at the level of *caletas*.

In addition to self-organization and empowerment, another specific point of interest maturing during this stage relates to the change of attitudes of fishers with respect to management and conservation of resources. Research in the fishing cove of Los Vilos, which included three syndicates, evidenced that 94% of surveyed fishers agreed or strongly agreed with the statement: 'Fishers have a duty to conserve marine resources for the next generation' and all agreed with the statement: 'Improving earnings from resources by improving quality'. In this way, fishers have a self-perception of their role in marine conservation that was generated through their experience with co-management. Despite these encouraging results, it is important to be cautious because the positive attitude towards conservation might be related to recent memories of intervention when conservation affinity was synonymous with gaining donor projects (Gelcich et al. 2005a). This especially took place during the development phase when community-based development and conservation projects of MEABRs were greatly supported and fishers learnt that cooperation can bring status and also additional sources of income from eager national or international non-governmental organizations (NGOs) or agencies.

During these last three years, the MEABR policy process has matured and self-management has proved to be effective in many regions of Chile. Achievements concerning policy, sustainability, fishers' knowledge, and inter-sector collaboration encourage stakeholders to think that the exclusive use associated to common property user rights has generated the necessary incentives and institutions. Nevertheless, not everything is as positive as it seems with this marine, common-property-based policy intervention as discussed below (see also Gelcich et al. 2005a; Gelcich et al. in press).

Problems with MEABR policy

The problems associated with open access and the traditional command and control approach to fisheries, led the search for MEABRs as a management alternative under which the responsibility for benthic resource sustainability is shared by those who have an interest in its success, including government and fishers. To date, perceptions of the success of the MEABR policy have been largely dependent on resource restoration data within MEABRs and the official statistics on MEABR adoption. The large number of MEABR applications has been taken as evidence that fishers are organizing and adopting livelihoods as non-migrating businessmen, which was one of the original policy aims. However, a review of MEABR application statistics and the accompanying official documentation does not reflect the current and future problems of the policy.

There is a need to have a better understanding of the effect of management decisions on the social, cultural, and economic pressures on the people being managed (Kaplan and McCay 2004). In Chile, the human dimension of the management process has not been given the attention it deserves and no studies have examined the social consequences of the policy. This is unfortunate, as the attitudes and perceptions of key stakeholders influence their engagement with policies and consequently should increasingly be fed into policy developments (Aipanjiguly et al. 2003). Primary analysis of human dimensions shows the existence of important sources of conflict with the MEABR implementation. Many of them deal with the fact that open-access fishing sites are becoming scarce and overexploited (Gelcich et al. 2005b). This has important consequences depending on the type of fisher engaged in MEABR. From a few case studies, such as El Quisco and Matanzas, we suspect that around only 20–30% of fishers' annual earnings come from MEABR, with the rest coming from finfish or other resources. Nevertheless, there is no data to prove this is the case throughout different regions of Chile. In Los Vilos, for example, diving in the increasingly scarce open-access areas is as much as ~65% of fishers' income. This is why fishers who have historically maintained a livelihood as divers say that this lifestyle is at risk.

For example, one fisher stated: 'This law didn't analyze the secondary effects. There is an indiscriminate extension of the areas. They (fisheries department) say that the sea cannot all be used by MEABR. But there is nearly nowhere to go and dive, everything is asked for and the little historical zones (open-access) left have collapsed... Divers have no where to dive....' (Fisher of Los Vilos, 2002 in Gelcich 2005). Fishers like the one quoted above define themselves as divers and, as such, they are reticent to adopt a livelihood as fin-fishers during the periods when MEABRs are closed for extractions. They therefore seek an optimization of MEABRs by including multispecies management plans. Additionally, due to the lack of open-access diving grounds, conflict between divers is rising, notably as a result of illegal poaching within the MEABRs. This is weakening social bonds between fishers and raises the costs of enforcing rights over MEABRs. This is a hard to solve problem especially if we recognize that fishers who are poaching within MEABRs feel it is their right to do so: '... They (unions who applied for MEABRs first) not only compete with us in the historical (open access) areas but they started extracting resources from these sites to repopulate their own management areas, it was like putting things in a bag, every organisation did this, and exploited historical sites. Therefore we stayed empty handed... we had to go and get what was ours inside their areas... ' (Fisher of Los Vilos, 2003 in Gelcich 2005).

In conclusion, we must not forget that a fine-tuning period to the policy may last several years and could be accompanied by conflict and low or negative financial returns due to poaching and market adjustments. This transition could have stronger effects in unions where occupational mobility of fishers is restricted to diving and to regions where *loco* sizes and abundances are historically lower. Therefore, new research regarding economic valuations, historical access rights, and the need for income diversification are necessary steps in order to achieve real practical benefits for conservation and fisher communities if MEABRs are to be successful.

Discussion

Perspectives for the future

MEABR policy has incorporated many of the classic common-property regime design principles (Ostrom 1990). In this section we discuss two ideas that could contribute to the long-term sustainability of benthic resource management through MEABRs. The first includes the need for further devolution of power to fisher communities as a requirement of a multispecies ecosystem approach within MEABRs and the second considers the need of marine protected areas within the MEABR network.

Ecosystem management

Scales for the implementation of operational and institutional management arrangements should correspond with the scales of the population movements and dynamics of the benthic stocks. In Chile, the boundaries selected for each MEABR are generally based on the physical structure of the coastline. Moreover, MEABRs were designed by giving high priority to the *loco*. Even though fishers, managers, and scientists could easily observe and understand the limits of *caletas*, these are not the boundaries defined by the life-history habits of benthic invertebrates, particularly of mobile species that offer a new challenge in the design and implementation of management tools (Fernández and Castilla 1997, 2000). This implies a trade-off between conflicting criteria given by the scale of implementation of a given area-specific management measure and the probability of response of the benthic species to fishing pressure (Castilla and Defeo 2001).

An ecosystem approach incorporating interspecific interactions and physical environmental influences would be more realistic in managing MEABRs than the currently used single-stock management approach in the sense that it considers competitive release or a diminution of predation effects on species with the highest value in the market (Botsford et al. 1997). Nevertheless, this holistic approach to restoring populations within MEABRs is difficult to implement, as fishing effort becomes more heterogeneous and therefore more difficult to link to specific stocks, as there is insufficient information about local ecosystems and it implies a large social and economic cost (Castilla and Defeo 2005). Therefore, ways forward to achieve resource sustainability should not be limited to a biophysical ecosystem management approach but should include incentives for the effective governance of resources.

In the Chilean MEABR experience there is still the need to include knowledge generated from the resource management experience, explicitly within future management practices, as an ever-evolving feedback process (Gelcich 2005). As suggested for other natural resource domains, research and development can no longer be the exclusive domain of natural/social scientists (Olsson and Folke 2001). Fishers should be able to adapt policy to their own conditions through experimentation. Researchers, both social and natural scientists, and managers could gain from fisher experimentation because they will observe the results of numerous experiments over a wide range of conditions between and within years, allowing them to generalize about outcomes of experiments (i.e. meta-analysis), and to develop or amend theory accordingly (Edwards-Jones 2001; Johannes 2002; Gelcich et al. in press). It should be highlighted, however, that manipulation of species abundance should be conducted with scientific advice because the complex dynamics in multispecies assemblages precludes a synthetic forecast of the ecological outcome of these manipulations (Castilla 2000).

In our view, devolution of further management responsibility to fisher communities that encourages further participation in the planning, experimentation, and surveillance of management measures is a promising short to mid-term solution to current problems with Chilean MEABRs and could eventually provide a starting point for socio-economic and biophysical approaches to ecosystem based management.

Marine protected area network

Castilla (2000) suggested that spatial linkages, promoting cross-fertilization between permanently closed, managed, and unmanaged areas may represent a front line example for the joint achievement of conservation and management goals. This spatial model, enhancing connectivity between different locations and environments along the Chilean coastline, is presently being developed. For instance, the first Chilean multiple-use marine and coastal protected area has recently been established in far southern Chile (Parque Francisco Coloane, Fernández and Castilla 2005) and the inauguration of two new marine and coastal protected areas will soon follow in northern and southern Chile. Their locations will be necessarily flanked by several MEABRs already established and in operation. Therefore, a natural connectivity between areas for management and conservation purposes should soon occur. It has already been demonstrated for northern and central Chile (Castilla and Rho 1997; Manríquez and Castilla 2001) that marine reserves and no-take areas act to replenish benthic resources, leading to increased larval production and also as 'seeding grounds' for nearby overexploited fishing grounds. This novel strategy may serve to abolish the traditional confrontations between conservation and management measures and better communicate pure management from a biological and fishery perspective with bio-economic and social science perspectives (Castilla 2000).

Conclusions

The rationale for introducing the co-management arrangement in the form of MEABR was the fact that these fisheries were almost overexploited. In this respect, co-management is a form of crisis management that anticipates sustainability, efficiency, and equity among the resources and its users. There is evidence that MEABRs have had several positive repercussions for marine benthic fisheries in Chile:

(1) MEABRs were successful in increasing bio-economic fishery indicators (benthic stock abundance and CPUE levels, unit prices, and individual sizes of targeted stocks) when compared with either past value during the declining phase of the fisheries or with open-access grounds (Castilla 1994; Castilla and Fernández 1998; Castilla et al. 1998).
(2) From a socio-economic point of view, MEABRs have initially increased economic returns received by many local communities (Castilla 1997; Castilla et al. 1998). The elucidation of ownership was critical: enhancement of high-value invertebrate stocks under a specified ownership regime through MEABRs allowed fishers to absorb an important share of the economic benefits. The legitimacy of co-management and the perception of ownership by small-scale fishers override the expectations of the benefits to be derived from shellfish extraction (Castilla 1994). Methods of restricting access to the benthic stocks, together with legislation to protect rights of fishing of those fishers that participate in the management plan (Castilla 1994; 1997) were successfully implemented and applied. The assignment of fishing grounds to well-defined groups of fishers also represents recognition of the role of small-scale communities in conservation and management (Castilla and Defeo 2001) and has given incentives for fishers' engagement in non-fisheries' business activities such as tourism and seafood restaurants.
(3) From a scientific point of view, improvement of information, including data from small-scale fishers, consolidated the management schemes even in data-poor situations when stock-assessment methodologies are usually difficult to apply. The cross-linkage between experimental management protocols and the active participation of fishers was successful in improving knowledge of fishery indicators. This helped identify and compare the conditions and indicator values in pre-policy phases to those achieved during the policy process (Fig. 2.2), thus suggesting values that could be followed to maintain stock indicators at desirable levels.

Concerning the management system, the legitimization of the participation of fishers in the management process was seen as the only way to promote compliance with regulations, thus attenuating the usually high enforcement and policing costs occurring in artisanal fisheries (Seijo et al. 1998). We are aware that the issue of limiting access to the fishery is still controversial under the philosophy that no one should be prevented from extracting resources (Mahon et al. 2003). Artisanal fisheries function as an economic buffer and as a safety valve for thousands of people moving in and out of the fisheries according to the opportunities in the national economies (Jul-Larsen et al. 2003). Limiting numbers of fishers through zoning of coastal areas can endanger the stability of this situation as well as the role that diver livelihoods play in society.

Despite the above criticism, history demonstrates that the implementation of co-management strategies together with an allocation of user rights have been necessary for benthic invertebrate management in Chile. We advocate the importance of maintaining policy process through further socio-economic and biophysical research aimed at feeding back to ecosystem management in order for marine benthic management policies to keep changing. Finally, we suggest that the combination of coastal MEABRs and no-take areas, including reserves and parks, may lead to a strategy where management and conservation practices can be merged, resulting in a novel and sustainable fishery conservation approach.

References

Aipanjiguly, S., Jacobson, S. and Flamm, R. (2003) Conserving manatees: knowledge, attitudes, and intentions of boaters in Tampa Bay, Florida. *Conservation Biology*, **17**: 1098–105.

Bernal, P., Oliva, D., Aliaga, B. and Morales, C. (1999) New regulations in Chilean fisheries and aquaculture: ITQ's and territorial user rights. *Ocean and Coastal Management*, **42**: 119–42.

Botsford, L.W., Castilla, J.C. and Peterson, C.H. (1997) The management of fisheries and marine ecosystems. *Science*, **277**: 509–15.

Bustamante, R. and Castilla, J.C. (1987) The shellfishery in Chile: an analysis of 26 years of landings (1960–1985). *Biología Pesquera (Chile)*, **16**: 79–97.

Castilla, J.C. (1988) Earthquake-caused coastal uplift and its effects on rocky intertidal kelp communities. *Science*, **242**: 440–3.

Castilla, J.C. (1989) Un seguimiento de la actividad pesquera artesanal en dos caletas de Chile central. *Memorias del Simposio Internacional de los Recursos Vivos y las Pesquerías en el Pacifico Sudeste, Chile. Viña del Mar, Chile, 9–13 May.*

Castilla, J.C. (1990) Clase magistral: importancia y proyección de la investigación de ciencias del mar en Chile. *Revista Biología Marina (Chile)*, **25**: 1–18.

Castilla, J.C. (1994) The Chilean small scale benthic shellfisheries and the institutionalization of new management practices. *Ecological International Bulletin*, **21**: 47–63.

Castilla, J.C. (1997) Environmental regulations and the hazards of risk assessment: attitudes in developing countries. *Human and Ecological Risk Assessment*, **3**: 659–64.

Castilla, J.C. (1999) Coastal marine communities: trends and perspectives from human-exclusion experiments. *Trends in Ecology and Evolution*, **14**: 280–3.

Castilla, J.C. (2000) Roles of experimental marine ecology in coastal management and conservation. *Journal of Experimental Marine Biology and Ecology*, **250**: 3–21.

Castilla, J.C. and Defeo, O. (2001) Latin American benthic shellfisheries: emphasis on co-management and experimental practices. *Reviews in Fish Biology and Fisheries*, **11**: 1–30.

Castilla, J.C. and Defeo, O. (2005) Paradigm shifts needed for world fisheries. *Science*, **309**: 1324–5.

Castilla, J.C. and Durán, L. (1985) Human exclusion from the rocky intertidal zone of central Chile: The effects on *Concholepas concholepas* (Gastropoda). *Oikos*, **45**: 391–9.

Castilla, J.C. and Fernández, M. (1998) Small-scale benthic fisheries in Chile: on co-management and sustainable use of benthic invertebrates. *Ecological Applications*, **8**: S124–32.

Castilla, J.C. and Rho, E. (1997) Conservación y repoblamiento en el litoral del norte de Chile: el caso de Minera Escondida en Punta Coloso, Antofagasta, Chile. *Estudios Oceanológicos (Chile)*, **16**: 51–66.

Castilla, J.C., Manríquez, P., Alvarado, J., Rosson, A., Pino, C., Espóz, C., Soto, R., Oliva, D. and Defeo, O. (1998) Artisanal 'Caletas' as units of production and co-managers of benthic invertebrates in Chile. In: G.S. Jaimieson and A. Campbell (Eds) *Proceedings of the North Pacific Symposium on Invertebrate Stock Assessment and Management. Canadian Special Publication in Fisheries and Aquatic Science*, **125**: 407–13.

Defeo, O. and Castilla, J.C. (2005) More than one bag for the world. Fishery crises and keys for co-management success in selected artisanal Latin American shell fisheries. *Reviews in Fish Biology and Fisheries*, **15**: 265–83.

Durán, L.R and Castilla, J.C. (1989) Variation and persistence of the middle rocky intertidal community of central Chile, with and without human harvesting. *Marine Biology*, **103**: 555–62.

Edwards-Jones, G. (2001) Should we engage in farmer-participatory research in the UK? *Outlook on Agriculture*, **30**: 129–36.

Fernández, M. and Castilla, J.C. (1997) The Chilean artisanal stone crab (*Homalaspis plana*) fishery: catch trends in open access zones and the effect of management areas in Central Chile. *Journal of Shellfish Research*, **16**: 371–7.

Fernández, M. and Castilla, J.C. (2000) Recruitment of *Homalaspis plana* in intertidal habitats of central Chile and implications for the current use of management and marine protected areas. *Marine Ecology Progress Series*, **208**: 157–70.

Fernández, M. and Castilla, J.C. (2005) Marine conservation in Chile: historical perspective, lessons and the future. *Conservation Biology*, **19** (6): 1752–62.

Gelcich, S. (2005) The Human Dimensions of Co-management in Chilean Coastal Fisheries. PhD thesis, University of Wales, Bangor.

Gelcich, S., Edwards-Jones, G., Kaiser, M.J. and Watson, E. (2005a) Using discourses for policy evaluation: the case of marine common property rights in Chile. *Society and Natural Resources*, **18**: 377–91.

Gelcich, S., Edwards-Jones, G. and Kaiser, M.J. (2005b) Importance of attitudinal differences among artisanal fishers towards comanagement and conservation of marine resources. *Conservation Biology*, **19**: 865–75.

Gelcich, S., Edwards-Jones, G., Kaiser, M. and Castilla, J.C. (in press) Co-management policy can reduce resilience in traditionally managed marine ecosystems. *Ecosystems*.

Jentoft, S., (1989) Fisheries co-management: delegating government responsibility to fishermen's organisations. *Marine Policy*, **13**: 137–54.

Johannes, R.E. (2002) The renaissance of community-based marine resource management in Oceania. *Annual Reviews in Ecology and Systematics*, **33**: 317–40.

Jul-Larsen. E., Kolding, J., Overå, R., Raakjær Nielsen, J. and Zwieten, P. (2003) *Management, Co-management or No Management? Major Dilemmas in Southern African Freshwater Fisheries. 1. Synthesis Report*. FAO Fisheries Technical Paper, 426/1.

Kaplan, I.M and McCay, B. (2004) Cooperative research, co-management and the social dimension of fisheries science and management. *Marine Policy*, **28**: 257–8.

Leiva, G. and Castilla, J.C. (2002) A review of the world gastropod fishery: evolution of catches, management and the Chilean experience. *Reviews in Fish Biology and Fisheries*, **11**: 283–300.

Ley de Pesca y Acuicultura (1991) Ministerio de Economía Fomento y Reconstrucción, Gobierno de Chile. Ley N°18.892 DS N°430.

Mahon, R., Almerigia, S., McConney, P., Parker, C. and Brewster, L. (2003) Participatory methodology used for sea urchin co-management in Barbados. *Ocean and Coastal Management*, **46**: 1–25.

Manríquez, P.H and Castilla, J.C. (2001) Significance of marine protected areas in central Chile as seeding grounds for the gastropod *Concholepas concholepas*. *Marine Ecology Progress Series*, **215**: 201–11.

Meltzoff, S., Lichtensztajn, Y. and Stotz, W. (2002) Competing visions for marine tenure and co-management: genesis of a marine management area system in Chile. *Coastal Management*, **30**: 85–99.

Myers, R.A. and Worm, B. (2003) Rapid worldwide depletion of predatory fish communities. *Nature*, **423**: 280–3.

Olsson, P. and Folke, C. (2001) Local ecological knowledge and institutional dynamics for ecosystem management: a study of Lake Racken watershed, Sweden. *Ecosystems*, **4**: 85–104.

Ostrom, E. (1990) *Governing the Commons: the Evolution of Institutions for Collective Action*. Cambridge University Press, Cambridge.

Palumbi, S. (2003) Population genetics, demographic connectivity, and the design of marine reserves. *Ecological Applications*, **13**: S146–58.

Pauly, D., Christensen, V., Dalsgaard, J., Froese, R. and Torres, F. (1998) Fishing down marine food webs. *Science*, **279**: 860–3.

Pauly, D., Christensen, V., Guenette, S., Pitcher, T., Sumaila, U., Walters, J., Watson, R. and Zeller, D. (2002) Towards sustainability in world fisheries. *Nature*, **418**: 689–95.

Pauly, D., Alder, J., Bennett, E, Christensen, V., Tyedmers, P. and Watson, R. (2003) The future for fisheries. *Science*, **302**: 1359–61.

Reyes, E. (1988) Nuevo colapso de la pesquería del *Loco*. *Chile Pesquero*, **47**: 41–4.

Roberts, C. and Polunin, N. (1993) Simple solutions to managing complex fisheries. *Ambio*, **22**: 363–8.

Roberts, C., Bohnsack, J., Gell, F., Hawkins, J. and Goodridge, R. (2001) Effects of marine reserves on adjacent fisheries. *Science*, **294**: 1920–3.

Roberts, C., McClean, C., Veron, J., Hawkins, J., Allen, G., McAllister, D., Mittermeier, C., Schueler, W., Spalding, M., Wells, F., Vynne, C. and Werner, T. (2002) Marine biodiversity hotspots and conservation priorities for tropical reefs. *Science*, **295**: 1280–4.

San Martin, G. (2001) Áreas marinas costeras con derechos de uso exclusivo destinadas al manejo y explotación de recursos bentónicos en Chile.*Taller Áreas de Manejo: Experiencias y Proyecciones, Puerto Montt 16–17 Agosto*.

Schurman, R. (1996) Snails, southern hake and sustainability: neo-liberalism and natural resource exports in Chile. *World Development*, **24**: 1695–709.

Seijo, J.C., Defeo, O. and Salas, S. (1998) *Fisheries Bioeconomics. Theory, Modelling and Management.* FAO Fisheries Technical Paper, 368.

Sernapesca (2004) Servicio Nacional de Pesca. http://sernapesca.cl

Subpesca (2000) *Áreas de Manejo y Explotación de Recursos Bentónicos.* Documento de Difusión 3. Subsecretaría de Pesca. http://www.subpesca.cl

Subpesca (2002) *Concepto de Areas de Manejo y Recursos Bentónicos.* Documento de Difusión 1. Subsecretaría de Pesca. http://www.subpesca.cl

Subpesca (2004) *Estadísticas Areas de Manejo.* http://www.subpesca.cl

Thorpe, A., Ibarra, A. and Reid, C. (1999) *The New Economic Model and Fisheries Development in Latin America.* CEMARE Research Paper, 141, University of Portsmouth.

Vial, J. (1991) Reflexiones en torno a la Ley de Pesca. *Colección Estudios CIEPLAN*, CIEPLAN Santiago de Chile.

Chapter 3

From Crisis to Institutional Sustainability in the Chilean Sea Urchin Fishery

Carlos A. Moreno, Nancy Barahona, Carlos Molinet, J.M. (Lobo) Orensanz, Ana M. Parma, and Alejandro Zuleta

Introduction

Artisanal and small-scale fisheries are frequently characterized as low-investment, part-time family operations, conducted by fishers who fish close to their villages mainly for subsistence, occasionally selling or bartering the surplus at local markets. Commercial diving for sea urchins in Chile is no doubt a small-scale fishery, if gauged by the size of the fishing units, and 'artisanal' according to the country's fisheries legislation. Yet, it could not be more different from the above stereotype. Above all, this small-scale fishery has astounding dimensions along several axes. For decades the world's largest sea urchin fishery, in 2002 it contributed approximately 65% of the world's supply (Andrew et al. 2002; Fig. 3.1). Spread along the entire 36 degrees of latitude spanned by the country, from 18° S (border with Peru) to 54° S (Cape Horn), this strictly commercial fishery fuels a multimillion export-oriented industry. Most of the landings originate in remote and inhospitable regions in the fjords of southern Chile, far away from the towns and villages where the thousands of participating fishers and their families live.

Idiosyncratic and defying categorization, this fishery presents daunting challenges for assessment and management. Its complexity has led to cycles of regulatory innovation and frustrations, and occasionally to political or legal stalemate and violent confrontations. Yet ultimately, conflicts forced the government, managers, fishers, processors, and scientists to sit together (whether they liked it or not) and cooperate to find alternatives for monitoring, assessment, and management. The result has been the first fisheries management plan produced in Chile, one that was elaborated in a truly participatory context and is backed by consensus from all the parties. Here we describe how this fishery, once regarded as hopeless, is reshaping itself and discuss the serious difficulties faced.

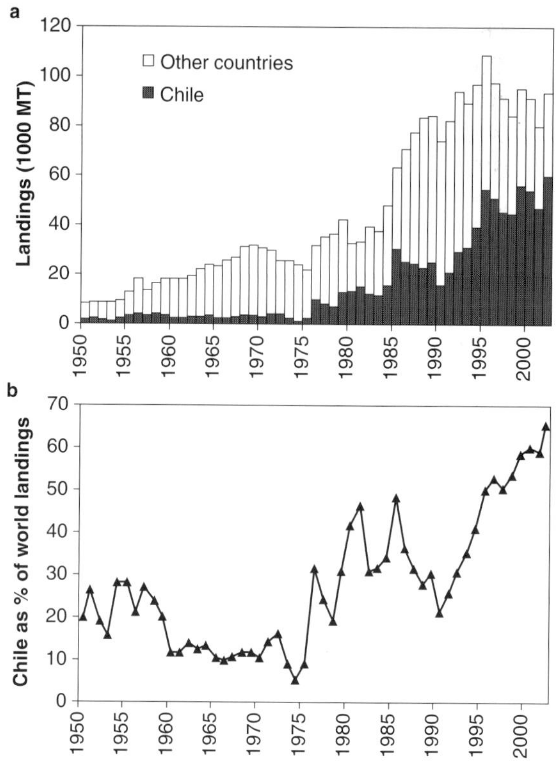

Figure 3.1 Sea urchin landings, from 1950–2002: (a) total world and Chilean landings in thousand metric tons; (b) Chilean landings as percentage of world landings.

The system

Chilean fisheries: political jurisdictions and management institutions

The Fisheries Act of 1991 (Ley General de Pesca y Acuicultura, LGPA, enacted on January 21, 1992) and its modifications, delineate the formal scaffolding of the management system. A centralized agency, the Undersecretary of Fisheries or Subpesca, has authority over the management of all fisheries, offshore and industrial as well as inshore and artisanal. Under its umbrella, the National Fisheries Service or Sernapesca, is in charge of enforcement and compiling landing statistics. New legislation would introduce important modifications, including decentralized management of benthic resources. The country is divided into 13 administrative regions (Fig. 3.2), organized by latitude from Region I in the north to Region XII in the south and a metropolitan region. Effective authority would be transferred to the zonal fishing councils (COZOPEs), where Regions X–XI

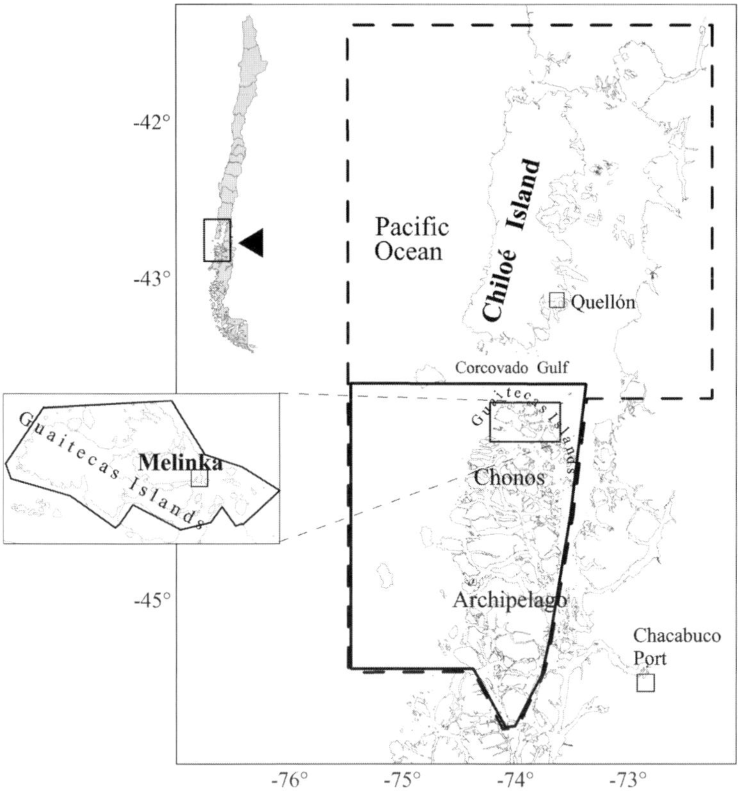

Figure 3.2 Area where the sea urchin fishery of Regions X–XI operates. Top left: map of Chilo with indication of the 13 administrative regions. Solid polygon: initial area of implementation of the management plan; dashed polygon: extended area of implementation. Detail: exclusion zone of the Guaitecas Archipelago, where the experimental fishery will be conducted.

addressed in this paper correspond to a single zone. Much of the management-oriented research is supported through a special fund, namely the Fisheries Research Fund (FIP), and the semi-governmental Institute for the Promotion and Development of Fisheries (IFOP) conducts a significant part of the work. IFOP has conducted a port-sampling program for the sea urchin fishery since 1984; monitoring consists of detailed interviews and records of the catch per trip, as well as sampling of size composition.

Access regimes

As defined in the Chilean legislation, artisanal boats are those smaller than 18 m and 50 gross tons. Artisanal fishers are the sole harvesters of benthic resources, and dredging and trawling are banned in inshore waters.

Management of benthic fisheries in Chile spans a range of access regimes, from 'open access' to individual quotas, to territorial use rights (Chapter 2). In a strict sense, there are no 'open-access' fisheries in Chile, as access is restricted even in non-regulated fisheries. Artisanal fishers are required to be registered in a regionalized National Registry, which has been closed since 1995, resulting in a *de facto* limited entry at the regional scale. Fishers can operate legally only in the region where they are registered or, in exceptional circumstances, in 'contiguous zones' of adjacent regions. Almost half of the commercial divers of the country are registered in Region X.

Individual quotas were introduced in the LGPA for resources designated as 'fully exploited' with the expectation of solving the problems created by open access. Those problems were made most patent by a crisis in the *loco* fishery, which was closed for four years, leaving a disastrous sequel of social and economic problems (Chapter 2). The quota regime consisted of a total allowable catch (TAC) established for each region based on estimates of resource abundance, split into individual non-transferable quotas among registered divers. As the direct assessment of benthic resources is not practical at that spatial scale, abundance was estimated indirectly using the size composition of the commercial catch.

Loco quotas were introduced in 1993 following closure and abandoned in 1999 after the system failed to regulate the harvests (González et al. 2006; Orensanz et al. 2005). The reasons were multiple. The assessment model, while internally consistent, was inadequate for sedentary resources where data are dominated by spatial heterogeneity at multiple scales (Orensanz and Jamieson 1998). Even more important, enforcement of individual quotas was ineffective in a system where tens of thousands of small boats operate along thousands of kilometers of coastline.

A new system based on small-scale territorial use rights for fishing (TURFs) and co-management was introduced along the central to north coasts, as described in Chapter 2. Local organizations of fishers can request exclusive fishing rights for benthic resources over stretches of intertidal or subtidal seabed. TURFs have been successful in the conservation and management of artisanal fisheries for valuable benthic species, including sea urchins, and in transforming the fortunes of nearshore fisheries (Chapter 2).

The exploitation of sea urchins within TURFs is based on a plan that requires fishers to evaluate their abundance within the areas and make projections of stock status with the help of a consultant. A harvest plan is agreed on that basis and fishers then decide how the quota is to be caught.

In addition to TURFs, the LGPA contemplates a system of two types of marine protected areas: parks and reserves. Marine parks are no-take zones established to preserve areas of special ecological significance or scientific interest. Reserves were conceived as reproductive refugia where fishing activities are allowed under certain circumstances. This distinction, clearly established in the LGPA, is however the subject of much confusion in other management contexts.

Southern Chile: a geographic labyrinth

The coasts of continental Chile are naturally split into two distinct realms. Northern and central Chile (Regions I–IX and the northern end of Region X) has a quasi-linear coast (Fig. 3.2). South of 42° S (most of Region X and Regions XI–XII) the coast is broken and convoluted, forming an intricate maze of islands, fjords, and channels. The fishery of Regions X–XI, addressed in this paper, operates over four latitudinal degrees of coast, from Chacao Channel (42° S), which separates the northern end of Chiloé Island from the continent, to Taitao Peninsula (~46° S), which effectively splits the inner waters of southern Chile, blocking navigation between Regions XI and XII. Most fishing grounds are located on the insular part: Chiloé Island (Region X), Guafo Island (Region XI) and the Guaitecas–Chonos Archipelago (Region XI).

The labyrinthic geography creates a great diversity of complex local circulation patterns. Maximum depth in the exploited fjords is in the order of 300 m. Surface temperature down to 5–10 m ranges from 8.5–9°C in winter to 13–15°C in summer but can be as low as 5°C near the mouth of the rivers, and as high as 18–20°C in wave-protected zones during the summer.

The Chilean sea urchin

Loxechinus albus is a relatively slow growing sea urchin that may live to be as old as 20 years and may reach a maximum diameter of 130 mm (Stotz et al. 1992; Gebauer and Moreno 1995; Vásquez 2001). It feeds mainly on drifting algal material (Castilla and Moreno 1982; Moreno and Vega 1988) and is common in intertidal pools on rocky shores and on subtidal reefs; maximum depth record is 340 m (Larraín 1975). A photographic study conducted in Regions X–XII indicated that the frequency of sea urchins drops with depth (~30% at 0–30 m, ~10% at 31–90 m and ~4% below 90 m, Inostroza et al. 1983). Estimated size at maturity is in the range of 35–50 mm (Bay-Schmith et al. 1981). The spawning season varies along the long coast of Chile; in the south, spawning appears to be restricted to late spring (Bay-Schmith et al. 1981). Larvae remain in the plankton for 28–34 days under laboratory conditions (González et al. 1987).

The sea urchin fishery of Regions X–XI

Fleet and fishing operations

The number of divers registered with SERNAPESCA (National Fisheries Service, Chile) to fish in Regions X–XI declined between 2000 and 2004 when only 22% of the registered divers were effectively engaged in the sea urchin fishery (Table 3.1). The diving population is aging, due in part to the moratoria on new entrants. For example, a sample of 63 divers interviewed had an experience of 10–30 years in the urchin fishery (Barahona et al. 2003).

Table 3.1 Number of registered divers and divers effectively fishing sea urchins in Regions X and XI (data from Aycon and Sernapesca).

	Registered divers (Sernapesca)		Divers effectively engaged in the sea urchin fishery (Aycon)	
Year	Region X	Region XI	Region X	Region XI
2000	5405	749		
2001				
2002			727	313
2003			664	296
2004	3731	514	639	288

Fishing units include vessels with a cabin or *lanchas* 12–15 m long (Fig. 3.3) and boats 8–10 m long. The sea urchin fishery has two distinct components, one supplying the local demand for fresh urchins and the other, which dominates the activity in quantitative terms, supplying the export-oriented processing plants (Barahona et al. 2003). Local consumers prefer large urchins – the larger the better, while the gonads of relatively small urchins, below the legal size limit, are more suitable for the Japanese market.

The fleet supplying the domestic market operates mostly from harbors from north and northeast Chiloé Island and adjacent areas (Fig. 3.2). Boats make daily trips to fishing grounds. Since large urchins tend to occur in deeper water, divers specializing in the domestic product make short dives at depths of 25–50 m (Fig. 3.4). Not surprisingly, diving-related health problems are common in these communities.

Figure 3.3 Boats of the sea urchin fishery fleet in the harbor at Ancud.

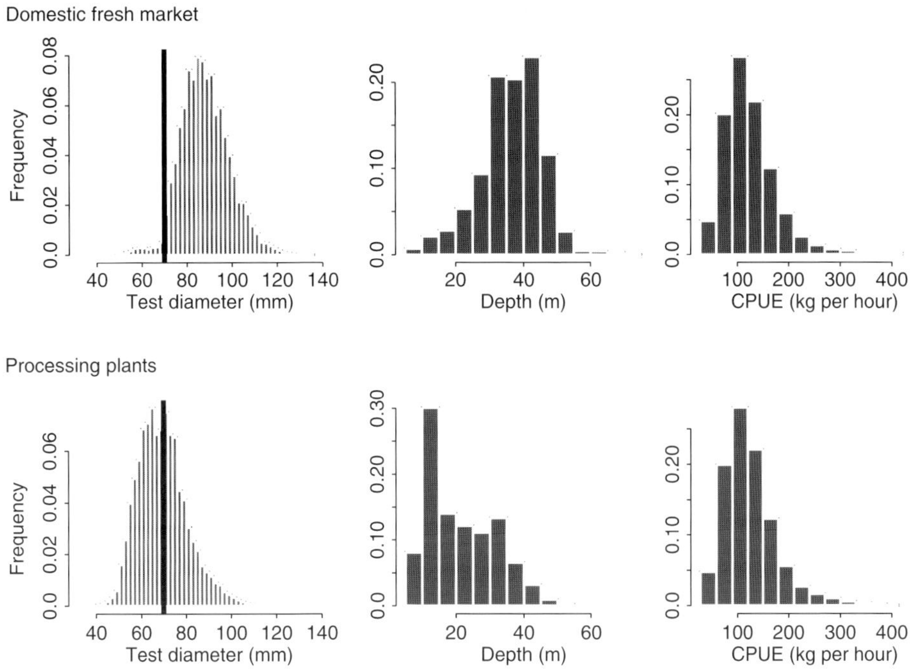

Figure 3.4 Size composition of the catch, diving depth, and catch per unit of effort (CPUE), 2001–2004. Top: fishery that supplies the domestic fresh market; bottom: export-oriented fishery. Vertical lines indicate the legal size limit.

Divers supplying the export-oriented industry, in contrast, target smaller urchins in shallower water (Fig. 3.4). Some of their vessels make daily trips from Quellón, Melinka, and other harbors (Fig. 3.2) but, since most fishing takes place in remote areas of Region XI, fishers cannot commute daily between the ports and the fishing grounds. In that case there is a different operational mode. Groups of 10–20 boats are organized into *faenas* by operators that tow or lead them to temporary camps, from where they fish in neighboring fishing areas. Fishing areas, known as *procedencias*, are identifiable by their vernacular names. Operators provide services through mother boats (*lanchas de acarreo*), which are typically 15–25 m long, and transport supplies from the nearest harbor to the *faena* and the catch from the *faena* to processing plants. Some operators work for a single processor, while others are independent. Most of the catch is landed in Quellón (Region X) where processing plants are located, and the catch landed at Melinka (Region XI) is principally also processed in Quellón (Fig. 3.2). There is growing fishing and processing activity centered in Chacabuco/Puerto Aysén (Region XI, Fig. 3.2).

Social matrix

The artisanal fisheries of central and northern Chile are organized around *caletas* that, in rural areas, are analogous to fishing villages (Chapter 2). *Caletas*, their nearby fishing grounds, and fishers' organizations constitute the ecological, social, and economic template of the system. This structure, however, is almost non-existent in much of the sea urchin fishery of Region XI. Fishers and their families reside in townships like Ancud, Quellón, Melinka and Puerto Chacabuco, far away from the most significant fishing grounds. Fishers' organizations are comparatively weak and there are other proximate players: the *Faena* operators and the processors, many of whom are organized into an association, namely Asociación de Industriales Ericeros de Chile. Fishers are organized into 'syndicates', 'associations' or 'cooperatives', and these community-level organizations are grouped into regional federations and national confederations. Organized artisanal fishers have shown considerable political muscle on several occasions.

Historical trends and stages in development

Statistical data on the Chilean sea urchin fishery go back to 1945. Annual landings were relatively small before 1975, never reaching 3000 metric tons (MT) (Fig. 3.1); most of the product was sold in local markets for fresh consumption. The fishery grew rapidly during the next decade as a result of increasing demand from Japan and the development of an export-oriented industry; landings reached 30 000 MT in 1986. At the beginning of the boom most of the production originated from Region X but yields started to decline there during the 1980s. Anecdotal reports point to the serial depletion of the traditional fishing grounds of Chiloé Island, starting from the north–northeast sector. Effort increased as divers attracted from other regions in search for *loco* entered the sea urchin fishery. As yields declined in Region X, the fleet was forced to gradually move southward to the Guaitecas and Guafo Islands at the northern end of Region XI. Processing plants, however, were always based in Quellón (Region X, Fig. 3.2).

As the fleet expanded further to the south into the remote and virtually unpopulated Chonos Archipelago, operations were gradually adjusted in order to deliver live sea urchins to the processing plants. The small town of Melinka, the only permanent settlement in the Guaitecas Islands at the northern end of Region XI (Fig. 3.2), became an intermediate station for ferrying fresh urchins to the processing plants in Quellón. Over the last two decades (1986–2004), landings from Regions X–XI have been relatively stable, fluctuating around 20 000 MT (Fig. 3.5). The demand for sea urchin roe went up in the Japanese market due to a strong yen. During the early 1990s the fishery expanded southward once again, this time into Region XII (Fig. 3.5); annual national landings doubled and have been maintained around 50 000 MT since then. Within Region XII, the focus of fishing has radiated out from landing points such as Puerto Natales and Punta Arenas to the more exposed margins of the archipelagos and channels. By 1999, fishing had extended to the furthermost reaches of the region.

Management and scientific advice before 2001

Before 2001, management of the sea urchin fishery followed the traditional finfish-inspired paradigm. For many years the fishery was open-access, the only control

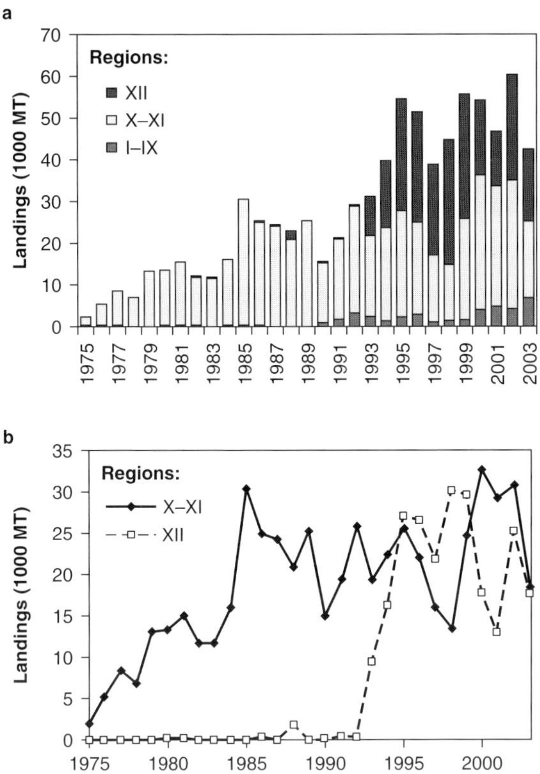

Figure 3.5 Landings by region, between 1970 and 2003, in thousand metric tons: (a) cumulative for Regions I–IX, cumulative for Regions X–XI, and Region XII; (b) Regions X–XI and XII.

being a minimum size limit of 10 cm dating back to 1934. Two important regulations were introduced in 1980: (1) a seasonal closure from 15 December to 1 March to protect the stocks' reproductive potential; and (2) a reduction of the legal size to 8 cm. In 1986, the seasonal closure was extended and the legal size was further reduced to 7 cm in order to accommodate the size standards required by the export market. This size limit is still well above the size at maturity.

Beginning in 1994, Subpesca started to explore the development of a management plan for the stocks of Regions X–XI. Problem identification was followed by preliminary simulation modeling aimed at the exploration of scenarios and management alternatives suggested by different stakeholders. The latter included processors, fishers, managers, and scientists, all of whom participated in two workshops in an attempt to initiate an exercise in adaptive management (Walters 1986). The main positive outcome of the exercise was the testing of a mechanism for communication between the stakeholders. Unfortunately, the fisheries authority could not implement the management options identified. The diagnostic was that the harvest of sea urchins below the legal size limit was growing and that the maximum annual surplus production estimated at ~20 000 MT was being fished (Moreno et al. 1991; Zuleta et al. 1994, 1995; Moreno and Zuleta 1996).

A three year decrease (1995–1998) in the catches (Fig 3.5), prompted concern about the status of the resource as well as apparent serial depletion of fishing grounds inferred from the southward displacement of the fishing fleet (Barahona and Jerez 1997). A perception of latitudinal serial depletion gave rise to fears of a collapse once the last unfished beds were reached. In 1999, managers considered bringing the fishery under the TAC regime described above. A size-based model similar to that developed for *loco* was used to assess total regional abundance of sea urchin, using size compositions of the landings. Lacking spatial structure, the model could not capture the strong spatial dynamics of the stock and the fleet, nor the impressive size variation and plasticity of sea urchins. Aware of these limitations, scientists abandoned the model-based assessments in 2001. Fear of an imminent crisis grew during 2001. Managers felt frustrated and confused without a TAC recommendation, without a management plan, and with no formal consultation process with fishers and processors. They immediately started a new quest for assessment and management alternatives.

The dynamics of depletion and recovery

As described earlier, the epicenter of fishing activity moved from central and northern to southern regions during the 1980s and 1990s (Fig. 3.5), raising fears of serial depletion on a grand geographical scale (Barahona and Jerez 1997). The pattern is, however, more complex than it appears at first sight. The decline of landings from Region X was real and in fact steeper than indicated by the landing statistics because many of the landings recorded in Quellón (Region X) originated in Region XI. Expansion of the fishery into Region XII, in contrast, was not accompanied by a decline of the combined catch from Regions X–XI, which has been relatively stable for two decades (Fig. 3.5). The drop between 1995 and 1998, interpreted initially as depletion (Barahona and Jerez 1997), was also attributed to a retraction of demand in the international market. The overall regional trends are indicative of staged geographical expansion rather than serial depletion.

Serial depletion, however, is still possible within a region even if not apparent in the aggregate fishery statistics. Indeed, serial depletion and poor performance of catch per unit of effort (CPUE) as an index of abundance have been blamed for the collapse of many benthic fisheries targeting sedentary resources (Orensanz et al. 1998). As fishers move on from depleted beds to new locations, CPUE remains stable while abundance falls, a process known as hyperstability (Hilborn and Walters 1992). Serial depletion may thus proceed, masked in the CPUE data, until all grounds are depleted and the fishery crashes.

Examination of the spatial allocation of effort by the sea urchin fleet indicates that although *procedencias* close to harbors, such as in the Guaitecas Islands, were harvested earlier and more intensively, the entire region has been harvested for many years (Figs. 3.6 and 3.7). IFOP's database contains more than 350 sea urchin *procedencias* within Regions X–XI, but more than 80% of the catch each year comes

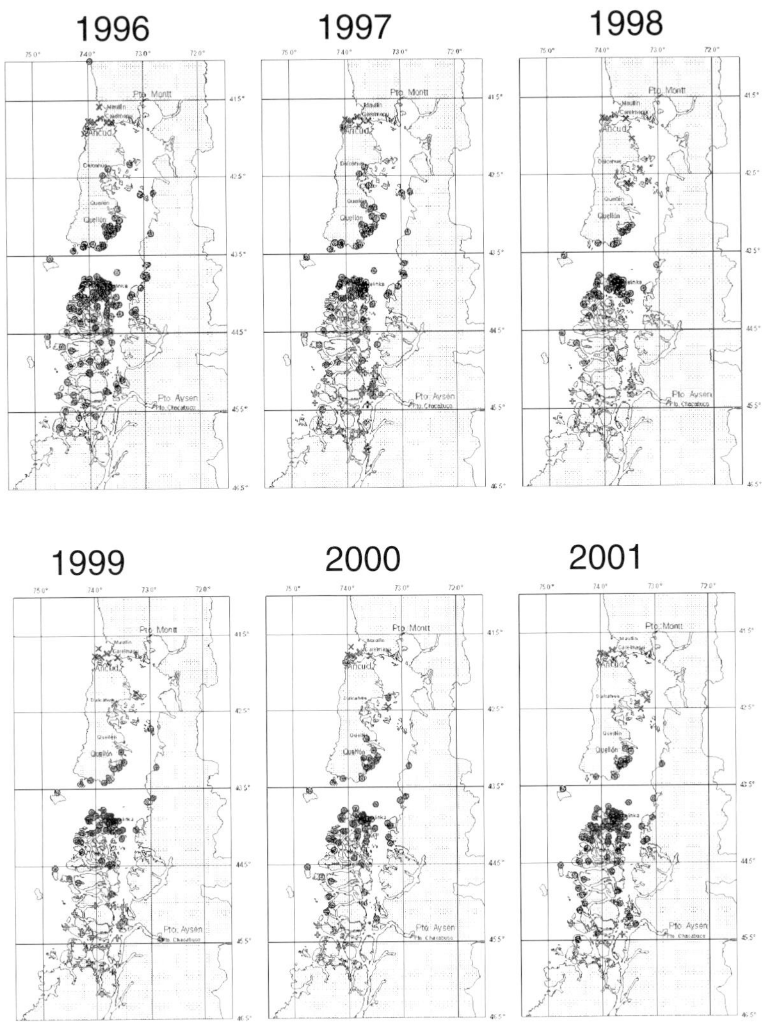

Figure 3.6 Geographical distribution of effort directed to sea urchins in Regions X–XI during the 1996–2001 fishing seasons (dots indicate locations or *procedencias*). Data from IFOP's monitoring program.

from fewer than 30. Spatial contractions and expansions in the spatial range of the fleet are well illustrated by concentration curves (Fig. 3.7), where cumulative catch is plotted as a function of locations ranked by catch. The steeper the curve, the more spatially concentrated the fishery. As an example, when the fishery started to operate from Puerto Chacabuco in the south of Region XI, the fleet spread over most of the Chonos Archipelago but then contracted to the most profitable locations (Fig. 3.6).

Examination of scatterplots of catch per trip over time and cumulative catch by season disaggregated by *procedencias* confirms a pattern recurrently described by processors, *faena* operators, and fishers. Most *procedencias* were visited intermittently over the 14 fishing seasons examined (Fig. 3.8). Many were abandoned

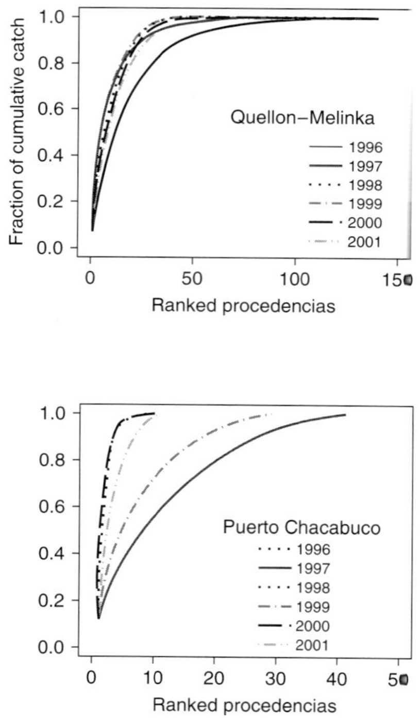

Figure 3.7 Concentration curves for the central and south sectors of Regions X–XI, 1997–2001 seasons. Note contractions and expansions in the south. Data from IFOP's monitoring program.

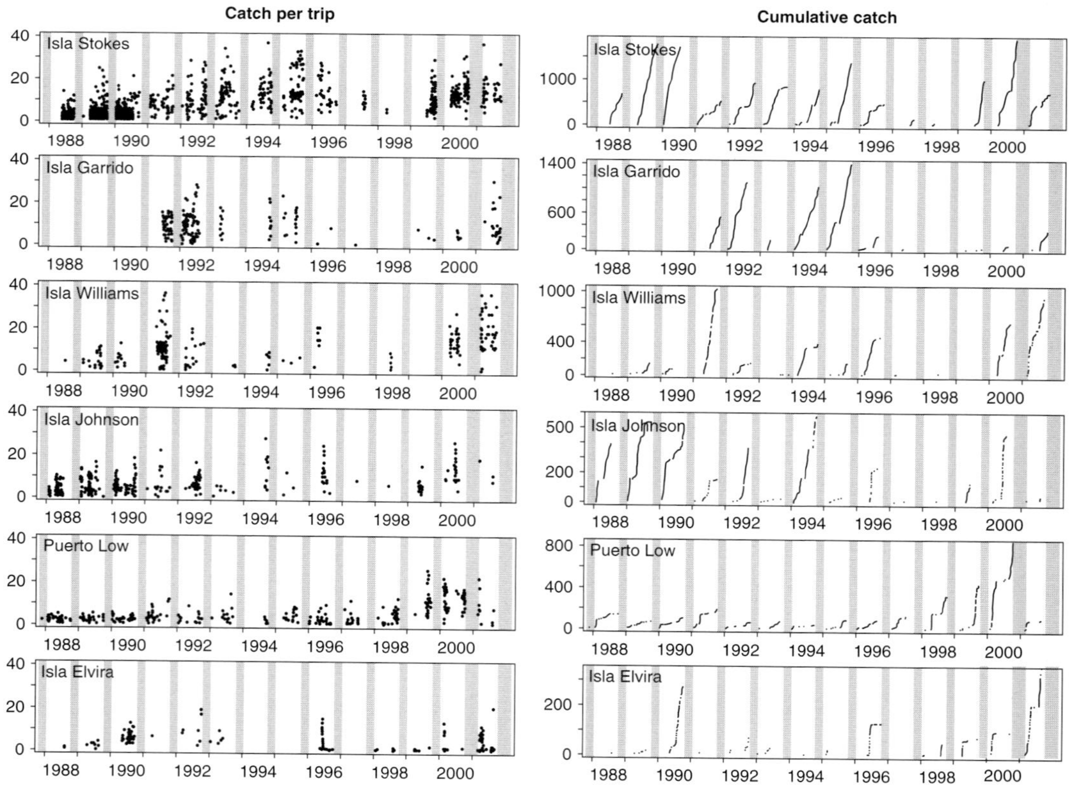

Figure 3.8 Catch per trip and cumulative total catch over the fishing seasons of 1988–2001 for six illustrative *procedencias* of Region XI. Shaded areas correspond to reproductive closures.

temporarily for a few years; some were harvested only at the beginning of the fishery, while others were not exploited until recent years. Intermediate-scale hyperstability is present, at least at the resolution of the available CPUE data: abandonment of a *procedencia* was never preceded by a measurable decline in catch per trip (Fig. 3.8). What the data show is rotation at the scale of the entire region, the complex result of the dynamics of localized events of recovery and depletion. This has happened over two decades (1985–2003) where the entire region has been accessible to the fleet, while annual landings have been remarkably stable.

The dynamics of local recovery in depleted locations is of central interest in the development of realistic management strategies, but is still poorly understood. Short-term recovery could be the result of sea urchins moving from refuge areas into depleted spots. Fishers insistently mention the existence of a vast refuge at depths beyond the reach of divers and the scarcity of exploratory surveys does not permit even a qualitative evaluation of this hypothesis. Inostroza and colleagues (1983) however, observed sea urchins down to a depth of 100 m. Fishers indicate that a common practice is to deplete a good *procedencia* at the beginning of the season, abandon it temporarily, and then return towards the end of the season. This behavior is easily confirmed by examination of IFOP's database (Fig. 3.8).

The picture that starts to emerge from a close look at the fishing process is not a simple one, and certainly not one of serial spatial depletion. The fleet has remarkable mobility and, as a result, the fishery expands and contracts over the entire range on a year-to-year basis. Fishers move on after depleting each bed, ranking locations according to roe quality (brown gonads have low or no commercial value), accessibility (distance and exposure), shell phenotype (long versus short spines), and prior knowledge. Locations are revisited after variable periods, ranging from intraseasonal to periods of several years. These data and results of interviews revealed that, at least within Region X–XI, the fleets rotate areas spontaneously rather than proceeding through a latitudinal depletion of the beds, as was earlier believed (Barahona et al. 2003).

A quest for alternative models

Conscious of the need for a new paradigm, in 2000 the fisheries authority started to look for better information that would support innovative management. The Fisheries Research Fund funded the first phase (2001–2002) of a project titled Biological Bases for the Rotation of Areas in the Sea Urchin Fishery of Regions X–XI that was conducted by IFOP (Barahona et al. 2003). The basic idea was to explore spatially-explicit strategies, such as rotation and reserves, that would be consistent with the nature of the fishing process. The project stressed joint work among scientists, managers, processors, and federations of artisanal fishers from the two regions. A formal interview program was conducted to gather information from industry managers, *faena* operators, and fishers.

The results revealed that: (1) there is spontaneous rotation of fishing areas at different scales; and (2) there are large areas that are not harvested because they are dominated by sea urchins with dark roes of low commercial quality. Experimental studies indicate that the main determinant of roe color is the algal composition of the diet (Bustos et al. 2004; Vidal 2004). These findings provide the basis for the design of a strategy that, by matching and formalizing fishers' behavior, has a chance of acceptance and compliance. The main elements of the plan discussed were: a formal rotation program; tracking of recovery rates of harvested plots in a network of observation sites used to fine-tune rotation times; a legal size that, while guaranteeing a significant reproductive contribution would fit the requirements of the market; and the creation of extensive reproductive refuges in areas where roe is of bad quality in terms of color but where sea urchins are otherwise physiologically and reproductively normal. Execution of the project was interrupted in 2002 after a serious crisis erupted.

Social turmoil, political complexities, and prospects

The 'contiguous zones' conflict

A serious social crisis erupted during 2001. As mentioned earlier, fishermen from Region X had traditionally fished in the largely unpopulated archipelagos of Region XI and their catch was ferried to Quellón (Region X) for processing. These trans-boundary operations were legally contemplated in a special proviso of the LGPA, known as 'contiguous zones'. Over the years, however, fishing activities gradually established local roots, and fishers registered in Region XI began exercising political pressure to curtail the intrusion of fishers from Region X. Local politicians also perceived the opportunity to promote investment in their own region. A contentious and at times violent confrontation originated, and management was crippled as a result of litigation and political squabbling (Barahona et al. 2003).

Negotiations were further complicated due to the fact that artisanal fishers' organizations from Region XI were not unified. The organizations from Aysén and Puerto Chacabuco, at the south of Region XI (Fig. 3.2), were controlled by long liners that target mostly southern hake (*Merluccius australis*), while fishers from Melinka were primarily commercial divers. Management of hake stocks has been convoluted and contentious, with artisanal long liners recurrently pushing for higher quota and/or a larger allocation. In the course of hake negotiations, leaders have sometimes manipulated conflicts in the sea urchin fishery in order to put indirect pressure on fisheries authorities, strengthening their bargaining position. In a sense, the management of sea urchin stocks has sometimes been held hostage to hake allocation negotiations.

The crisis precipitated when fishers from Region XI took legal action against a decree from Subpesca that extended the contiguous zones proviso to allow the 2001 season to proceed. The court ruled in their favor, effectively crippling the normal development of fishing activities. Turmoil and violent confrontations erupted in normally quiet fishing villages. The crisis grew so serious that a high-level meeting was called at the Presidential Palace in Santiago. An agreement, named Acuerdo de La Moneda, was reached between Subpesca, the Ministry of Internal Affairs, the governments of Regions X and XI, and leaders from fishers' organizations from the two regions. It contemplated: (1) an experimental fishing season; (2) input controls to put a cap on the number of users; and (3) the definition of 'areas of exclusion', closed to fishers from Region X (COZOPE 2005). In spite of its broad support, the agreement that was made into a decree by Subpesca on 15 October 2001 was challenged again, this time by an organization of commercial divers from Melinka (Region XI). The case was, however, dismissed by an appeals court in February 2002, paving the way for a new stage in the development of the fishery.

Participatory management

The 2001 crisis made it clear that the traditional command-and-control management approach was no longer an option for this complex fishery. Subpesca consequently invited a participatory process to develop a management plan, an approach contemplated in the LGPA, but never implemented in Chilean fisheries. The zonal authority (COZOPE) organized a participatory forum that brought together all the parties including fishers from the two regions, processors, *faena* operators, managers, and scientists. The process was initially unruly and difficult but in less than a year it led to the creation of a management commission, formally known as CoMPeB X–XI that stands for Comisión de Manejo de las Pesquerías Bentónicas de las Regiones X y XI, with representatives of all the interested parties and co-chaired by the zonal fisheries director and an independent scientist. In addition, a technical advisory team, led by scientists from Universidad Austral, was appointed to provide advice on strictly technical matters. In 2003, the advisory team produced the technical elements of a management plan but discussions during 2003–2004 were difficult and protracted. A special program for controlling the access was put in place during 2002–2004 but the climate of uncertainty during this period reflected in poor performance: the catch increased above the upper bounds considered adequate based on historical records and the portion of the catch below the size limit exceeded the historical range (Fig. 3.9).

The plan was finally approved by consensus and promulgated through an executive resolution from Subpesca in February 2005. This is the first formal management plan in the history of Chilean fisheries management and is framed in the letter and spirit of the new fisheries legislation. Seizing the momentum, COZOPE expanded the scope of the management plan to include other benthic fisheries for invertebrates and algae. The management plan specifies objectives and goals, area of application (Fig. 3.2), actions in the short-term (2005–2006) and mid- to long-term (2007–2010), and includes indicators (Table 3.2; Fig. 3.9), feedback mechanisms

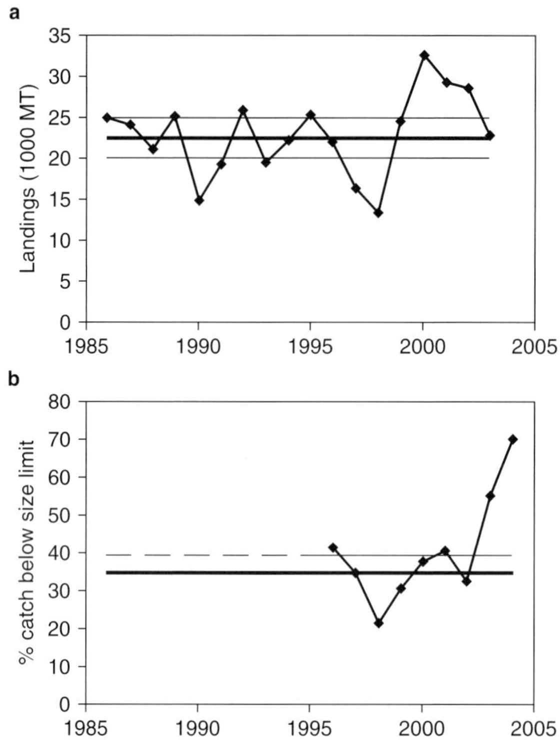

Figure 3.9 Indicators of performance for Regions X and XI: (a) annual landings compared to reference levels based on historical records; thick and thin lines indicate reference levels (data from Sernapesca and Aycon); (b) fraction of the commercial catch below the legal size limit, and maximum reference bounds (data from IFOP).

(Fig. 3.10), and norms for the functioning of the CoMPeB and the technical advisory team (COZOPE 2005).

Management controls include guideline harvest levels and a minimum legal size. In the mid- to long-term, the management plan contemplates two very significant developments: a rotational experimental fishery and reproductive reserves. Both require the committed and effective participation of the users. Different stakeholders have advocated some form of rotational strategy (Bórquez 2002) that would build from the spontaneous rotation that appears to guide spatial effort allocation in Region XI.

Fishery scientists are increasingly valuing and exploring this old empirical practice for the case of benthic resources. The main piece of information required for the design of a rotational strategy is the recovery rate of locally-depleted areas, which has to be investigated at the relevant spatial scale of the fishing grounds. Hence the need for an experimental fishery conducted in a small sector, such as the Guaitecas Archipelago (Fig. 3.2). The implementation of such a pilot plan would also assist in the development of monitoring protocols to be conducted with the

Table 3.2 Criteria and indicators selected to gauge performance of the management plan.

Criteria	Indicator	Structure	Reference point
Performance of the fishery	Landings	By fleet By size By area	Historical levels
	Effort (divers day^{-1})	By area By port of origin By *faena*	Historical levels Shifts in the spatial allocation of effort
Biological sustainability	Density (network of observation stations)	Pre-established geographical locations	Changes in density over time
	CPUE	By area	Historical levels
	Size structure: fraction above legal size, size frequency distribution, mean size	By area By depth	Historical levels
	Color of the gonads	By area By depth	Percentage processed of grade D
Economical sustainability	Contribution to GNP Value of the exports Ex-vessel price	Global By product By landing port	Historical levels Relation between ex-vessel price and export value
Social sustainability	Employment (number of workers)	Subsectorial	Historical levels
Institutional sustainability	Compliance (violations and/or accusations)	Subsectorial	100% compliance
	Conflicts (number)	Subsectorial	0

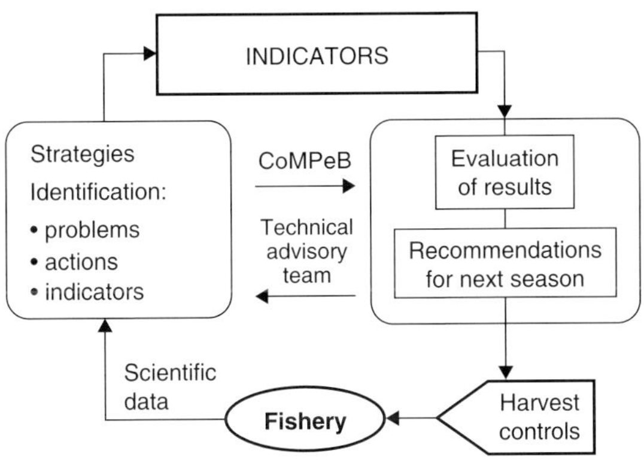

Figure 3.10 The management cycle as conceived by the management plan involves feedback through a series of indicators of system status, which are used to adjust harvest controls for each fishing season. Decisions are made by a participatory forum (CoMPeB) with technical advice from the technical advisory team.

active participation of fishers' organizations. In the long-term, lessons learned from the experimental fishery will be considered to adjust the management of the entire fishery. Given the known longevity of sea urchins it is anticipated that the results of the plan will not be visible in less than six years.

Any spatially-explicit harvest strategy must include, as an insurance policy, the implementation of reproductive refuges. This particular form of marine protected area has received much attention in recent years (Clark 1996; Lauck 1996; Houde et al. 2000). In the specific case of sea urchins, Rogers-Bennett et al. (1995) attended to natural circumstances that provide *de facto* refuges and Quinn et al. (1993) investigated variations in design through computer simulation. As described later, empirical knowledge of the fishers is providing unexpected angles for the design of reproductive refuges.

Limited entry, clear entry rules, and monitoring of effort, all implemented with the participation of fishers' federations, are a first step in the direction of incentives for fishers' responsible behavior. In this fishery, however, there are other proximate players. Processors depend on a sustainable supply of sea urchin roe and often operate their own *faenas*, are well organized, and have a substantial role in the participatory management of the fishery. For them, counting on a steady and orderly supply is a most important incentive. The convergence of incentives, participation of all the stakeholders, and innovative science are at the core of collective optimism about the future of a fishery that three years ago seemed doomed to institutional collapse.

Revamping the monitoring and knowledge support

Two main projects are now under way to support the implementation of the new management plan. In 2002, fishers' organizations from Region XI approached Aysén Consultores Ltda., a local consulting firm that had implemented a rather sophisticated and effective system to control access in the hake fishery. The goal was to provide support for the implementation of actions defined in the Acuerdo de La Moneda over a three-year period. Organizations from Region X concurred, requesting the assistance of a local NGO, namely Fundación Chinquihue, to act as technical counterpart on their behalf. Basic elements of the project, approved by Subpesca, were to monitor access of fishers from both regions to the fishing grounds of Region XI using an electronic system to identify and track individual fishers and to geo-reference the *procedencias* identified by fishers (Aycon 2002). The project was conducted between 2002 and 2004, and provisions have been made for it to continue until 2006. In 2003, a catch of 18 312 MT was recorded from the contiguous zone: 67% from Region XI, landed by 1136 divers of whom 74% were registered in Region X (Aycon 2003; Table 3.3). In addition, 900 MT were landed from areas outside the contiguous zone, bringing up the total for Regions X–XI to 19 212 MT. The catch was processed in 35 plants, three of which processed 50% of the total. The program provided the first hard information to identify divers

Table 3.3 Numbers of divers registered in Regions X and XI that operated in the two regions in 2003 (data from Aycon 2003).

Region of origin	Region of operation	Number of divers	Catch (MT)
X	X	178	2340
	XI	664	11 016
XI	XI	294	4944
Total		1136	18 300

according to area of operation and region of registration, an issue that had been a subject of contentious debate.

A second research project, conducted by Centro de Estudios Pesqueros at the Universidad Austral and IFOP (CEPES 2004) is aimed at designing a spatially-explicit management strategy for the experimental rotational fishery. Specifically, this includes mapping of the area where the fishery operates, and developing a spatially-explicit simulation model (Valero et al. 2004) to be used for testing alternative management procedures (Butterworth and Punt 1999). Possible harvest controls and monitoring plans will be discussed in workshops involving the stakeholders. The objective is to identify those enforcement and management alternatives that perform well given various scenarios and uncertainties.

A special effort is being made to gather and organize the empirical ecological knowledge of the fishers. This is critical because the area where the fishery operates is extensive and difficult to access, making surveys very costly or logistically impracticable. One of the first results, of unexpected significance for the design of reserves, was the identification of extensive areas dominated by brown low-quality roe (Fig. 3.11); fishers may more easily accept these areas as closures due to their low commercial interest. Fishers and processors also pointed to two shell phenotypes that have not been documented in the scientific literature. Short and long spine varieties have different acceptance and economic values and also show well-defined patterns of geographic distribution. A similar phenomenon has been described in California, with similar implications for conservation (Rogers-Bennett et al. 1995).

New challenges

The challenges ahead are many and not minor. The most complex issues are related to the zoning of uses, particularly the rapid development of salmon aquaculture facilities seen as a major threat by fishers. The plume of residues generated by salmon pens can affect significant areas of seabed, smothering algae and degrading the quality of urchin roe.

Another zoning problem relates to the fast growth of TURF requests by fishers' organizations in a geographic setting where the viability of this system is still untested. Requests of TURFs within the experimental fishery area create serious

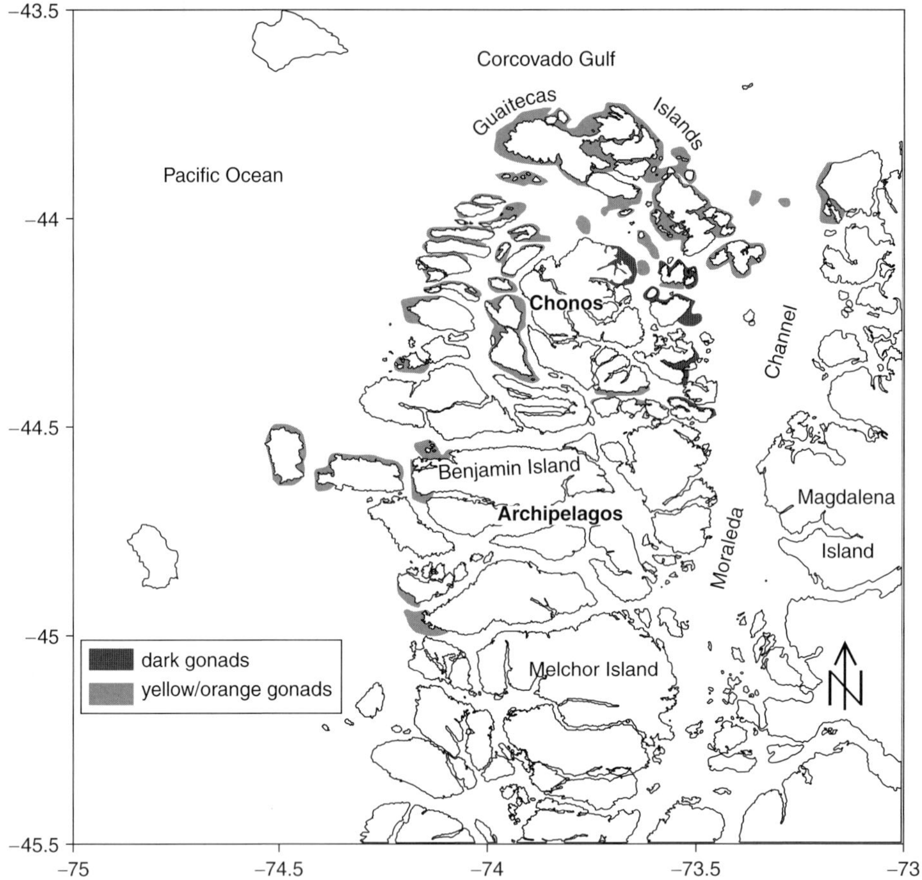

Figure 3.11 Geographic distribution of gonad color phenotypes in the Guaitecas and Chonos archipelagos. Information gathered from artisanal fishers through an interview program.

equity problems. By February 2003 the number of TURFs requested in Regions X and XI was 330 and 175 respectively. Sixty-two were located within the area of the management plan, although only 39 had declared sea urchin as a primary target species (most expressed interest in *loco*).

Besides salmon farms and TURFs, zoning has to consider the implementation of marine parks and reserves, including reproductive refuges for sea urchin. It is clear that the management of sea urchin and other coastal fisheries will be increasingly embedded in a complex matrix of spatial uses, requiring an integrated approach to the management of the coastal zone.

The scientific and technical information available to address these challenges is far from satisfactory. In the specific case of the sea urchin fishery, little is known about the processes involved in recruitment. Larval dispersal and related patterns of connectivity among subpopulations, including deep-water populations, are important in the design of spatial harvest schemes. These factors are not documented and knowledge about circulation in the complex labyrinth of southern Chile is still rudimentary. Post-settlement processes associated with the recovery of depleted grounds have never been studied, although they are a dominant factor for sustainability under a rotational harvest strategy. Restocking with juveniles produced in hatcheries has been recurrently advocated as a possible way to supplement natural recruitment. Although the technical protocol for the mass production of juveniles is well established (Bustos and Olave 2001), the practicality and economical feasibility of restocking Chilean sea urchin beds has not been convincingly demonstrated.

Also unknown are the possible ecological cascading effects of the massive removal of a major herbivore from extensive subtidal areas. Although dramatic effects due to the removal of sea urchins have been documented in the northern hemisphere (Lawrence 1975), these processes remain little known for the Chilean sea urchin and some results are contradictory: while it has been often held that effects may be weaker because the Chilean species feeds mostly on drifting algae (Vásquez et al. 1981, 1984; Castilla and Moreno 1982; Moreno and Sutherland 1982; Dayton 1985), recent studies suggest stronger than expected ecological effects of grazing (Buschmann et al. 2004).

Threats to sustainability are not just biological but also social, economic, and institutional (Charles 2001). Institutional sustainability was at the core of the crisis that brought this fishery to the brink of collapse in 2001–2002. Although major progress has been made, the new institutional framework is largely untested. The main challenge is now to streamline participatory management institutions that will provide the appropriate context for addressing future emerging challenges in this fishery.

Summary

The sea urchin fishery of southern Chile contributes more than half of the world's supply of sea urchin roe. The fishery provides for the livelihood of thousands of artisanal divers and their families and fuels a significant export-oriented industry. By the year 2000 there was a feeling of imminent crisis. An apparent north to south serial depletion of fishing grounds gave rise to fears of a collapse once the fleet reached the last unfished southernmost beds. Provision of scientific advice was limited and irregular. Stocks had been assessed using models that did not account for the spatial dynamics of the fishery. These assessments were later abandoned as scientists concluded that models were inappropriate. Managers felt frustrated and confused without a management plan or formal consultation process with fishers and processors. To make things worse, a contentious confrontation erupted between fishers of two adjacent regions in 2001.

In 2002, the regional government led an effort to bring all the parties together to discuss the future of the fishery. The process was initially difficult but in less than a year it led to the formalization of a participatory forum (CoMPeB) with representatives of the fishers from the two regions, processors, managers and scientists, and a technical advisory team led by scientists. The advisory team produced a formal management plan that was approved by the CoMPeB. The plan contemplates short, medium, and long-term goals, as well as clear indicators of success. The parties have accepted the implementation of a rotational experimental fishery and the creation of reproductive reserves, and traditional knowledge has been fundamental for the preliminary selection of the latter. Top-down decision-making has lent its way to an effectively participatory form of governance. The challenges ahead are many, and not minor. Yet, the convergence of incentives, participation of all the stakeholders, and innovative science are at the core of collective optimism about the future of a fishery that three years ago seemed doomed to institutional collapse.

References

Andrew, N., Agatsuma, Y., Ballesteros, E., Bazhin, A., Creaser, E., Barnes, D., Botsford, L., Bradbury, A., Campbell, A., Dixon, J., Einnarsson, S., Gerring, P., Hebert, K., Hunter, M., Hur, S.B., Johnson, C., Juinio-Meñez, M.A., Kalvass, P., Miller, R., Moreno, C., Palleiro, J., Rivas, D., Robinson, S., Schroeter, S., Steneck, R., Vadas, R., Woodby, D. and Xiaoqi, Z. (2002) Status and management of world sea urchin fisheries. *Oceanography and Marine Biology Annual Review*, **40**: 343–425.

Aycon (Aysén Consultores Limitada) (2002) *Localización, prospección y estructura de talla del recurso erizo* (Loxechinus albus), *racionalización y control del esfuerzo de pesca artesanal y control de los desembarques en aguas interiores de la X y XI región.* Propuesta Técnica para una 'Pesca de investigación'.

Aycon (2003) *Resumen consolidado pesca de investigacón recurso erizo X y XI Regón, año 2003.* Documento preparado para la Mesa Técnica encargada del diseño del plan de manejo del recurso erizo.

Barahona, N. and Jerez, G. (1997) La pesquería del erizo (*Loxechinus albus*) en las regiones X a XII de Chile: 13 años de historia (1985–1997). *XIII Congreso de Ciencias del Mar (Chile).*

Barahona, N., Orensanz, J.M., Parma, A., Jerez, G., Romero, C., Miranda, H., Zuleta, A., Catasti, V. and Gálvez, P. (2003) *Bases biológicas para rotación de áreas en el recurso erizo.* Informe final. Fondo de Investigación Pesquera, Proyecto FIP No 2000-18. Instituto de Fomento Pesquero (IFOP), Valparaíso (Chile). http://www.fip.cl

Bay-Schmith, E., Werlinger, C. and Silva, S. (1981) *Ciclo anual de reproducción del recurso* Loxechinus albus *entre la X y XI Región.* Proyecto de Investigación. Subpesca, Informe Final. Universidad de Concepción (Chile).

Bórquez, J. (2002) *Organización de la extracción para la preservación mediano y largo plazo del recurso erizo.* A proposal prepared for consideration by the Asociación de Industriales Ericeros, Quellón (Chile).

Buschmann, A., García, C., Espinoza, R., Filún, L. and Vásquez, J. (2004) Sea urchin (*Loxechinus albus*) and kelp (*Macrocystis pyrifera*) in protected areas in southern Chile. In: J.M. Lawrence and O. Guzmán (Eds) *Sea Urchins – Fisheries and Ecology*,

Proceedings of the International Conference of Sea-Urchin Fisheries and Aquaculture. DEStech Publ., Pennsylvania.

Bustos, E. and Olave, S. (2001) *Manual: cultivo del erizo* (Loxechinus albus). Instituto de Fomento Pesquero (IFOP), División Acuicultura (Chile).

Bustos, R., Rivas, P., Romo, C., Guzmán, O. and Guisado, C. (2004) Color improvement of brown gonads of the sea urchin *Loxechinus albus* by artificial feed. Abstract in: J.M. Lawrence and O. Guzmán (Eds) *Sea Urchins – Fisheries and Ecology, Proceedings of the International Conference of Sea-Urchin Fisheries and Aquaculture.* DEStech Publ., Pennsylvania.

Butterworth, D.S. and Punt, A.E. (1999) Experiences in the evaluation and implementation of management procedures. *ICES Journal of Marine Science*, **56**: 985–98.

Castilla, J.C. and Moreno, C. (1982) Sea urchins and *Macrocystis pyrifera*: experimental test of their ecological relations in southern Chile. In: J.M. Lawrence (Ed.) *International Echinoderms Conference.* A.A. Balkema, Rotterdam.

CEPES (Centro de Estudios Pesqueros) (2004) *Bases biológicas para la rotación de áreas del ecurso erizo – Fase II.* Propuesta Técnica, Proyecto FIP No 2003–13.

Charles, A. (2001) *Sustainable Fishery Systems.* Blackwell Science, Oxford.

Clark, C.W. (1996) Marine reserves and the precautionary management of fisheries. *Ecological Applications*, **6**: 369–70.

COZOPE (Consejo Zonal de Pesca X–XI Regiones) (2005) *Plan de Manejo Pesquerías Bentónicas Zona Contigua Regiones X–XI.*

Dayton, P.K. (1985) The structure and regulation of some South American kelp communities. *Ecological Monographs*, **69**: 219–50.

Gebauer, P. and Moreno, C.A. (1995) Experimental validation of the growth rings of *Loxechinus albus* (Molina, 1782) in southern Chile (Echinodermata: Echinoidea). *Fisheries Research*, **21**: 423–35.

González, H.E., Castilla, J.C. and Guisado, C. (1987) Effect of larval diet and rearing temperature on metamorphosis and juvenile survival of the sea urchin *Loxechinus albus* (Molina, 1782) (Echinodermata: Echinoidea). *Journal of Shellfish Research*, **6**: 109–15.

González, J., Stotz, W., Garrido, J., Orensanz, J.M., Parma, A.M., Tapia, C. and Zuleta, A. (2006) The Chilean TURF system: how is it performing in the case of the loco fishery? *Bulletin of Marine Science* (in press).

Hilborn, R. and Walters, C.J. (1992) *Quantitative Fisheries Stock Assessment: Choice, Dynamics and Uncertainty.* Chapman and Hall, New York.

Houde, E., Coleman, F.C., Dayton, P., Fluharty, D., Kelleher, G., Palumbi, S., Parma, A.M., Pimm, S., Roberts, C., Smith, S., Somero, G., Stoffle, R. and Wilén, J. (2000) *Marine Protected Areas: Tools for Sustaining Ocean Ecosystems.* Ocean Studies Board, National Research Council. National Academy Press, Washington.

Inostroza, F, Robotham, H., Salas, R. and Serey, I. (1983) *Cartas de distribución de los recursos bento-demersales de las aguas interiores de la X, XI y XII región. Zona 'I', 'II', 'III' y 'IV'.* Informe Final. Instituto de Fomento Pesquero, Valparaíso (Chile).

Larraín, A.P. (1975) Los equinoideos regulares fósiles y recientes de Chile. *Gayana Zoología (Chile)*, **35**: 1–188.

Lauck, T. (1996) Uncertainty in fisheries management. In: D.V. Gordon and G.R. Munro (Eds) *Fisheries and Uncertainty – A Precautionary Approach to Resource Management.* University of Calgary Press, Calgary.

Lawrence, J.M. (1975) On the relationships between marine plants and sea urchins. *Oceanography and Marine Biology Annual Review*, **13**: 213–86.

Moreno, C.A. and Sutherland, J.P. (1982) Physical and biological processes in a *Macrocystis pyrifera* community near Valdivia. *Oecologia*, **55**: 1–6.

Moreno, C. and Vega, R. (1988) Valor científico de las Reservas Marinas Costeras: un ejemplo de estudio ecológico en poblaciones de *Loxechinus albus* (Molina). *Informe UNESCO, Ciencias del Mar*, **47**: 124–34.

Moreno, C.A. and Zuleta, A. (1996) *Investigación complemento pesquería erizo 1995*. Informe Final prepared for Subpesca, Universidad Austral de Chile.

Moreno, C.A., Rivas, D., Zuleta, A., Vergara, L. and Jeréz, G. (1991) *INVES: Estimación captura total permisible recurso erizo 1989*. Prepared for Subpesca, Universidad Austral de Chile.

Orensanz, J.M. and Jamieson, G.S. (1998) The assessment and management of spatially structured stocks. *Proceedings of the North Pacific Symposium on Invertebrate Stock Assessment and Management. Canadian Special Publications in Fisheries and Aquatic Sciences*, **125**: 441–59.

Orensanz, J.M., Armstrong, J., Armstrong, D.A. and Hilborn, R. (1998) Crustacean resources are vulnerable to serial depletion – the multifaceted decline of shrimp and crab fisheries in the greater Gulf of Alaska. *Reviews in Fish Biology and Fisheries*, **8**: 117–76.

Orensanz, J.M., Parma, A.M., Jerez, G., Barahona, N., Montecinos, M. and Elías, I. (2005) What are the key elements for the sustainability of 'S-fisheries'? Insights from South America. *Bulletin of Marine Science*, **76**: 527–56.

Quinn, J.F., Wing, S.R. and Botsford, L.W. (1993) Harvest refugia in marine invertebrate fisheries: models and applications to the red sea urchin, *Strongylocentrotus franciscanus*. *American Zoologist*, **33**: 537–50.

Rogers-Bennett, L., Bennett, W.A., Fastenau., H.C. and Dewees, C.M. (1995) Spatial variation in red sea urchin reproduction and morphology: implications for harvest refugia. *Ecological Applications*, **5**: 1171–80.

Stotz, W., González, S. and López, C. (1992) Siembra experimental del erizo rojo *Loxechinus albus* (Molina) en la costa expuesta del centro – norte de Chile: efectos del erizo negro *Tetrapygus niger* (Molina) sobre la permanencia y crecimiento de juveniles. *Investigación Pesquera (Chile)*, **37**: 107–17.

Valero, J.L., Parma, A.M., Ernst, B., Orensanz, J.M. and Hilborn, R. (2004) *MetaPesca: a Simulation Model to Explore Management Options in Benthic Invertebrate Fisheries*. World Fisheries Congress, Vancouver, Canada.

Vásquez, J.A. (2001) Ecology of *Loxechinus albus*. In: J.M. Lawrence (Ed.) *Edible Sea Urchins – Biology and Ecology*. Elsevier Science, Amsterdam.

Vásquez, J.A., Castilla, J.C. and Santelices, B. (1981) Repartición de recursos por erizos en praderas de *Macrocysytis pyrifera* de Pto. Toro, Isla Navarino, Chile. In: *Pontificia Universidad Católica de Chile, Informe Final*, prepared for Armada de Chile, Santiago (Chile).

Vásquez, J.A., Castilla, J.C. and Santelices, B. (1984) Distributional patterns and diets of four species of sea urchins in giant kelp forest (*Macrocystis pyrifera)* of Puerto Toro, Navarino Island, Chile. *Marine Ecology Progress Series*, **19**: 55–63.

Vidal, G.B. (2004) Use of artificial diets in the culture of the sea urchin *Loxechinus albus*. In: J.M. Lawrence and O. Guzmán (Eds) *Sea Urchins – Fisheries and Ecology*,

Proceedings of the International Conference of Sea-Urchin Fisheries and Aquaculture. DEStech Publ., Pennsylvania.

Walters, C.J. (1986) *Adaptive Management of Renewable Resources*. Macmillan, New York.

Zuleta., A., Moreno, C.A. and Rubilar, P. (1994) *INVES: evaluación de stocks y monitoreo de la pesquería del erizo 1993*. Informe Final Contrato BIP No. 200311748, Subpesca, Valparaiso, Chile.

Zuleta, A., Moreno, C.A., Vergara, L. and Rubilar, P. (1995) *INVES: evaluación de stocks y modelo de manejo del recurso erizo*. Contrato BIP No. 20039061. Subpesca, Valparaiso, Chile.

Chapter 4

Achievements and Setbacks in the Commercial Diving Fishery of San José Gulf, Argentine Patagonia

J.M. (Lobo) Orensanz, Ana M. Parma, Néstor Ciocco, and Ana Cinti

'We should not speak of successes, just of achievements.'

– Oscar Avilez, artisanal fishers' leader from Chile. From a presentation made at the Biennial Conference of the International Association for the Study of Common Property, Oaxaca, Mexico, August 2004.

Introduction

The management of natural resources is highly demanding of scientific support and therefore applied scientists often become involved in the management process, whether as providers of technical advice for decision-making or as scientists-turned-managers. But scientists, including ourselves, are accustomed to the directed progress of knowledge: we know the process of research and discovery can be meandering, but a sense of progress is always there. It is not surprising therefore if scientists become frustrated with management as it moves forward and backwards, and breakthroughs are too often followed by setbacks. Furthermore, management decisions are not based solely on scientific support, with its emphasis on ecological sustainability, but also have to include societal values and agendas that are inevitably exposed to pressure from politicians, power brokers, and lobbies. Scientists involved with fisheries management tend to be pessimistic, as they perceive this compromise as irrational, setbacks as lack of success, and progress as insufficient.

In this brief essay we discuss a series of 'rounds' in the 40 year history of a small-scale scallop fishery from Patagonia: two were won, two lost, and round five is on. In the terms of Oscar Avilez, fisher and friend, this is not the history of a definitive success, but one of hard-to-attain significant achievements, and some setbacks. On balance, we argue, the achievements set the compass in the direction of success.

Settings

The gulfs of Northern Argentine Patagonia

Patagonia, at the southern end of the Americas, is a vast and sparsely-populated region shared by Chile on the west and Argentina on the east (Fig. 4.1a). On the Atlantic side, the Argentine Patagonian coast spreads over 15 degrees of latitude, from the mouth of the Colorado River (approximately 40° S) to the Beagle Channel (55° S). These coasts were pronounced 'wretched and useless' by Darwin (1962), who sailed along as a scientist on board HMS *Beagle* in 1834. He found them describable 'only by negative characters; without habitations, without water, without trees, without mountains'. Yet, 'in calling up images of the past,' he wrote, 'I find that the plains of Patagonia frequently cross before my eyes . . . why have these arid wastes taken so firm a hold of my memory?' and went on to conclude that '. . . it must be owing to the free scope given to the imagination'. Nowadays, 170 years later, most of the coast continues to be unpopulated, inaccessible, and inhospitable, dotted by just a dozen or so urban areas.

The operation of artisanal or small-scale coastal fisheries is made difficult by exposure and lack of shelter from the strong prevailing winds, blowing mostly from the southwest. Favorable conditions are found in a few areas, most noticeably the three North Patagonian gulfs: Nuevo, San José, and San Matías (Fig. 4.1b). Several commercial small-scale fisheries operate in the region, gathering shellfish and algae along the intertidal zone, beach seining and long lining for fish, dredging, and commercial diving for shellfish (Table 4.1). Among the latter, the most significant is the Tehuelche scallop (*Aequipecten tehuelchus*). Approximately 200 fishers are currently active in the artisanal fisheries around the Valdés Peninsula, which include San José, Nuevo, and the southern part of San Matías gulfs.

Here we focus primarily on the scallop fishery of San José Gulf, a roughly oval-shaped basin (814 km^2) connected to San Matías Gulf through a narrow (9 km) but relatively deep (80 m) mouth (Fig. 4.1c). The bottom below 35 m is a flat muddy plain. Circulation is driven by tides that range from 2.9 to 8.7 m between mean neap and spring tides. Strong winds (15 km/h average) blow predominantly from the SW. There are no estuaries or significant human settlements along the coast; average annual precipitation is 180 mm. Surface water temperature ranges seasonally from 9–17°C. The gulf is divided into an east and west half by a recurrent front (Fig. 4.1c). To the west, tidal flows produce a complex eddy system (Amoroso 2004). As compared to the west, water in the east has a longer residence time and is therefore warmer in summer but colder in winter. The spring phytoplankton bloom starts by early October; a second, smaller bloom occurs in late February. Average gross and net annual production have been estimated at 223 and 161 g C/m^2/yr (Charpy-Roubaud et al. 1982).

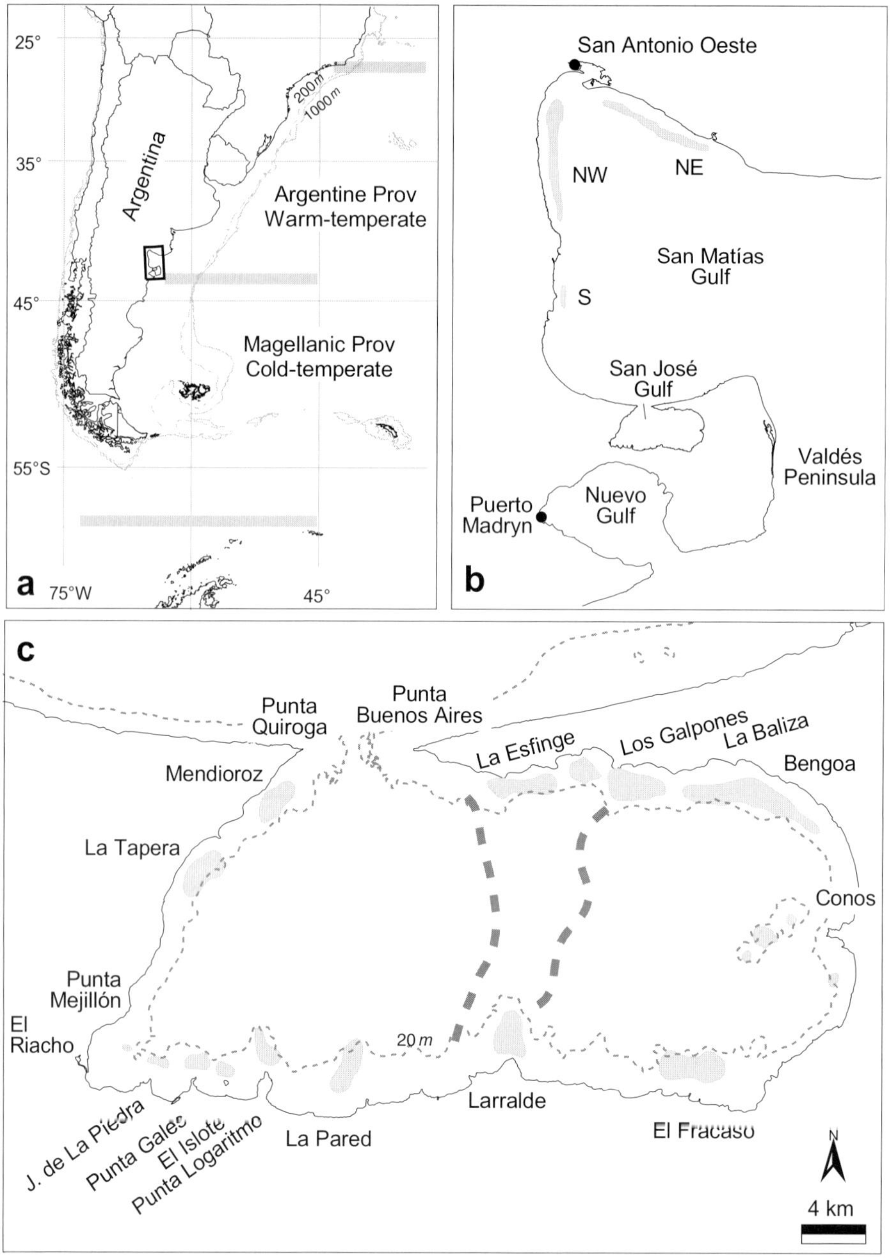

Figure 4.1 (a) The southwestern Atlantic, showing major biogeographic domains. (b) The North-Patagonian gulfs (area shown as a small rectangle in (a)). Shaded areas indicate scallop-fishing grounds in San Matías Gulf. (c) San José Gulf. Light dashed line indicates the 20 m isobath (at low tide). Shaded areas correspond to historical scallop grounds. Heavy dashed lines indicate the approximate range of the front (see text) (modified from Ciocco et al. 2005).

Table 4.1 Main benthic resources harvested by artisanal fishers in San José Gulf (Argentine Patagonia).

Group	Common name	Scientific name
Gastropod mollusks		
	Caracol	*Odontocymbiola magellanica*
	Caracol picante	*Buccinanops gradatum*
	Caracol chico	*Buccinanops globosum*
Bivalve mollusks		
	Ostra puelche (puelche oyster)	*Ostrea puelchana*
	Cholga	*Aulacomya ater*
	Mejillón (blue mussel)	*Mytilus edulis platensis*
	Vieira tehuelche (scallop)	*Aequipecten tehuelchus*
	Navaja	*Ensis macha*
	Almeja (hardshell clam)	*Ameghinomya antiqua*
	Panopea (geoduck)	*Panopea abbreviata*
Cephalopod mollusks		
	Pulpito (octopus)	*Octopus tehuelchus*
	Pulpo Colorado (octopus)	*Enteroctopus megalocyathus*
Brachyuran crustaceans (true crabs)		
	Cangrejo nadador	*Ovalipes trimaculatus*
	Cangrejo buey	*Platyxanthus patagonicus*

A sensitive region for marine conservation

San José Gulf is at the boundary between the cold- and warm-temperate domains of the southwestern Atlantic, denoted in the biogeographic literature as, respectively, the Magellanic and Argentine biogeographic provinces (Fig. 4.1a). Cold-temperate species predominate in the central and colder basin, and warm-temperate species in the intertidal and shallow subtidal zones. The gulf and neighboring coasts around Valdés Peninsula are ecologically very sensitive due to the presence of large colonies of marine birds and mammals, notably sea lions and elephant seals, but mostly because they are the mating and calving areas of the southern right whale (*Eubalaena australis*) (Campagna and Lichter 1996). Back in 1974, San José Gulf was consecrated by Chubut Province as a provincial marine park (Ciocco 1995), the first of its kind in the southwest Atlantic. UNESCO declared the entire region of Valdés Peninsula and adjacent coastal areas as a Natural World Heritage Area in 1999. A management plan was developed with participation of all stakeholders (DGCAP 1998) and has served as a framework for the implementation of specific conservation-oriented actions.

The Tehuelche scallop

The Tehuelche scallop is endemic to the Argentine Biogeographic Province, ranging in nearshore areas of the shelf from Rio de Janeiro (Brazil) to Camarones (Fig. 4.1a). The maximum recorded shell height is 102 mm, but specimens larger than 90 mm are uncommon; maximum estimated longevity is 11 years, but survival beyond age eight is exceptional. Analysis of stomach contents indicate that re-suspended microphytobenthos is the main component of the diet, the rest being phytoplankton. It is a simultaneous hermaphrodite; spawning has been observed in late spring and late summer, with considerable geographic and year-to-year variation. Larvae stay in the water column for approximately 20 days.

The San José Gulf stock is structured as a metapopulation, with local subpopulations interconnected through larval dispersal (Fig. 4.1c) (Orensanz 1986). Results from a three year study (Orensanz 1986) suggest that at least in the best-studied ground of San Román, growth is density-dependent, with effects being significant above a local biomass of 1 kg/m^2, where the maximum observed biomass was 1.5 kg/m^2. Calculations based on primary productivity (Charpy-Roubaud et al. 1982) and two independent population dynamics studies (Orensanz 1986; Ciocco 1991) yielded very consistent results on the production of meat, which is in the range of 360–380 g/m^2/yr. Recruitment of juveniles also appeared to show compensatory density-dependence. The commercial size of 60 mm is reached at an age of 1.5 to 2.5 years. We refer the reader to Orensanz et al. (1991a) and Ciocco and colleagues (2005) for further information and references on the biology and ecology of the species.

Achievements, setbacks, and prospects

Round 1: Boom and bust of the San Matías Gulf Scallop Fishery (1968–1972)

A decline in scallop landings from the Georges Bank, Canada during the late 1960s (Caddy and Lord 1971) sent brokers looking for alternative products. To seize the opportunity, many small-scale scallop fisheries sprouted around the world, among them a dredge fishery for the Tehuelche scallop in San Matías Gulf, Argentine Patagonia (Orensanz et al. 1991a; Ciocco et al. 2005). Starting in 1969, a fleet converged from other harbors of the country on San Antonio Oeste, a small sleepy town in the province of Río Negro, northern Patagonia (Fig. 4.1b). The wooden boats, 12–22 m long, towed a heavy dredge from the side (3 m × 2.5 m, Fig. 4.2a).

Scallop beds were located at depths greater than 20 m along the northwest coasts of San Matías Gulf (Fig. 4.1b), under the jurisdiction of Río Negro Province. Exploitation had been very light before 1969, when one or two boats supplied a small domestic market. Before the scallop boom, the substratum of the grounds was a

a

b

Figure 4.2 Fishing units. (a) A dredger arrives loaded with unsorted scallop bags to San Antonio Oeste, 1970. Notice the design and size of the dredge (photograph by Dr S.R. Olivier). (b) One of the fishing units that operate in the San José Gulf commercial diving fishery.

mixture of sandy or gravel sediment covered by abundant shell fragments that had accumulated over time. Empty shells provided a suitable substratum for scallop larval settlement and early benthic survival. Four years of intensive dredging, between 1969 and 1972, unselectively removed large and small scallops, also scouring the top shelly substratum of the beds. The catch was not sorted on board but transported in bulk to port. Small scallops and shell hash were discarded. Thousands of tons of shelly material were used during that period to fill lowlands around San Antonio and to construct loading docks. It has been calculated that during a single fishing season, each spot of scallop bed was dredged, on average, seven times (Iribarne et al. 1991). Sequential depletion of the beds proceeded quickly.

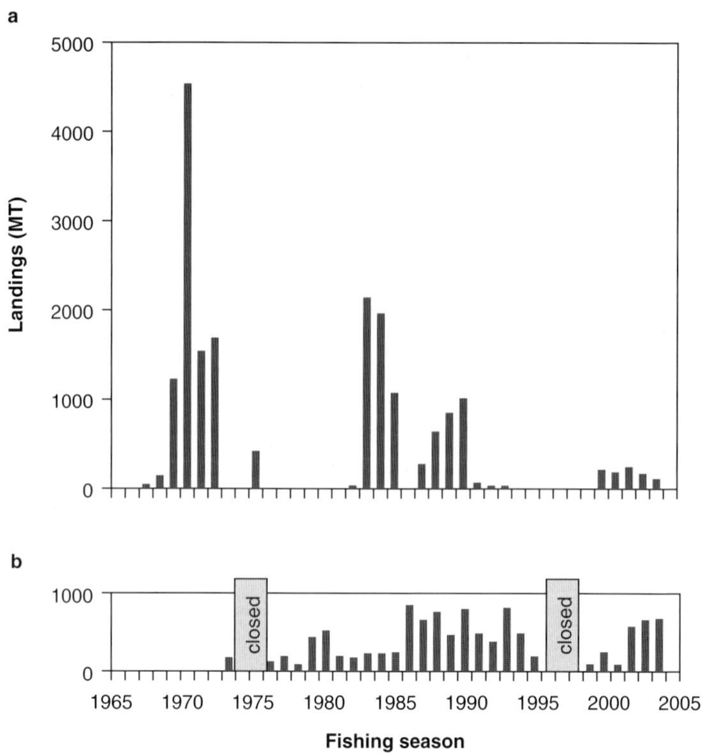

Figure 4.3 Historical series of landings. (a) San Matías Gulf, Río Negro Province, mostly based on dredging. (b) San José Gulf, Chubut Province, based on commercial diving.

The scallop boom was followed by a decade (1972–1981) of virtually no recruitment. In the ensuing two decades there were three harvesting pulses, all supported by sporadic pulses of recruitment, and none of them comparable to the boom years (Fig. 4.3a). Scallop stocks provide many examples of spasmodic dynamics (Caddy and Gulland 1983; Orensanz et al. 1991b, 2006), and it has been suggested that spasmodic recruitment during the three decades that followed the boom of the San Matías Gulf scallop fishery reveals the 'natural' dynamics of the population, recruitment hypothetically being controlled by unidentified environmental conditions. However, the age distribution of the catch (Olivier et al. 1970; Olivier and Capítoli 1980) and of a sample retrieved from the shell dumps (Parma and Orensanz unpublished data) indicate that recruitment was rather regular, not spasmodic, during the time that led to the accumulation of the biomass removed during the boom.

In retrospect, several pieces of information suggest that intensive dredging between 1969 and 1972 irreversibly changed the substratum of the beds and the dynamics of the stock. The fact that only four pulses of recruitment occurred during the 35 years that followed the boom of the fishery suggests the possibility of recruitment overfishing. The scallop boom also changed the social dynamics of a closely-knit community by creating a transient climate of prosperity that vanished with the scallop beds. After the fishery was formally closed in 1972, brokers, processors, and skippers started looking for new grounds. Round 1 was lost.

Round 2: From dredging to commercial diving (the 1970s)

The existence of attractive scallop beds in San José Gulf, under the jurisdiction of the Chubut Province, had been well known to a few skippers since at least 1968. By 1971, up to five dredgers from San Antonio and Rawson fished in the gulf, but operations were irregular due to lack of logistical facilities; the catch was processed in makeshift camps on the beach. The local market was supplied by the collection of scallops stranded on the beach as a consequence of frequent windstorms. However, after the collapse of the San Matías fishery, with good markets and well-established marketing connections, the area suddenly became attractive for commercial dredging. A few boats started to operate more regularly in San José Gulf during the early 1970s, landing the catch at El Riacho (Fig. 4.1c).

In 1971, a group of commercial divers arrived from Uruguay on board a now legendary 'black sailboat', looking for adventure and opportunity. The Picallo brothers, Santiago and Zeito, led the team and claimed that commercial diving for scallops was a habitat-friendly and economically-viable alternative to dredging. Their idea was hard to sell, as commercial diving was virtually unknown in Argentina. Some sport divers occasionally sold their catch locally, but this was a small and irregular activity. The Picallo brothers teamed up with a few local fishers who contributed their boats and rigged them for commercial diving. In February 1973 they were joined by a small group of scientists to conduct a survey of the scallop grounds and investigate the merits of commercial diving (Olivier et al. 1974). Scientists advised provincial managers to ban the use of dredges, and started a research project in November 1974 with the intent of providing better scientific support and, hopefully, identifying a sustainable management option. Concerned because of the collapse of the scallop fishery in the adjacent Río Negro Province and the pressure to allow dredging, the fisheries authority of Chubut Province closed all fishing activities for two years (1974–1975). The commercial diving fishery was formally opened in 1976 and dredging has been effectively banned ever since then.

The fishing units initially geared up for commercial diving were three small wooden side-trawlers, namely *San José Chico*, *Virgen de Luján*, and *Ría de Vigo*, equipped with large compressors to supply up to seven hookah divers. Air supply systems were rudimentary. The operation became commercially viable once the divers designed a bag, locally known as *salabardo*, that allowed them to use both hands when fishing. This modest technological development paved the way for an entirely new way of harvesting the Tehuelche scallop. Fishers quickly tried several types of vessels and gear. The converted trawlers were abandoned because the simultaneous operation of multiple divers was impractical. In 1977, Luigin Isola, a local entrepreneur and boat-maker, developed the basic fishing unit still in use today

(Fig. 4.2b). It consists of an 8 m long boat built of wood or fiberglass, powered by a 35–110 hp outboard motor (Ciocco 1995; Ciocco et al. 2005). Boats are launched from sandy beaches with the help of tractors (Fig. 4.2b).

The fight to keep the ban on dredging in San José Gulf was long, protracted, and coincided with a politically-difficult period: the repressive military dictatorship that ruled Argentina between 1976 and 1983. Pressure from the dredging fleets based in San Antonio Oeste and Rawson was strong and was supported by some local power brokers. At one point, a well-respected skipper entered a highly publicized hunger strike to protest against the ban. To further complicate things, during the late 1970s the ecological effects of dredging and trawling were still far from being an issue of scientific interest, managerial attention, or concern to conservationists. References in the international literature to support the ban were virtually non-existent, in contrast with the numerous articles and reviews published on that subject over the last 15 years.

Under ideal circumstances a controlled experiment should have been conducted to investigate the effects of dredging, but there was an immediate urge to provide a tangible rationale and to make a decision. A short-term project was conducted during May 1976 using a dredging boat geared for both commercial diving and dredging, a sort of spirited contest where an old-time skipper and his crew, a bunch of seasoned divers and a small team of young scientists tried to demonstrate their points. Divers documented photographically the ecological effect of the heavy dredges (Fig. 4.4): scouring of the ground, silting, and incidental mortality of various organisms (Fig. 4.5). Scientists confirmed the complete lack of selectivity of the gear (Fig. 4.6), and the skipper searched in vain for deep grounds that, being beyond the reach of divers, could justify the use of dredges. Pictures taken during the survey (Figs 4.4 and 4.5) provided the powerful and tangible testimony that tilted the balance in favor of the ban.

At the same time, a project conducted between 1974 and 1977 (Orensanz 1986) provided support for management. It was shown that the legal size limit of 60 mm, which had been established years earlier with no sound rationale, was very reasonable: it allowed scallops to go through one or two reproductive seasons, and corresponded to the minimum muscle size required by exporters. Divers quickly developed a search image for commercial size, the result being very high selectivity (Fig. 4.6). Divers concentrated in very dense patches, with a threshold density around 30 scallops/m^2, and did not have to harvest deeper than 10 m to conduct a profitable activity. It was expected that thinning by divers would release intraspecific competition for food. Scallops below legal size, and those living at densities below the threshold of commercial profitability or in beds deeper than the normal operational range of divers, provided a sizeable reproductive reserve. The commercial diving fishery, it was concluded, had a great potential for self-regulation as long as there were no increases in price that would make deeper or sparser beds worth targeting (Orensanz 1986).

Figure 4.4 Experimental dredging in San José Gulf, May 1976. This picture gives a sense of the weight of a loaded dredge. Skipper of the FV *Adelante Boca Juniors* (Boca Juniors is a popular Argentine soccer team) was Dn José Fondacaro.

Collaboration between scientists and a group of innovative fishers showed that habitat-friendly commercial diving was feasible, and resulted in the first and still the only dredging and trawling-free area of the southwestern Atlantic. Round two was won.

Round 3: No controls or incentives (the 1980s and 1990s)

Following the decline and fall of the military dictatorship, developments during the 1980s and 1990s were punctuated by political and economic crises. The scientific team that had conducted the scallop research project had been disbanded in 1978 as part of a purge of the country's scientific community, and it was not until the mid-1980s that a new cohort of young scientists was in place to conduct research on the coastal resources. The San José Gulf commercial diving fishery grew with little regulation apart from the legal size, while institutions, including fisheries management and science, were rebuilt, although often distracted by more pressing matters. There was a license program, but no cap on the number of licenses. By 1992 there were more than 30 teams in operation (Fig. 4.7). In the absence of incentives for conservation, including secure access rights, a 'race for fish' resulted from a 'tragedy of the commons' situation.

a b c

Figure 4.5 The scallop dredge in action. (a) View of a dense unfished bed. (b) Scouring of the bottom. (c) Siltation. These are part of a series of pictures that tilted the balance in favor of the banning of dredges three decades ago (photographs by Zeito Picallo).

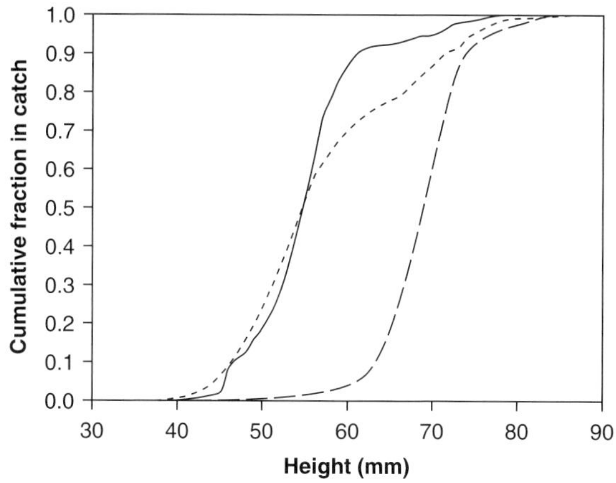

Figure 4.6 Cumulative size-frequency distributions of experimental fishing on board the *Adelante Boca Juniors*, May 1976. Solid line: dredge; dotted line: scientific diver attempting to pick every scallop along his path; dashed line: commercial diver.

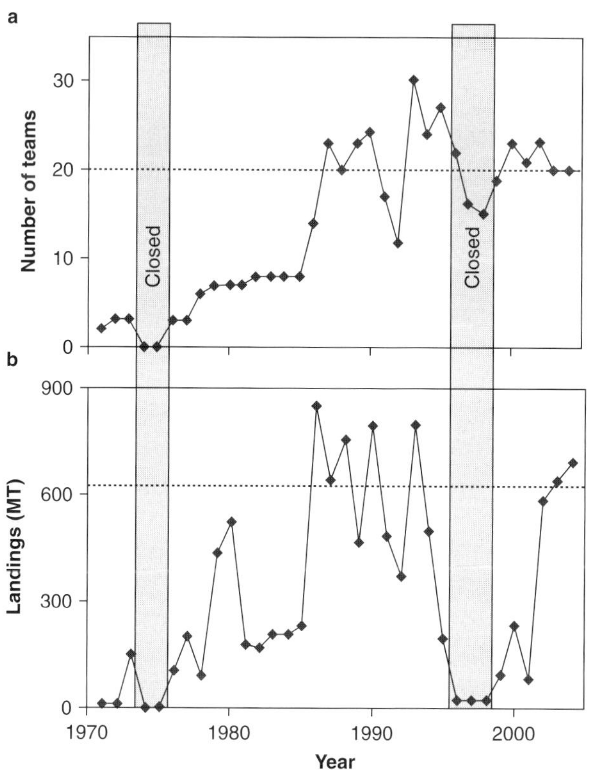

Figure 4.7 San José Gulf, commercial diving fishery, 1970–2004. (a) Size of the fishing force (number of teams). (b) Landings (MT, closed periods indicated). Dotted lines indicate precautionary levels (20 teams, 600 tons/yr).

Reported annual landings, which underestimate the real catch, were generally above 600 tons during the period 1987–1995 (Fig. 4.7). To finally precipitate a collapse, a severe economic crisis erupted in 1989; by the middle of the year, inflation was running at a daily rate of 3%, the relative value of the *peso* fell to historically low levels, and the democratic government that replaced the dictatorship collapsed. Since the scallop fishery was export-oriented, the relative value of the product skyrocketed. The threshold density level for profitable operation went down, and divers risked working in deeper water. Commercial divers are habitat-friendly compared to dredges, but they are also tremendously efficient at searching and depleting highly valuable resources. In addition, a new market developed for small-sized scallops, which were used to decorate cans of shellfish stew. All the factors that had previously contributed to maintain a reproductive stock suddenly evaporated.

A time series of CPUE assembled by Ciocco (1995) for the period 1976–1994 using data from various sources shows a declining trend, with a historical low in 1994 (Fig. 4.8). Information collected through interviews with old and active fishers confirmed the trend, even if an apparent subjective positive bias increased with the time lapsed (Fig. 4.8). The provincial fisheries administration, concerned by reports from the field, contracted with the National Patagonic Center (a branch of the National Council for Scientific and Technical Research) to survey the grounds in December 1995 and November 1996 (Ciocco et al. 1996, 1997, Fig. 4.9a).

The surveys were done by divers counting scallops along transects, following an adaptive design (Thompson 1994). Their results revealed a grim reality: the scallops left could scarcely support a single team during the fishing season (Ciocco

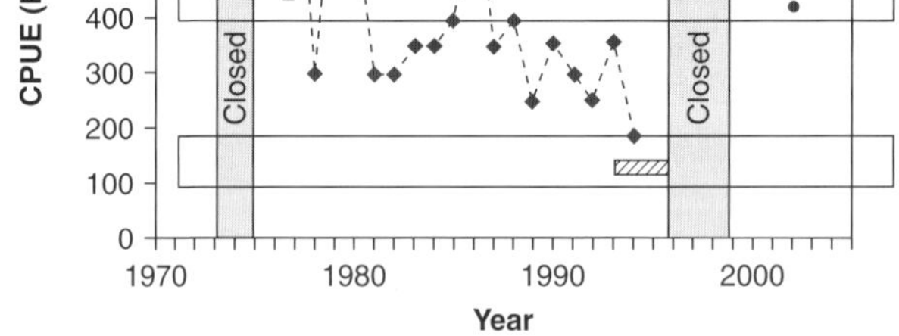

Figure 4.8 San José Gulf, commercial diving fishery, 1970–2004. Catch per unit of effort (CPUE), expressed as kg/diver/day. Connected diamonds: series compiled from various sources by Ciocco (1995). (■) Data from Orensanz (1986). (•) Data from logbook program (2002–2003). Cross-hatched rectangles: ranges reconstructed from interviews; older values (early 1970s) are likely to reflect maxima rather than averages due to a positive subjective bias. Horizontal rectangles indicate average 'good times' CPUE (500 kg/diver/day) and the low level that preceded the closure.

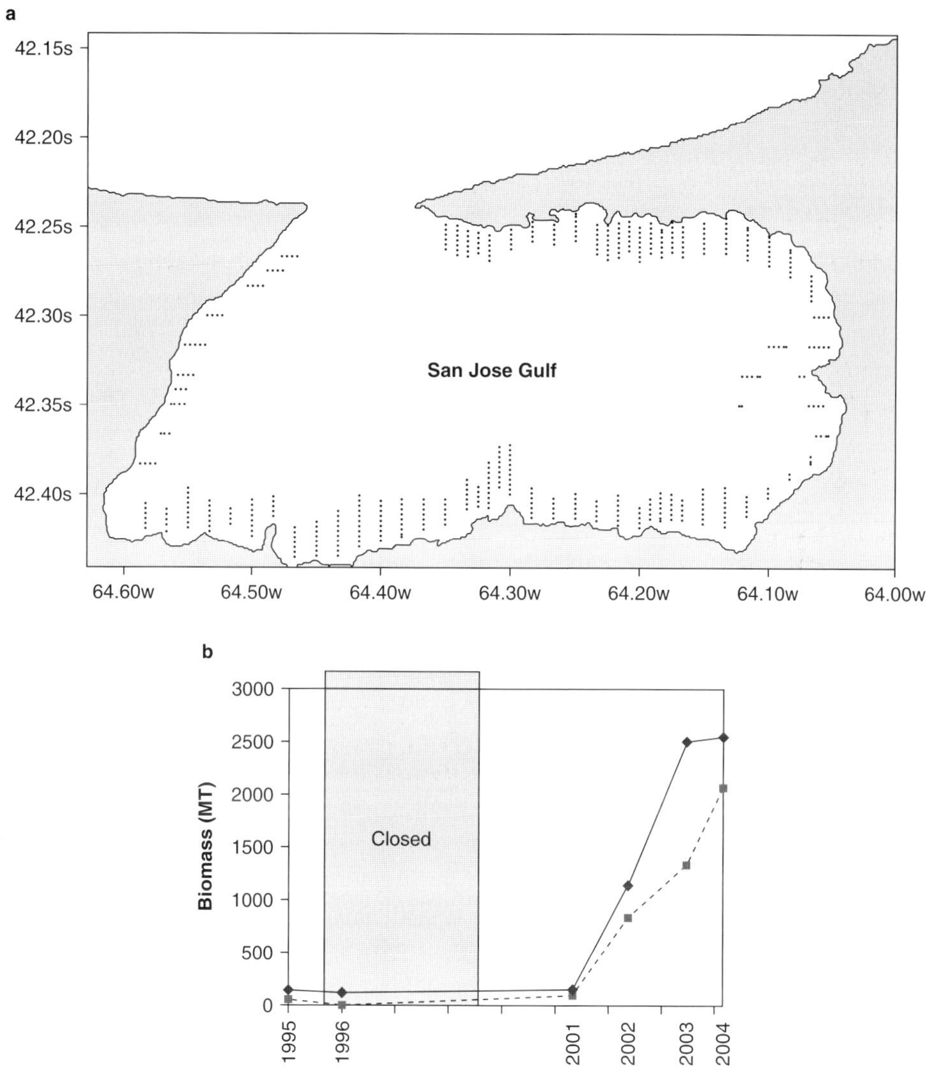

Figure 4.9 Recovery of the San José Gulf scallop stock. (a) Survey design. (b) Estimated abundance in six surveys conducted between 1995 and 2004. Solid line: total biomass; dashed line: biomass of sub-legal stock.

and Orensanz 1997). Besides, the concentration profile had changed (Orensanz et al. 1998), partly as a result of divers targeting the densest patches (Ciocco and Orensanz 1997; Orensanz et al. 2006) – the few scallops left were dispersed at very low densities. The fishery was completely closed for three years between 1996 and 1998. Only 15 of the fishing units continued operating during those years, targeting other resources, mostly the southern ribbed mussel or *cholga* (*Aulacomya ater*), or poaching the few remaining scallops. *Cholga* beds, which take a long time to develop, were severely depleted during that period, and have not recovered. In the absence of monitoring, appropriate incentives, and efficient controls, a fishery that had survived for two decades, in spite of little regulation, collapsed, with devastating consequences for fishers and their families. Round 3 was lost.

Round 4: Recovery, participatory management, and incentives (2000–2004)

During the late 1990s, a crisis hit the Argentine industrial fishery, whose backbone is the Argentine hake (*Merluccius hubbsi*). The provincial administration, under pressure to provide alternatives, reopened the scallop fishery in 1999 and 2000 with fixed monthly quota allocations to license holders (10 in 1999 and 19 in 2000); an additional group of teams operated illegally (9 in 1999 and 4 in 2000). Divers soon noticed that the quota were out of proportion with availability: they were unable to reach them even though they were running the risks of fishing at 30 m depth or more. A key element for solving this crisis was the Association of Artisanal Fishers of Puerto Madryn (APAPM), an organization formed in 1993, as the first symptoms of an impending crisis became evident. By 2000 the membership of APAPM included approximately 60% of the fishers, had formal legal status, and played an increasingly proactive role (Santa Ana 2001). It was significant that their charter has a strong commitment to sustainability.

The APAPM made a formal presentation to the provincial fisheries authority questioning the unrealistically high catch allocation for the 2000 fishing season. Managers, uncertain about how to proceed, approached CENPAT (Centro Nacional Patagónico) looking for scientific advice, but scallop beds had not been surveyed in years and thus scientists were unable to recommend a quota. It soon became apparent that because of the uncertainties surrounding the fishery, there was need for discussion of management issues among scientists, managers, and stakeholders. In 2001, the provincial fisheries authority convened a technical advisory board for the commercial fishery of San José Gulf, integrated by technical staff from the administration, scientists from CENPAT and representatives of APAPM (Parma et al. 2003; Orensanz et al. 2005). In 2005, the advisory board was expanded to incorporate representatives of the entity that administers the Natural Protected Area Peninsula Valdés, and the provincial authority of tourism and protected areas that has a management mandate. Participation of different sectors in the advisory board was the first structured co-management experiment in the history of Argentine fisheries.

In November 2000, a group of 15 permit holders, all of whom had survived the three year scallop closure, produced an important document where they proposed a well fleshed-out program that would grant long-term secure rights to a limited number of fishing units and introduce incentives for conservation. Based on the fishers' proposal, the advisory board evaluated the prospects of a limited-entry program (Orensanz et al. 2003). Although the dynamics of the exploited scallop metapopulation is not well understood (Amoroso 2004), historical data and experience suggest that the number of permits should not exceed 20, and the annual scallop catch should have a precautionary cap of 600 tons (Fig. 4.7). This advice from the board resulted in a moratorium on the number of permits.

Since 2001, scientists from CENPAT, with the collaboration of fishers from APAPM and support from the administration, have conducted annual surveys of the scallop beds leading to annual catch quota recommendations. Data show a sustained rebuild of the stock starting in 2001 (Fig. 4.9b). Since 2002 the fishery has been thriving again. Round 4 was won.

Round 5: New challenges (2005–?)

Recovery of the resource has brought pressure to increase the number of licenses, posing the risk of a ratchet effect (Hennessey and Healey 2000) – it is easy to increase the number of permits when the stock goes up, but difficult to cut them back when the stock goes down. Some managers in the provincial fisheries authority have, however, suggested that the number of permits could be adjusted to track stock abundance. Besides the feared ratchet effect, one well-documented problem with this policy is that it creates a disincentive for conservation when the resource is at a low level. Fishers that stay in the fishery have no incentives to be conservative if the rebuilt stocks are to be shared with new entrants (Hilborn et al. 2005). In addition, one cannot expect an honest collaboration in data gathering and monitoring when the number of fishing permits is pending on the assessments.

Limited-entry programs are usually preceded by a moratorium (Charles 2001), but the transition between the two is a delicate matter. Functional limited-entry systems offer appropriate incentives for responsible fishing (Heizer 2000), provided they are accompanied by effective disincentives for non-compliance, generally in the form of effective enforcement and penalties. The latter may prove to be the Achilles heel of the San José Gulf commercial diving fishery. Conditions for enforcement are favorable – it is supported by most fishers – transit of shellfish from landing sites to processing plants must go through a bottleneck that is the single road that runs through the isthmus (Fig. 4.1b), and the region is a protected area. However, legal constraints, lack of coordination between agencies, and a weak judiciary system conspire to make enforcement ineffective. This became clear during the 2004 season, when, according to available evidence, there was an unprecedented gap between real and declared landings.

Equity or social sustainability issues also require serious attention. Currently, there are no clear rules for the cancellation of a permit in the case of non-compliance, nor a system for ranking prospective permit-holders waiting to take the place of dismissed non-compliers. There is, however, agreement on general criteria: years and continuity in the fishery, compliance with regulations, and being the son of a retiring fisher. Also, there is agreement on conditions that have been a point of contention in other limited-entry systems: permits should not be transferable, permit-holders should be physically engaged in fishing activities, and a one permit per permit-holder policy.

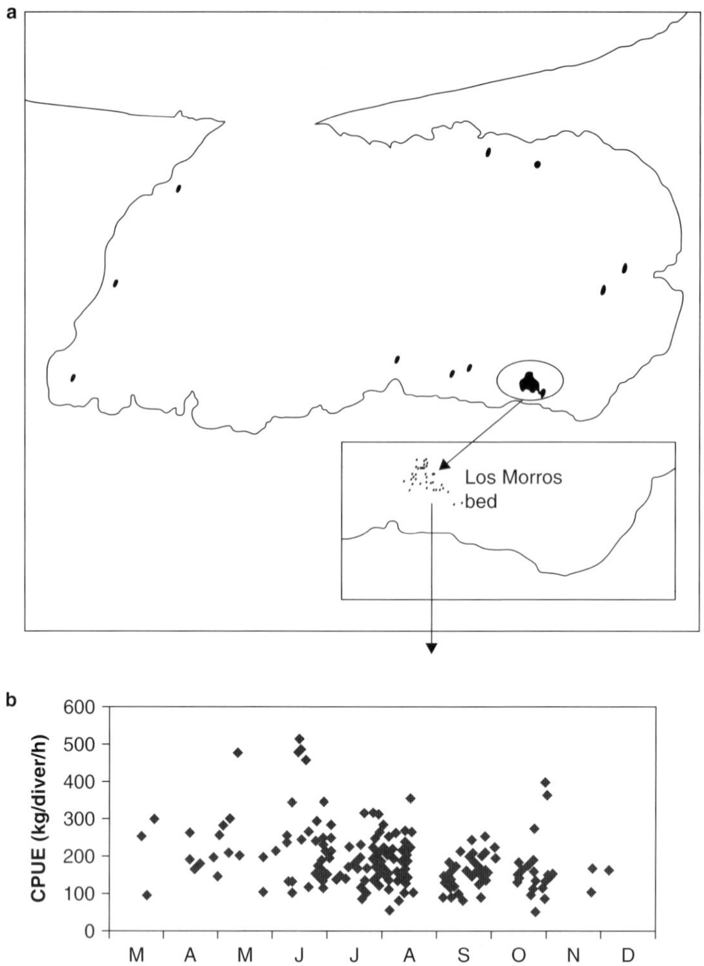

Figure 4.10 San José Gulf commercial diving fishery, 2002. Results from the logbook program. (a) Spatial allocation of fishing effort. Notice the concentration of effort in a single fishing ground (Los Morros). (b) Trends in CPUE (kg/diver/h) in the Los Morros fishing ground.

It has become clear that the technical support for management must be upgraded, particularly with regards to the fishing process. A voluntary logbook program was started in 2002 with the active participation of several fishers under a confidentiality agreement (Cinti et al. 2003). The value of this information is well illustrated with data from the 2002 season (Fig. 4.10), when most of the effort concentrated in a single bed, namely Los Morros. Fine spatial resolution made it possible to understand fishers' behavior and to interpret CPUE data (Fig. 4.10b) at an appropriate scale and attention to sources of variation. Round 5 is on, and the clues required for winning are becoming clear.

Summary

A decline in scallop supplies from the Georges Bank during the late 1960s created a market for alternative products. Many small-scale scallop fisheries sprouted to seize the opportunity, among them a dredge fishery for the Tehuelche scallop in San Matías Gulf, Argentine Patagonia. After the fishery collapsed in 1971, dredgers turned their attention to dense grounds discovered in San José Gulf, an ecologically sensitive area that had just been designated as a provincial marine park. Factual evidence and common sense suggested that the collapse of the San Matías fishery was due to the use of a fishing gear that was unselective and disruptive of the benthic ecosystem.

Collaboration between scientists and a group of innovative fishers showed that habitat-friendly commercial diving was feasible, and this resulted in the first commercial diving fishery of Argentina. Subsequently, the fishery was neglected and grew unregulated during a period of political and economical uncertainty. The commercial diving scallop fishery collapsed in 1996 and was closed for three years. The fishers that survived the crisis developed a substantive proposal for recovery, its backbone being a limited-entry program that would introduce appropriate incentives and put a cap on effort. A co-management program, involving scientists, fishers, and managers was put together to provide guidance. The fishery subsequently recovered, and is thriving again. Recovery of the resource brought pressure to increase the number of licenses, posing the risk of a ratchet effect on effort. A new round is on, but this time fishers, scientists, and managers have learned important lessons and know how to work together. The direction for continued achievement is now clearer.

References

Amoroso, R. (2004) Heterogeneidad espacial en la dinámica de la metapoblación de vieira tehuelche (*Aequipecten tehuelchus*) del Golfo San José. Thesis (licenciatura degree), Universidad Nacional de la Patagonia, Puerto Madryn, Argentina.

Caddy, J.F. and Gulland, J.A., (1983) Historical patterns of fish stocks. *Marine Policy*, **7**: 267–78.

Caddy, J.F. and Lord, E.I. (1971) High price of scallop landings conceals decline of offshore stocks. *Fisheries of Canada*, **23**: 3–7.

Campagna, C. and Lichter, A. (1996) *Las ballenas de la Patagonia*. Emecé Editores, Buenos Aires.

Charles, A. (2001) *Sustainable Fishery Systems*. Blackwell Science, Oxford.

Charpy-Roubaud, C.J., Charpy, L.J. and Maestrini, S.Y. (1982) Fertilité des eaux cótieres nord-patagoniques: facteurs limitant la production du phytoplancton et potentialités d'exploitation mytilicole. *Oceanologica Acta*, **5**: 179–88.

Cinti, A., Parma, A.M. and Orensanz, J.M. (2003) *Seguimiento de la pesca de vieiras en el Golfo San José durante la Temporada 2002. Recomendaciones para el monitoreo y control de la pesquería.* Technical Advisory Board for the Management of the Commercial Diving Fishery, Chubut Province, Document 7.

Ciocco, N.F. (1991) Differences in individual growth rate among scallop (*Chlamys tehuelcha* (d'Orb.)) populations from San José Gulf (Argentina). *Fisheries Research*, **12**: 31–42.

Ciocco, N.F. (1995) *La marisquería mediante buceo en el Golfo San José (Chubut, Argentina).* Plan de Manejo Integrado de la Zona Costera Patagónica (PNUD-GEF), Informe Técnico 2.

Ciocco, N.F. and Orensanz, J.M. (1997) *Collapse of the Tehuelche Scallop* (Aequipecten tehuelchus) *Fishery from San José Gulf (Argentina).* 11th International Pectinid Workshop, La Paz, Mexico.

Ciocco, N.F., Gosztonyi, A.E., Galvan, D., Monsalve, M.A., Diaz, M.A., Vera, R., Ibañez, J., Ascorti, J., Signorelli, J.C. and Berón, J.C. (1996) *La vieira tehuelche del golfo San José: primeros resultados de la campaña de relevamiento SANJO/95.* LAPEMAR, Centro Nacional Patagónico, Puerto Madryn, Argentina, Technical Report 1.

Ciocco, N.F., Monsalve, M.A., Diaz, M.A., Vera, R., Signorelli, J.C. and Diaz, O. (1997) *La vieira tehuelche del golfo San José: primeros resultados de la campaña de relevamiento SANJO/96.* LAPEMAR, Centro Nacional Patagónico, Puerto Madryn, Argentina, Technical Report 2.

Ciocco, N.F., Lasta, M.L., Narvarte, M., Bremec, C., Bogazzi, E., Valero, J. and Orensanz, J.M. (2005) Argentina. In: S. Shumway (Ed.) *Scallops: Biology, Ecology and Aquaculture* (2nd edn). Elsevier Publishers, Amsterdam.

Darwin, C. (1962) *The Voyage of the Beagle.* Doubleday Anchor Book and The American Museum of Natural History (New York). Based on the 1860 version, the last one edited by Darwin (the diaries were first published in 1839).

DGCAP (Dirección General de Conservación de Areas Protegidas) (1998) *Plan de manejo 'Sistema Península Valdés'.* Gobierno de la Provincia del Chubut, Secretaría de Turismo y Areas Protegidas, Rawson, Argentina.

Heizer, S. (2000) The commercial geoduck (*Panopea abrupta*) fishery in British Columbia, Canada – an operational perspective of a limited entry fishery with individual quotas. In: R. Shotton (Ed.) *Use of Property Rights in Fisheries Management.* FAO Fisheries Technical Paper 404/2.

Hennessey, T. and Healey, M. (2000) Ludwig's ratchet and the collapse of New England groundfish stocks. *Coastal Management*, **28:**187–213.

Hilborn, R., Orensanz, J.M. and Parma, A.M. (2005) Institutions, incentives and the future of fisheries. *Philosophical Transactions of the Royal Society B*, **360:** 47–57.

Iribarne, O., Lasta, M., Vacas, H., Parma, A.M. and Pascual, M.S. (1991) Assessment of abundance, gear efficiency and disturbance in a scallop dredge fishery: results of a depletion experiment. In: S.E. Shumway and P.A. Sandifer (Eds) *An International Compendium of Scallop Biology and Culture: A Tribute to James Mason. Selected Papers from the VII International Pectinid Workshop. World Aquaculture Workshops, No. 1.* World Aquaculture Society.

Olivier, S.R., and Capítoli, R. (1980) Edad y crecimiento en *Chlamys tehuelcha* (d'Orbigny) (Mollusca, Pelecypoda, Pectinidae) del golfo San Matías (Pcia. de Río Negro, Argentina). *Anales del Centro de Ciencias del Mar y Limnología, UNAM (Mexico)*, **7**: 129–40.

Olivier, S.R., Christiansen, H.E. and Capítoli, R. (1970) *Notas preliminares sobre la vieira tehuelche del golfo San Matías (Prov. de Río Negro).* Proyecto de Desarrollo Pesquero, UNDP/FAO, Mar del Plata, Argentina, Documento Informativo.

Olivier, S.R., Orensanz, J.M., Quesada, L.A. and Capítoli, R. (1974) *Estado actual de las poblaciones de vieira tehuelche (Chlamys tehuelcha) del golfo San Jose.* Centro Nacional Patagónico, Puerto Madryn, Argentina, Informe Técnico 3.

Orensanz, J.M. (1986) Size, environment and density: the regulation of a scallop stock and its management implications. *Canadian Special Publications in Fish and Aquatic Sciences*, **92**: 195–227.

Orensanz, J.M., Pascual, M.S. and Fernández, M. (1991a) Biology and fisheries of the scallops from the southwest Atlantic Ocean. In: S. Shumway (Ed.) *Scallops: Biology, Ecology and Aquaculture* (2nd edn). Elsevier Publishers, Amsterdam.

Orensanz, J.M., Parma, A.M. and Iribarne, O. (1991b) Population dynamics and management of natural stocks. In: S. Shumway (Ed.) *Scallops: Biology, Ecology and Aquaculture* (2nd edn). Elsevier Publishers, Amsterdam.

Orensanz, J.M., Parma, A.M. and Hall, M.A. (1998) The analysis of concentration and crowding in shellfish research. *Canadian Special Publications in Fishery and Aquatic Sciences*, **125**: 143–57.

Orensanz, J.M., Parma, A.M., Ciocco, N.F. and Cinti, A. (2003) *Programa de entrada limitad para la pesca comercial de mariscos mediante buceo en el Golfo San José.* Technical Advisory Board for the Management of the Commercial Diving Fishery, Chubut Province, Document 9.

Orensanz, J.M., Parma, A.M., Jerez, G., Barahona, N., Montecinos, M. and Elías, I. (2005) What are the key elements for the sustainability of 'S-fisheries'? Insights from South America. *Bulletin of Marine Science*, **76**: 527–56.

Orensanz, J.M., Parma, A.M., Turk, T. and Valero, J. (2006) Population dynamics, assessment and management. In: S. Shumway (Ed.) *Scallops: Biology, Ecology and Aquaculture* (2nd edn). Elsevier Publishers, Amsterdam.

Parma, A.M., Orensanz, J.M., Elías, I. and Jerez, G. (2003) Diving for shellfish and data: incentives for the participation of fishers in the monitoring and management of artisanal fisheries around southern South America. In: S.J. Newman, D.J. Gaughan, G. Jackson, M.C. Mackie, B. Molony, J. St John and P. Kaiola (Eds) *Towards Sustainability of Data-limited Multi-sector Fisheries.* Department of Fisheries of Western Australia, Fisheries Occasional Publications No. 5.

Santa Ana, C. (2001) *Relevamiento del Sector Pesquero Artesanal en los Golfos Nuevo, San José y sur del San Matías.* Pew Fellows in Marine Conservation Program and CENPAT (Puerto Madryn). Database (ACCESS) and associated documentation.

Thompson, S.K. (1994) *Sampling*. John Wiley and Sons, New York.

Chapter 5
Evolution and Lessons from Fisheries and Coastal Management in the Philippines

Alan T. White, Edgardo Gomez, Angel C. Alcala, and Garry Russ

Introduction

Over the last three decades, Philippine institutions and communities have been experimenting with new approaches to coastal management to try to stem the increasing tide of destruction to coastal habitats and the decline of fisheries. Many attempts at coastal and fisheries management have been conducted, ranging from broad-area management planning for whole bays to small community-based marine protected area (MPA) projects (Courtney et al. 2000). The impetus driving coastal and fisheries management projects and MPAs is attributed to the interest of local and national government through devolution of authority under national policies to promote protection and management of coastal resources. Municipal, city, and provincial governments have thus become the key players in the implementation of integrated coastal management (ICM) and MPA projects in association with national government, non-governmental organizations (NGOs), people's organizations, research institutions, and multilateral and bilateral donor organizations, employing different strategies and approaches (White et al. 2002, 2005a). The many useful lessons being learned in the context of ICM programs are summarized in this chapter.

Declining fisheries, mangrove and coral reef destruction, and poverty among coastal communities are issues of primary concern. Fisheries-related food production in the Philippines has been static for the last 10 years despite an increased number and tonnage of commercial vessels, increased number of municipal fishers, and increased coverage of fishponds (Bernacsek 1996; BFAR 1997; Courtney et al. 1999). Municipal fish catch has been on a steady decline, accelerated by the use of illegal fishing practices, such as dynamite and small-mesh fishing nets, as well as overfishing (Fig. 5.1). Philippine fisheries generally reflect world trends of declining catches and fishing down the marine food web, such that top predators

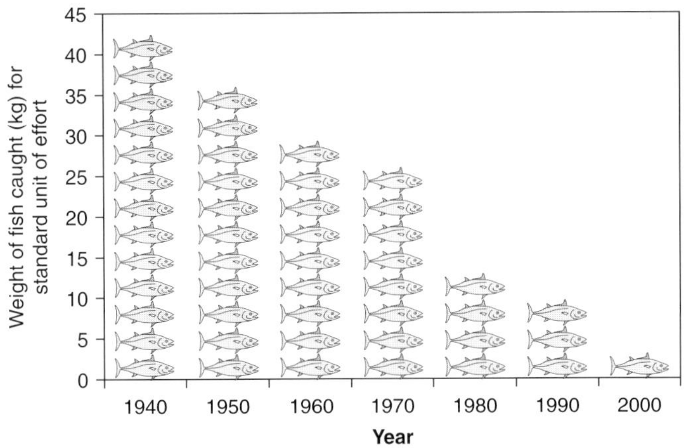

Figure 5.1 Decline in catch per unit effort since the 1940s for fishers using hook and line from six provinces in the Philippines (Green et al. 2003).

such as sharks, groupers, and others are increasingly scarce in Philippine catches (Pauly et al. 1998, 2002; Green at al. 2003). It is currently estimated that the 26 000 km^2 of coral reefs (with only about 5% in excellent condition) contribute at least US$1.35 billion annually to the economy (Gomez et al. 1994; White and Cruz-Trinidad 1998; Burke et al. 2002). It is suggested that this estimate could be significantly increased with improved coastal management efforts (White and Cruz-Trinidad 1998).

Factors contributing to this decline are intertwined and not easily isolated for the purposes of management. Nevertheless, the primary issues and conflicts of interest are: various kinds of pollution; illegal and destructive fishing practices; overfishing due to an open-access fishery regime throughout the country; a lack of planning and control of development in the shoreline and beach areas; increasing poverty among coastal dwellers; a rapidly growing population; and variable political will to address the problems (Courtney et al. 2000; NSO 2000; White and De Leon 2004). A force driving many of the issues is a lack of economic alternatives that would reduce the dependency of coastal dwellers on their natural resource base (Green et al. 2003).

Thus, a diversity of human activities are eroding the natural resource base along the 18 000 km[1] coastline of the Philippines. The intensity of these impacts is threatening the area's potential for future sustainable use. The lack of control of most coastal development and its impacts are indicative of what will occur if stronger and more effective institutions and procedures for coastal and fisheries management are

[1] 18 000 km is used here as the standard coastline of the Philippines based on the coastline of the main islands because this has been commonly cited. New estimates are higher, such as that used by Burke et al. (2002) of 33 900 km based on the perimeters of all islands calculated using digital maps.

not put in place. The challenges are of such magnitude that Philippine institutions are now responding with more concern and integrated approaches than they did in the past. The history of this context and its evolution follows.

Evolution of coastal and fisheries management

The movement towards improved coastal and fisheries management in the Philippines has been influenced through programs and various projects that have tested and refined the practice and implementation (Courtney and White 2000; White et al. 2005b). The evolution of coastal management in the country has approximated a pattern of five stages suggested by Sorensen and colleagues (1990).

(1) Incipient awareness: the need for a coastal program becomes known as signs of coastal resource degradation worsens (1970s and 1980s).
(2) Growing awareness: the need for a program is heightened through national conferences, workshops, or hearings convened by government, academe, or environmental groups (1980s and early 1990s).
(3) National study: heightened awareness resulting from conferences, international assistance, national studies, and policy recommendations (1990s to present).
(4) New program creation: studies on the coastal situation and its management lead to new ICM programs and institutional arrangements (late 1990s to present).
(5) Program development, implementation and evaluation: policies, laws, and programs are implemented and evaluated (starting to occur in 2000 to present).

Important events influencing or affecting the development of coastal and fisheries management is shown in Table 5.1. A legal and jurisdictional transition from central to local government responsibility for coastal management is also reflected in the local government roles (Fig. 5.2).

Two major forces have influenced the development of coastal management in the country since the early 1980s (Courtney and White 2000). The devolution of authority from central to local governments (municipal, city, and provincial) is the first major influence (Fig. 5.2). Second, a series of donor-assisted projects have resulted in a number of experiments in coastal resource management (CRM) (Table 5.1). These projects, working with coastal communities, have focused on nearshore fisheries and coastal habitat management (White and Lopez 1991; Ferrer et al. 1996; Christie and White 1997). In addition, the various efforts made to initiate coastal management have been supported and nurtured by a variety of institutions, including government, non-governmental organizations (NGOs), people's organizations, and research institutions with the support, ideas, and differing strategies of bilateral and multilateral donors.

The decentralization of responsibilities to local governments in the Philippines in the early 1990s, although hailed as a solution to many of the nation's woes, created significant challenges. In the case of coastal management, few coastal municipal

Table 5.1 Important events, laws, and projects in the evolution of coastal and fisheries management. Sources: MCDP (1986), Alcala (1988), White and Lopez (1991), Chua and Scura (1992), Calumpong (1996), Pomeroy and Carlos (1997), Arquiza and White (1999), NIPAP (1999), Courtney and White (2000), DENR et al. (2001), White et al. (2002), and White et al. (2005a, b).

1932	Fisheries Act gives most management responsibility to central government
1940	Hundred Islands, Lingayen Gulf, established as National Park
1930–1960s	Resources considered unlimited in supply not requiring management
1960–1970s	Robust expansion and development in fisheries and aquaculture
1974	First working municipal marine reserve established by Oslob Municipality and Silliman University at Sumilon Island, Cebu
1975	Fisheries decree promoted the optimal exploitation and use of fisheries
1975	Forestry Code established the need to protect mangrove forests
1976	Environmental Impact System established
1976	National Mangrove Committee established
1976	Commercial fishing limited to areas beyond 7 km of the shoreline
1976–1980	Wide-scale survey of coral reefs by University of the Philippines Marine Science Institute and Silliman University Marine Laboratory
1978	Coral gathering is limited to scientific research
1978	Marine Parks Task Force created to recommend sites for marine parks
1978	The Philippine Extended Economic Zone established
1979	Coastal Zone Management Committee with 22 agencies formed
1981	Philippines becomes signatory to CITES
1984–1992	Central Visayas Regional Project begins community-based ICM
1985–1986	Marine Conservation and Development Program of Silliman University supported by USAID establishes Apo, Pamilacan, and Balicasag islands and Carbin Reef as no-take marine reserves
1986	*Muro-ami* and *Kayakas* fishing methods banned in Philippine waters
1986–1992	First bay-wide management program began in Lingayen Gulf
1987	Bureau of Fisheries and Aquatic Resources moves from the Ministry of Natural Resources to the Department of Agriculture
1988	First national marine park established at Tubbataha Reefs, Sulu Sea
1990–1997	Fishery Sector Program of DA–BFAR initiates bay-wide management
1991	Local Government Code devolves responsibilities to local governments
1992	Philippine Council for Sustainable Development created
1992	Philippines becomes a signatory to Agenda 21
1992	National Integrated Protected Areas System (NIPAS) Act passed
1993	Coastal Environment Program of DENR established
1995	Fisheries and Aquatic Resources Management Councils authorized
1995	First International Coral Reef Initiative Meeting held at Silliman University
1996–2004	Coastal Resource Management Project of DENR and USAID implements ICM
1998	Fisheries Code reinforces the roles of local government in management
1998	First national coastal Mayors' conference held to discuss ICM issues
1998–2005	Fisheries Resource Management Project builds on lessons of FSP for bay-wide coastal management
1999	May proclaimed the Month of the Ocean in Philippines
2000	DA and DENR sign Memorandum on implementation of Fisheries Code
2001	More than 100 municipalities and cities allocate budget for ICM
2001	Southern Mindanao Integrated Coastal Zone Project starts
2002	National Coastal Management Policy reviewed at national level
2003	Philippine Sustainable Archipelagic Development Framework drafted
2004	National Coastal Management Policy pending signing as Executive Order
2004	First Coastal Zone Philippines Symposium conducted in Cebu City
2004–2008	Fisheries Improved for Sustainable Harvests Project of BFAR supported by USAID initiates sustainable harvest regimes in four project sites

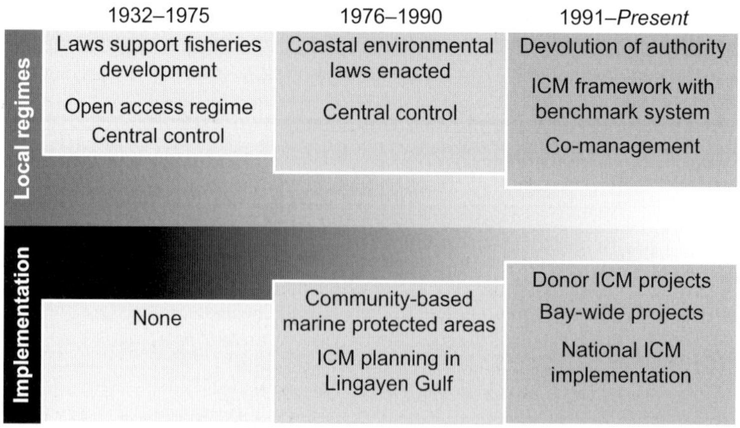

Figure 5.2 Transition of authority for coastal management from central to local government and implementation through time (from DENR et al. 2001 – Vol. 2).

or city governments in the country had the capacity to manage their natural resources. They generally lacked trained personnel, budget, planning capacity, and technical knowledge. In spite of these limitations, the motivation among local governments to manage their resources rapidly increased as they realized the seriousness of the problem and what they would lose if no action was taken (White and Cruz-Trinidad 1998). Thus, the opportunity to improve ICM in the country was seen as tremendous, given the 832 coastal municipalities bordering 18 000 km of coastline.

The current state of coastal and fisheries management in the country depends largely on decentralized authority and various ICM projects of the last 30 years (Table 5.1). These projects, supported by the Asian Development Bank, the United States Agency for International Development (USAID), the World Bank, and others, through the Philippine government, have provided key lessons for improving the state of ICM in the country and providing a sound foundation for the design of future ICM efforts. Such projects, often addressing a broad range of issues, have also included an emphasis on MPAs (White et al. 2005a).

A key lesson generated by the various ICM projects is that it is extremely difficult to plan and implement successful coastal and fisheries management without a multisector approach. Such programs must have sufficient support from the national and local government and its partners, and a strong level of acceptance among the resource-dependent communities. Most successful coastal and fisheries programs in the Philippines are still localized where the geographic scope is small and the number of stakeholders limited. Nevertheless, this is changing as more multi-municipal or city and bay-wide resource management plans are being developed and implemented that address a wide range of issues leading to improved management.

Legal and policy framework supporting decentralization

The legal and policy framework for coastal and fisheries management as well as for the establishment and management of MPAs in the Philippines, is found in the Local Government Code (LGC) of 1991, the Fisheries Code of 1998 and pertaining only to national MPAs, the National Integrated Protected Areas System (NIPAS) Act of 1992. The LGC and the Fisheries Code provide for the policy and institutional framework for carrying out ICM at the local level in the Philippines. This shift to decentralized control over the management of resources encouraged the active participation of local governments and communities.

In the formulation, planning, and implementation of ICM plans, the municipal or city governments are the primary agencies involved. There are many municipal ordinances supporting CRM implementation and MPA establishment within a *barangay* (smallest political unit in Philippines) or multiple *barangay* area. In the case of MPAs, most are established by municipal or city government ordinance. All other government agencies at the national, regional, and provincial level are tasked to provide technical, financial, and resource assistance to these local governments in the implementation of their ICM plans and enforcement of regulations in their municipal waters (Fig. 5.3). The laws and institutions that support implementation of ICM as well as the establishment of MPAs are described in more detail elsewhere (DENR et al. 2001 – Vols 2 and 5; White et al. 2002).

The Fisheries Code likewise provides for inter-local government cooperation for the management of contiguous fisheries and aquatic resources that straddle several municipalities, cities, or provinces. This management strategy is enhanced with the creation of Integrated Fisheries and Aquatic Resource Management Councils (IFARMC) composed of representatives from the local governments, non-governmental organizations, and private and fisher sectors including youth and

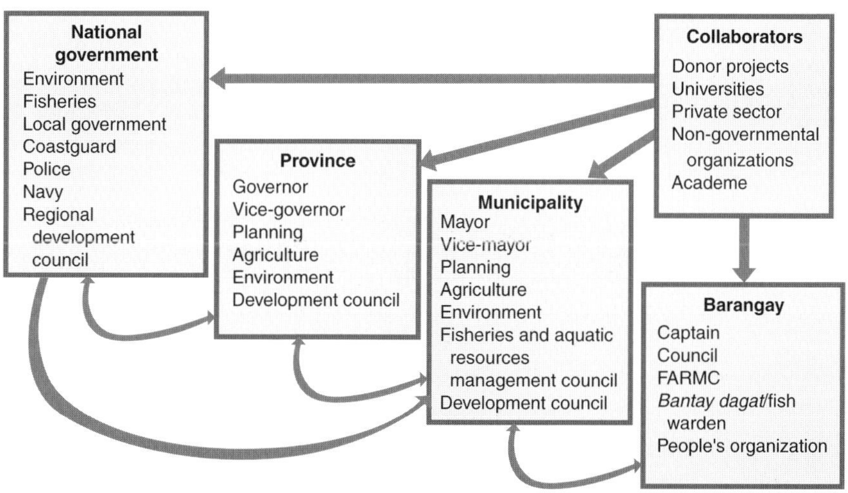

Figure 5.3 Institutions affecting coastal management and their interactions in the Philippines.

women. The IFARMC functions as a recommending body of municipal or city legislative councils for fishery management plans and related regulations. The Bureau of Fisheries and Aquatic Resources (BFAR) of the Department of Agriculture provides guidance for fisheries management within municipal or city jurisdictions. BFAR is also responsible for overall policies for fisheries management in the country and for the direct management of all fisheries that occur outside of the 15 km jurisdictional limit of local governments. The Department of Environment and Natural Resources (DENR) also has jurisdiction over coastal resources through its mandate to protect coastal habitats and to prevent pollution and to set up marine protected areas. The need for DENR and BFAR to coordinate their mandates and management efforts is great and is unfortunately not always attained.

Contribution of community-based marine protected areas

One theme of the three case studies in this chapter is that regardless of how MPAs are established, local support systems need to be in place and functional. Thus, a common thread is the importance of being part of a larger ICM system beyond the immediate community-based MPA. This larger ICM system is the local government planning and implementation framework elaborated below after first exploring community-based MPAs.

The first MPA in the country established as a fish sanctuary was in 1974 on Sumilon Island, Cebu under the guidance of the Silliman University Marine Laboratory (Alcala 1981). Sumilon Island fish sanctuary is often cited as the reason why coral reef fish sanctuaries contribute to improved reef fisheries management (White 1987). This experiment in reef management, that stopped all fishing on a portion of the Sumilon Island reef for about 10 years, allowed researchers to collect substantial data on the effects of such management on the coral reef and its related fisheries (Alcala 1988; Alcala and Russ 1990). The benefits provided compelling evidence for fish biomass spillover from no-take areas. Such evidence has been important for convincing scientists, reef managers, and fishers that fish sanctuaries improve reef fisheries while benefiting fishers in the area (Russ and Alcala 1996; Russ et al. 2003, 2004).

Between 1974 and the present, many similar municipal marine fish sanctuaries or MPAs have been established following the lead of Sumilon and the associated Apo Island (Pajaro et al. 1999; White et al. 2002) (Fig. 5.4). Other well-managed and documented MPAs in terms of their benefits for fisheries and tourism include: Balicasag and Pamilacan Islands, Bohol; Mabini, Batangas; and San Salvador Island, Zambales (Fig. 5.4) (White 1988a, b, 1989, 1996; White et al. 1994; Christie et al. 1999; Kuperan et al. 1999; Green et al. 2002). These examples have followed a general model whereby the portion of a fringing coral reef adjacent to an island or mainland is set aside in a 'no-take' or 'sanctuary' zone and where the area outside of this no-take zone is called a traditional fishing zone or in international terms, a buffer

Figure 5.4 Locations of important MPAs and ICM projects in the Philippines.

zone. Activities that do not damage the coral reef in any way such as traditional fishing methods are usually allowed within the buffer zone. Within the no-take sanctuary zone, entry for swimming and diving is normally permitted although some managers prohibit any form of entry (White 1988a, b, 1989; White and Vogt 2000).

Recent studies have not only indicated the beneficial effects of fish sanctuaries on fishery yields and protecting the coral reef, but also that people participating in such management efforts gain in a variety of ways related to food security, cash income from tourism, and pride in their protection role (White et al. 1994; Vogt 1997; Katon et al. 1999). A salient characteristic of all successful MPA projects in the country is the strong involvement of communities and the local government in planning and enforcement (White and Vogt 2000; DENR et al. 2001 – Vol. 5). This involvement builds the confidence of people to manage their own resources and encourages long-lasting outcomes. Thus, the success of Philippine MPAs hinges on two crucial actors: the (local and national) government and strong stakeholder community involvement.

However, in the context of these apparently successful models, there are problems. In a survey of approximately 439 MPAs, only about 44 MPAs were fully enforced (Pajaro et al. 1999). In a 2003 survey of 54 established and functioning MPAs, about 40% were managed in a sustainable manner (CCEF 2003; White et al. 2004). The area covered by the 54 MPAs reported covers about 810 km^2 of mostly coral reef habitat. Thus, the cumulative effects of the existing MPAs, assuming full implementation, would contribute significantly to the sustainability of the country's coastal ecosystems.

In spite of all the activity to install MPAs, there are only a few that are sustainable in their own right. Contrary to the popular opinion that they are an easy and successful ICM intervention, they are often in need of more assistance. Plus, the benefits from the successful and well-managed MPAs are often being lost to the drain of overfishing, pollution, climate change, and other human pressures in surrounding areas, outside of the controlled MPA or sanctuary (Christie et al. 2002). Nevertheless, they can be the centerpieces of a broader coastal management effort since they are site-specific and display tangible benefits through a locally-controlled mandate. In this context, three case studies follow that highlight a range of benefits and management approaches.

Case 1: Supporting fisheries at Sumilon and Apo islands

Research carried out over the past 28 years at Sumilon Island, and 22 years at Apo Island, both in the Central Visayas (Fig. 5.4), has provided some of the most compelling evidence available worldwide for the utility of no-take marine reserves as fisheries management tools. Abundances of fish targeted by fisheries have been monitored regularly inside and outside no-take reserves at each island since 1983 (Russ and Alcala 2003; Russ et al. 2004; Alcala et al. 2005). In addition, estimates of total annual fishery yields have been made eight times at Sumilon Island between 1976 and 2001 and seven times at Apo Island between 1981 and 2000 (Alcala 1981, 1988; Alcala and Russ 1990; Maypa et al. 2002; Russ et al. 2004). This extensive research addresses four key questions:

(1) Do no-take marine reserves increase the biomass of fish targeted by fisheries?
(2) Do no-take reserves display spillover (net export of adult fish biomass)?
(3) Do no-take reserves reduce, maintain or enhance local fishery yields?
(4) What is the role of local primary stakeholders (fisher communities and local government units) in the establishment and management of no-take marine reserves?

No-take marine reserves were established on the western side of Sumilon Island in 1974 and on the southeast side of Apo Island in 1982 that was legally designated in 1986 by municipal ordinance. Sumilon is a coralline island of 0.23 km^2, surrounded by a fringing coral reef of 0.5 km^2 that has a maximum depth of 40 m. Apo Island

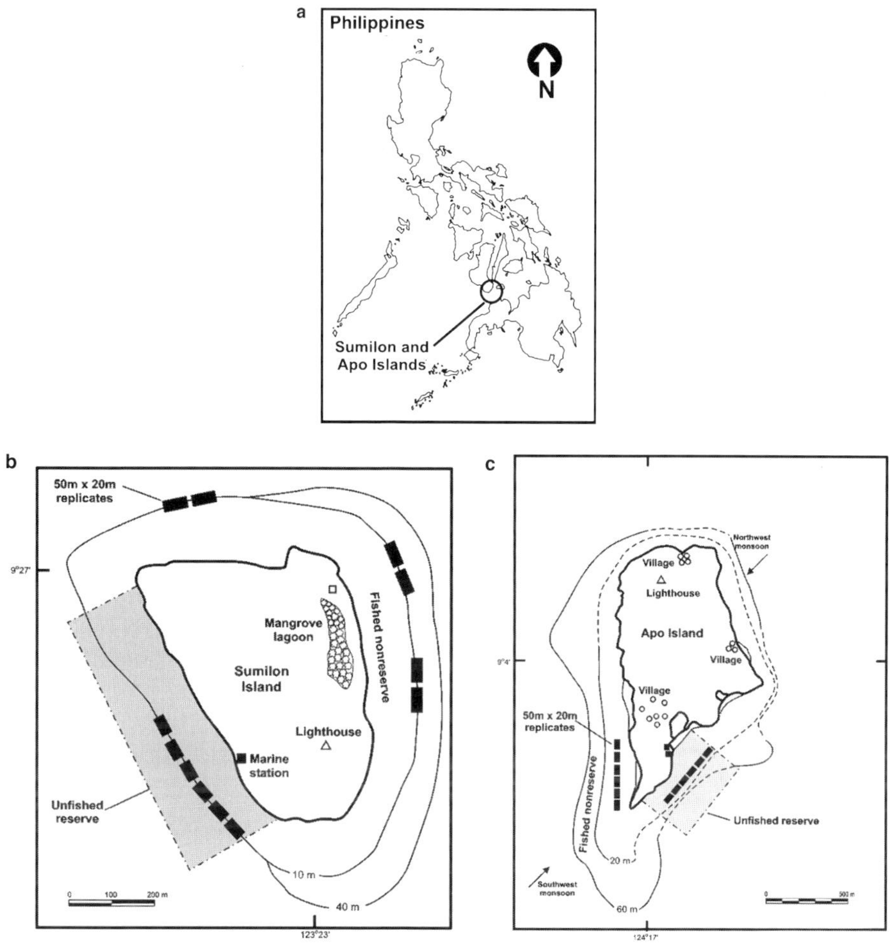

Figure 5.5 Sumilon and Apo Islands showing unfished reserve areas (shaded) and the fished non-reserve areas at each island. The six 50 m × 20 m replicate underwater visual census plots (black rectangles) monitored from 1983–2003 at each sampling site are shown.

is a mainland island of 0.7 km^2 surrounded by 1.1 km^2 of fringing coral reef to 60 m depth (0.7 km^2 to 20 m depth). The fisheries around each island use hook and line, traps, gill nets, and spears. Sumilon has no permanent residents, but is fished regularly by about 100 municipal fishers from the nearby Cebu Island. Apo Island has about 700 permanent residents, with about 100 municipal fishers. The Sumilon reserve is a 0.75 km section (25% of the coral reef area) (Fig. 5.5). The area of the reserve to 500 m offshore is 37.5 ha. The Apo reserve is a 0.45 km section (about 10% of the coral reef area) (Fig. 5.5). The area of the reserve to 500 m from shore is 22.5 ha.

The management histories of Sumilon and Apo reserves are given in Russ and Alcala (1999) and White and colleagues (2002). Both reserves were initially established under municipal ordinance. Day-to-day management was the responsibility of the local communities. However, Sumilon reserve has had a complex history of management from 1974 to 2005. Full no-take (1974–1984, 1987–1991) and partial no-take (all fishing except hook and line banned – 1995–2004) status was applied at Sumilon reserve in the hope of maintaining local fisheries and enhancing tourism at the island. No-take status at Sumilon was maintained in 1974–1983 under an agreement between the local municipal government led by a conservation-minded mayor and Silliman University. It was removed when a newly-elected mayor did not support the reserve concept (1984–1985), and due to poor communication of objectives from local mayors to fishers (1992–1994) (Russ and Alcala 1999). Full no-take status has again been re-established at Sumilon in 2005 by the local government. In contrast, the no-take status of the Apo reserve has been maintained from 1982–2005 due to strong support from the organized and empowered community.

Effects on fish biomass

The biomass of target fish (five fish families accounting for 92.3% of the fisheries yield at Sumilon; four families accounting for 75.6% of the fisheries yield at Apo) increased inside the no-take reserves by 3.1 times over nine years of no-take protection at Sumilon and by 4.6 times over 18 years of no-take protection at Apo (Alcala et al. 2005). The biomass of these fish did not increase outside each reserve. The biomass of large predatory fish, highly-favored targets of reef fisheries, increased by almost 12 times over nine years of no-take protection at Sumilon, and by 17.3 times over 18 years of no-take protection at Apo (Russ and Alcala 2003).

Spillover of fish

Both large predatory fish, also surgeonfish and jacks (accounting for 40–75% of the yield at Apo) increased substantially (by factors of 17 and 3 respectively) inside the Apo reserve over 18 years of protection. Outside the reserve, the biomass of these fish increased significantly closer to (200–250 m) than further away from (250–500 m) the reserve boundary over time (Russ and Alcala 1996; Russ et al. 2004). In addition, hook and line catch rates of surgeonfish were significantly higher closer to (within 200 m) than further away from the reserve (Russ et al 2004). In fact, 63% of the hook and line catch records of the surgeonfish *Naso vlamingii* in 2000–2001 were recorded within 200 m either side of the reserve, in just 11% of the reef fishing area (Russ et al. 2003). Combined, this is some of the best evidence that reserves can display spillover, and benefit local fisheries.

Effects on fisheries yields

The studies of the effects of no-take reserves on local fish yields at Sumilon and Apo islands are long-term manipulative experiments. Because data on reserve biomass and fish yields were collected over almost two decades, it was possible to investigate the relationships between the duration of reserve protection and both the fish biomass inside the reserves and local fish yields outside the reserves. At Sumilon, there were positive correlations between years of reserve protection and both targeted fish biomass in the reserve and catch outside the reserve, although the latter relationship was not statistically significant (Alcala et al. 2005). At Apo, there was a strong positive relationship between years of reserve protection and targeted fish biomass in the reserve, but not between years of reserve protection and catch outside the reserve (Alcala et al. 2005). These results, plus spillover evidence, suggest that marine reserves may help maintain, or even enhance, local fishery yields, in the long term.

Russ and colleagues (2004) reported on the potential fishery effects of the no-take reserve and a 1986 Marine Management Plan at Apo Island. They showed that the benefits over the 20 year period were: a slightly higher catch of surgeonfish and jacks; increased hook and line catch rate; and a reduction in hook and line fishing effort. The fishery and tourism benefits generated by the reserve enhanced the living standard of the island community.

The role of the local community and local government

Through social and community processes, the local fisher community was educated and empowered while the local government (village and town) learned to protect and manage their marine resources. Apo marine reserve, an early community-based MPA established in the country, is now a classic example of a highly successful community-based coral reef fishery and marine biodiversity conservation project operating under the policy and legal framework of the Local Government Code of 1991 and the Fisheries Code of 1998. This success is due to the collaborative partnership among an organized fisher community, a local government, and an NGO academic institution, serving primarily as technical and social facilitator-adviser (Alcala and Russ 2000, 2001).

Important lessons learned from the experiences at Sumilon and Apo Islands are:

- Strong local community support can protect no-take marine reserves over time.
- No-take marine reserves can increase the biomass of fish targeted by fisheries.
- No-take reserves can display some spillover, even though the effect may be small, and for coral reef fish, occur over a spatial scale of hundreds of meters, not kilometers. Such an effect, although small, can play a very direct role in developing strong local support for no-take reserves.
- No-take reserves can help maintain or even enhance local fishery yields. A result from these studies is that the removal of 25% (Sumilon) and 10% (Apo) of the available fishing area did not reduce total fish yield at either island over time.
- No-take reserves can generate substantial alternative incomes for local communities, particularly through tourism.

Case 2: Supporting fisher livelihood through tourism at Gilutongan Island, Cebu

The 15 ha Gilutongan Marine Sanctuary, just 5 km off the southern coast of Cordova Municipality, Cebu, in central Philippines, is fast becoming another favorite dive destination in the country. In 2004, the sanctuary realized earnings of over two million pesos (US$36 000) in users' fees from snorkelers and divers visiting the area.

The results of a reef monitoring effort conducted at the Gilutongan sanctuary in mid-2003 showed stable and improving living coral cover and a significant two-fold increase in the density of reef fish over the baseline monitoring data of 1998 (Nanola et al. 2004). In 1998, there were very few target reef species observed within the sanctuary and in the immediate vicinity, and the abundance of fish was very low. A steady increase in the number of species and total abundance has been recorded since. With the coral reef inside the sanctuary showing marked increases in fish population and improvements in live coral cover, it continues to attract an increasing number of divers and swimmers.

The local government of Cordova and resident fishers conduct a biannual reef monitoring activity in coordination with the Department of Environment and Natural Resources (DENR), the Bureau of Fisheries and Aquatic Resources, University of San Carlos – Marine Biology Section, and non-governmental organizations. This reef monitoring is now initiated and coordinated by the local government while data analyses are assisted by participating academic and NGO partners.

In the 1970s, Gilutongan was part of the hotbed of illegal fishers in the area around Mactan Island, Cebu. Blast and cyanide fishing, identified as highly destructive means of extracting from the sea, were killing not only the fish, but also the vulnerable reefs. The vicious cycle of poverty was grinding: a growing number of fishermen, saddled with obligations to their growing families, were trying to get more and more fish out of the sea through destructive methods. The resulting habitat degradation, however, caused the dwindling of fish stocks and pushed the Gilutongan Island residents even deeper into poverty (Sotto et al. 2001).

The reef sanctuary was established in 1991, but was not properly managed until 1999, with the creation of the Gilutongan Marine Sanctuary Board. Technical assistance was provided by the Coastal Resource Management Project (CRMP) of the DENR with funding assistance from USAID in the re-establishment of the Gilutongan Marine Sanctuary. At that time the *barangay* community, the municipal government and CRMP worked together to formulate a three year marine sanctuary management plan. The management plan, transformed into a municipal ordinance, specified management procedures, rules, and established the user fee system for visitors to the sanctuary.

Because Gilutongan Island is only about 20 minutes from Mactan, the resort center of Cebu and one of the prime Philippine diving destinations, a high volume of divers and tourists for Gilutongan's development as a destination was assured.

However, the Marine Sanctuary Board had to devise a scheme whereby utilization of the sanctuary would contribute to its conservation and management. The fee system was established in 2000. By the end of 2001, the sanctuary had 'earned' a million pesos.

Users' fees of 50 pesos (US$1) per diver are collected by the municipal government and are shared with the community in the island *barangay*. In addition, divers pay a premium for taking photographs or video footage underwater. With its 70% share, the Municipality of Cordova has constructed an US$8000 guardhouse at the sanctuary site and funds the reef monitoring activities. The municipal government also appropriates funds for more information, education, and communication activities, the salary of a project director, supplies and the maintenance of the guardhouse, anchor buoys, and two coastal law enforcement patrol boats stationed at the site. The *barangay* appropriates from it a 30% share of the revenue, which funds medicines and alternative livelihood for the immediate community. It also provides information, education and communication, and honoraria for law enforcers and monitoring personnel.

Local-level initiatives like Gilutongan Island are paving the way to help address the problem of degraded coastal habitats, declining fish catch, and the economic over-dependence of human communities on fishing. In addition, Gilutongan has provided benefits and lessons beyond the value of marine sanctuaries enhancing marine habitats. These include: (1) generating a significant revenue for sanctuary maintenance and to benefit poor resident fishers through a user fee and revenue sharing scheme; (2) assisting members in managing a marine sanctuary with technical assistance from NGOs, national agency, or university; and (3) an example that local municipal governments can achieve effective marine conservation in an affordable and sustainable manner. Although Gilutongan Marine Sanctuary is working well and providing numerous benefits, it is small, crowded, and already threatened with overuse. Thus, more such areas under protection are needed in the vicinity of Cebu to absorb the volume of divers and address the magnitude of area-wide problems.

Case 3: Supporting reef restoration, Hundred Islands National Park, Luzon

The Hundred Islands National Park (HINP) was established in 1940, long before the concept of marine protected areas became common in the Philippines. Hence, the focus of concern at that time was the islands and adjacent mainland shore, not the submerged coral reefs (NEDA 1992). The park, in the City of Alaminos, northern Luzon on the Lingayen Gulf, derives its name from the 107 islands, islets, and rocks of uplifted Karst topography within its boundaries (Fig. 5.4). Most islands have vegetation cover and in years past supported healthy fringing coral reefs.

The national Philippine Tourism Authority (PTA) managed Hundred Islands until the passage of the NIPAS Act in 1992 and then in 2005 management was transferred to the local city government. After that law came into effect, a Protected Area Management Board (PAMB) was created, chaired by the Regional Director of the DENR, with representatives from various national and local government agencies, as well as people's organizations. For practical reasons, however, it is still the PTA that exercises day-to-day management of the park. This situation did not work well until local government became more directly involved.

The park is highly accessible since the town of Alaminos is serviced by many buses traveling on good highways from Metro Manila and other parts of Luzon. This attribute is good and bad. The park could have a tremendous cultural, scientific, and educational value. However, the volume of visitors also makes the park subject to insults such as pollution, sedimentation and destructive fishing, among others.

Recognizing the stressed condition of the park's coral reefs, the PTA agreed in 2000 to a multi-year initiative of the Marine Science Institute of the University of the Philippines (UPMSI) to initiate restoration and restocking efforts. The objective was to enhance the attractiveness of the marine component of the park by reintroducing giant clams to various reefs and also to transplant corals at experimental sites in the northern, central, and southern sectors of the park. The islands are strung out from north to south on an average of five to six abreast, along a sedimentation gradient. The north is open to the clean waters of the gulf while the south is heavily silted from shore-based activities including fishponds. There is also an east–west gradient, with the east further from the mainland's influence but more wind and wave affected. Given this situation, two experimental sites for coral transplantation were established in each of the three sectors, on different islands, and giant clam restocking was undertaken in the northern sector.

The initial efforts in giant clam restocking actually commenced in the mid-1980s, necessitated, in part, by the overflow of giant clam stocks at the primary ocean nursery of the UPMSI in Bolinao town, 40 km north. Small numbers of the smooth clam *Tridacna derasa* and the horse hoof clam *Hippopus hippopus* were brought into the HINP together with large numbers of the true giant clam *Tridacna gigas.* Some of the aggregations of the latter species developed into 'coral reef communities', with the grown clams providing substrate, shelter, and rugosity on an otherwise rubble and rock reef flat. This area has been dynamited more than once by illegal fishers.

Corals and other invertebrates colonized many clamshells, as did various species of algae. These observations led to the expansion of the effort to include coral transplantation activities to enhance the environment of the park. As previously mentioned, six sites were identified in different sectors of the park and coral fragments were transplanted at a density of three fist-sized branches per m^2. In total, 10 000 coral fragments were transplanted using several methods. The most common was to cement the branches to the substrate and the least common way was to tie the branches to dead coral colonies. A third method was to tie coral fragments to marble tiles that served as substratum. In summary, after two years of observations, the survival of monitored transplants ranged from 50–89%. Of the corals used, the genera *Pavona* and *Porites* fared best followed by *Montipora* and *Acropora*.

One issue often raised regarding restoration or rehabilitation is the alleged negative effects of gathering coral fragments. To reduce these effects, the first option was to make use of naturally-occurring fragments, such as those broken by wave action. Second, small to fist-sized branches were carefully detached from donor colonies, usually in the order of 10–15% of the parent stock. The donor colonies were monitored throughout the experimental period and did not show any negative trends or death. In many cases, the donor colonies actually showed healthy recovery.

The next issue is whether this type of intervention is useful. A priori, yes, just as reforestation accelerates the recovery of a forest. Empirically, however, the jury is still out. In the two year observation period of the experiments, the fish life did not show consistently or statistically higher recruitment. Fishers have, however, used dynamite to catch fish where the clams have been deployed, and this might be seen as positive proof that the interventions were working, but also indicates that the local community and government enforcement of the laws is still lacking.

The management system for the HINP is still weak but improving. The PTA, a national agency, gets little cooperation from the local government which is critical to improvement. This highlights a basic lesson about coastal management in the Philippines, which is that local governments with jurisdiction over municipal waters, that sometimes include national parks such as HINP, must be included in the management system. In this case, it behooves the PTA to engage the town in management through sharing of benefits from the Park. The value of experimental intervention for reef restoration cannot be fully appreciated until these management issues are resolved.

Local governments and integrated coastal and fisheries management

The many MPA projects have increasingly revealed the need for more integrated forms of coastal management. Although the MPAs are seen as important interventions to protect coral reefs and to enhance nearshore fisheries, they only address a small portion of the coastal management issues in the country and sometimes inadvertently detract from the support required for broader management for fisheries, including pollution prevention, shoreline development, mangrove conservation, and others. In addition, the integration of reproductive health programs within ICM projects is beginning to slow population growth, a critical need given the dependence of people on limited resources.

This larger ICM system referred to earlier is the local government planning and implementation framework (Fig. 5.6). The MPA case studies above are developed within this framework through local municipal government and community support. An evolving ICM framework is highlighted below.

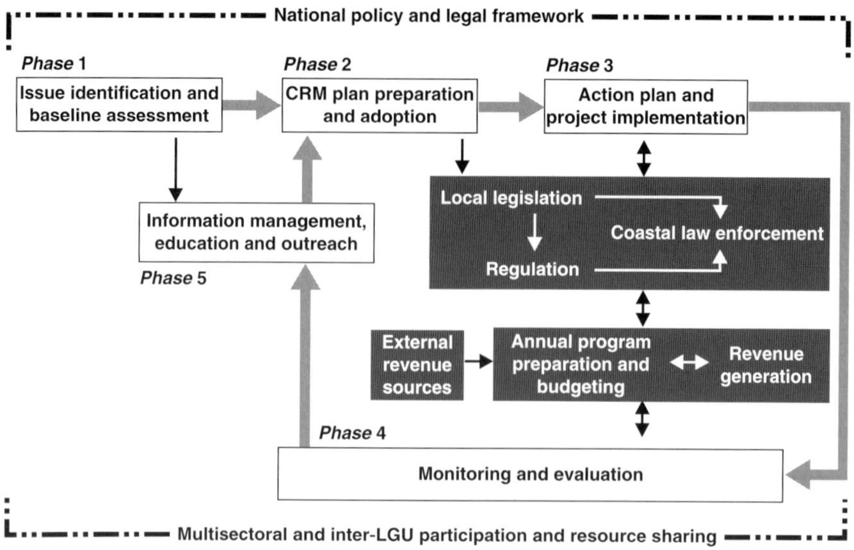

Figure 5.6 Five phase ICM planning process adapted for Philippine local government (DENR et al. 2001).

A benchmark system for local governments

A system of ICM benchmarks is being adopted as the national ICM strategy and policy framework through the DENR and the League of Municipalities of the Philippines. This benchmark system is now being utilized in more than 100 coastal municipalities (covering about 1/6 of the coastline) and cities through the assistance of the Coastal Resource Management Project of DENR since 1996 (DENR–CMMO 2003). Its purpose is to standardize the approaches and results of each local government in their coastal and fisheries management efforts so that they can follow a common monitoring and evaluation framework. It is a simple system whereby each local government adopts a planning framework and process for addressing their CRM needs, so that together with the national government they can assess the extent and effectiveness of their management interventions.

The basic ICM benchmarks required to achieve the first level are: (1) conduct a coastal resource assessment; (2) adopt a multi-year CRM plan; (3) establish coastal resources management organizations; (4) allocate a CRM annual budget; and (5) implement at least two best practices as appropriate. Common best practices include: (a) delineating and enforcing municipal water boundaries; (b) planning and implementing coastal zoning; (c) implementing specific fisheries management interventions; (d) having functioning marine protected areas; (e) implementing mangrove management; (f) implementing solid waste management; (g) developing coastal environment friendly enterprises; (h) passing legislation as required; (i) enforcing coastal laws; and (j) addressing other needs in the area.

Municipalities and cities are thus the primary units of government for ICM in the Philippines. Indeed, the size of a typical municipality or city that covers 20 km or more of shoreline and has marine jurisdiction to 15 km offshore, is large enough to effectively address most of the area's resource management problems. Local governments are confronted with multifaceted problems in their coastal areas and, therefore, the ICM undertaken by this scale of government can respond to local conditions. Development of ICM is at least beginning to follow a process and to adhere to a consistent set of benchmarks and criteria for evaluation (DENR–CMMO 2003).

The process of establishing MPAs is usually part of a broad community-based resource management program within the local government. A local or national NGO or a local university, as in the case of Silliman University and the formation of the Sumilon, Apo, and Gilutongan reserves, may facilitate this process. Being part of a larger ICM program raises awareness about the need for an MPA and also emphasizes the need for broad area plans where fisheries management and MPAs are integrated. The broad planning phases also provide guidance for MPA establishment by encouraging a planning process that leads to implementation through a participatory process (Fig. 5.6). In addition, a MPA management rating system and database is being adopted by most of the assisting organizations where the national government provides guidance for effective management and sharing information in a standard format. This system is helping to standardize the measurement of management performance among a diverse range of MPAs (White et al. 2004).

Boundary delineation for fisheries

A major issue triggered by the Fisheries Code of 1998, was the delineation of municipal boundaries for fisheries so that local governments knew exactly the areas for which they had responsibility. The delineation of the 15 km boundaries has implications for where commercial fishers (boats of more than 3 tons) are allowed to fish. Because of the increasing documentation of overfishing in nearshore waters and beyond, it became of paramount importance to formally establish the municipal boundaries so that these waters are reserved for small-scale fishers. The problem arose regarding how the boundaries are determined. For example, what landmass, the islands or mainland shore, was to be used as a baseline for determining the boundary. This question was initially settled in 2001 and then unsettled in 2002 by a Department of Justice letter stating that the order making the delineation was illegal. Nevertheless, local governments started and continued to delineate their municipal waters. As delineation has proceeded, more municipal and city governments have begun to enforce the 15 km limit and to ensure that these waters are reserved for the municipal small-scale fishers. This trend is continuing and promises to spread nationwide as part of local government ICM efforts. The weak link is that few municipalities have the capacity to patrol all their waters. In spite of this weakness, it is a healthy sign that open access is being restricted.

Local government law enforcement

In response to the need for improved capacity for coastal law enforcement by local governments, several experiments in collaborative law enforcement efforts have prospered. In the Central Visayas Region 7, for example, the Coastal Law Enforcement Alliance for Region 7 was formed in 2000. This alliance comprised all national agencies with some mandate over coastal law enforcement together with the four provincial governments. The alliance operates at both the regional or sub-regional scales to address increasing problems with illegal commercial fishing within municipal waters. It also provides logistical and technical support for single or clusters of municipal governments that want to improve their fisheries law enforcement. The real financing for most enforcement comes from the municipal and city governments themselves, but they need technical guidance and mentoring to encourage their work.

Presently, several clusters of municipal governments in Cebu and Bohol Provinces have created a joint patrol of their waters working together with the Region 7 support group. One group used a boat provided by the DENR with fuel purchased by the municipalities, and personnel from both local government, the Philippine National Police and the Philippine Coast Guard. Such operations have arrested and fined commercial fishing boats that operated illegally and these incidences serve as an important warning. Although this is only a beginning in a country with more than 800 coastal municipalities, it is setting a precedent for effective fisheries patrolling and management through local government initiatives jointly with key national agencies. Support for improved MPA surveillance is also being provided through these patrolling operations.

Co-management of MPAs with local governments and NGOs

Collaboration is an important factor in the success of the MPAs described in the case studies. Although legally ordained by the municipal or city government, the actual implementation brings together a number of actors that make them effective and sustainable entities over time (White et al. 2002). The *barangay* community of Apo Island, under the municipality of Dauin, has benefited greatly from its long-term association and mentoring of Silliman University. Key individuals at Silliman have maintained research and conservation interests in Apo Island. This involvement has aided the conservation of the island tremendously. However, day-to-day management is really done by the community leaders and the Protected Area Management Board, some of whom reside on the island. Gilutongan Island Marine Sanctuary is another example of collaboration where the key actors are the municipal government, diving tourism operators, an NGO – the Coastal Conservation and Education Foundation – and originally the USAID-supported Coastal Resource Management Project. Several very active community members have also been critical in providing vigilant enforcement in an area prone to illegal fishing at any time of day.

The private sector in the Philippines is playing an increasingly important role in the success of MPAs and ICM in general. As the examples illustrate, academic institutions such as Silliman University, NGOs such as the Coastal Conservation and Education Foundation, WWF (formerly known as the World Wildlife Fund), and others, including the tourism business community are contributing in appropriate ways. They provide technical guidance, mentoring to the communities, funding, and often bring tourists that are willing to pay entrance fees or purchase local products. These sectors also sit on management committees, thus collaborative management is indeed becoming the norm in the country.

Key lessons

A major message emerging from the experience of coastal management in the Philippines is that generalization cannot easily describe what is being learned. This is because the successes all have their own identity and to a limited degree, each community and local government must navigate their own course before their program becomes sustainable. This observation is borne out from the highly participatory process that each management body engages in before field implementation begins. This also reflects the relatively decentralized legal and institutional setting of the Philippines where local municipal and city governments have full authority over their coastal areas and resources. The only exception is when an area is within a national protected area and even here, there is a high level of local control because the management council is mostly comprised of local stakeholders. Thus, despite or, more likely, because of the highly participatory process of planning and implementation, many areas are coming under sustainable use regimes. These managed areas, such as Gilutongan, Sumilon, and Apo islands and many others, see improvements in the condition of their coral reefs and fisheries while socio-economic benefits to local stakeholders have improved.

An important common ingredient in the successfully managed areas in the Philippines is that multisector collaboration is always present. Such collaboration ranges from the fishing community to the national government while key players are more often local, rather than national. Catalytic academic organizations, NGOs and members of the business sector often make the difference between success and failure in localized management. The research efforts of several universities have, on the one hand, documented the real results of MPAs while at the same time serving as mentors for concerned communities and local governments. NGOs often bring technical assistance and funding to start projects that would never be initiated or pushed in the proper direction without their support. Thus, although each project may have a different mix of partners, there are always partners that have learned how to work together effectively without usurping local authority.

The final and perhaps most important conclusion is that ICM regimes are showing higher levels of success than those that simply focus on one issue or management approach. The local governments that plan for their entire area and then prioritize issues and interventions are doing better and moving toward sustainable management. The planning process helps them allocate budgets and to integrate conservation and even human reproductive health into their normal service. Integration across sectors, within and outside of coastal communities, and with a variety of field-tested interventions is a key to building management regimes that will endure over time, at least in the Philippines.

References

Alcala, A.C. (1981) Fish yields of coral reefs of Sumilon Island, central Philippines. *National Research Council Philippines Research Bulletin*, **36**: 1–7.

Alcala, A.C. (1988) Effects of marine reserves on coral fish abundances and yields of Philippine coral reefs. *Ambio*, **17**: 194–9.

Alcala, A.C. and Russ, G.R. (1990) A direct test of the effects of protective management on abundance and yield of tropical marine resources. *Journal du Conseil International pour l'Exploration de la Mer*, **47**: 40–7.

Alcala, A.C. and Russ, G.R. (2000) Role of socioeconomic factors in coral reef protection and management. *Proceedings 9th International Coral Reef Symposium, Bali, Indonesia.* **1**: 29–32.

Alcala, A.C. and Russ, G.R. (2001) Local partnership for coastal and marine resource management: the case of Selinog Island. *Tropical Coasts*, **8**: 4–9.

Alcala, A.C., Russ, G.R., Maypa, A.P. and Calumpong, H.P. (2005) A long-term, spatially replicated, experimental test of the effect of marine reserves on local fish yields. *Canadian Journal of Fisheries and Aquatic Sciences*, **62**: 98–108.

Arquiza, Y.D. and White, A.T. (1999) *Tales from Tubbataha, Natural History, Resource Use and Conservation of the Tubbataha Reefs, Palawan, Philippines* (2nd edn). Sulu Fund for Marine Conservation Foundation and Bookmark, Inc., Manila.

Bernacsek, G. (1996) *The Role of Fisheries in Food Security in the Philippines: A Perspective Study for the Fisheries Sector to the Year 2010.* Paper presented during the 2nd National Fisheries Policy Planning Conference, Puerto Azul, Cavite, Philippines.

BFAR (Bureau of Fish and Aquatic Resources) (1997) *1996 Philippine Profile*. Department of Agriculture, Manila.

Burke, L., Selig, E. and Spalding, M. (2002) *Reefs at Risk in Southeast Asia*. World Resources Institute, Washington.

Calumpong, H.P. (1996) The Central Visayas Regional Project: lessons learned. *Tambuli*, **1**: 12–17.

CCEF (Coastal Conservation and Education Foundation, Inc.) (2003) *Marine Protected Area Report Guide and Database*. Coastal Conservation and Education Foundation, Inc. and partners, Cebu City, Philippines.

Christie, P. and White, A.T. (1997) Trends in development of coastal area management in tropical countries: from central to community orientation. *Coastal Management*, **25**: 155–81.

Christie, P., Buhat, D., Garces, L.R. and White, A.T. (1999) *The Challenges and Rewards of Community-based Coastal Resources Management: San Salvador Island, Philippines.* Sunny Press, New York.

Christie, P., White, A.T. and Deguit, E. (2002) Starting point or solution? Community-based marine protected areas in the Philippines. *Journal of Environmental Management*, **66**: 441–54.

Chua, T.-E. and Scura, L.F. (Eds) (1992) Integrative framework and methods for Coastal Area Management. *ICLARM Conference Proceedings, Manila, Philippines.* 37.

Courtney, C.A. and White, A.T. (2000) Integrated Coastal Management in the Philippines: testing new paradigms. *Coastal Management*, **28**: 39–53.

Courtney, C.A., Atchue, J.A., III, Carreon, M., White, A.T., Smith, R.P., Deguit, E., Sievert, T. and Navarro, R. (1999) *Coastal Resource Management for Food Security*. Coastal Resource Management Project, Cebu City, Philippines.

Courtney, C.A., White, A.T. and Anglo, E. (2000) *Philippine Case Study: Managing Coastal Resources – Drawing Lessons and Directions from the Philippine Experience.* Sustainable Development Report, Asian Development Bank.

DENR–CMMO (Department of Environment and Natural Resources–Coastal and Marine Management Office) (2003) *Monitoring and Evaluating Municipal/City Plans and Programs for Coastal Resource Management.* Coastal Resource Management Project of Department of Environment and Natural Resources, Cebu City, Philippines.

DENR, DA–BFAR and DILG (Department of Environment and Natural Resources, Department of Agriculture–Bureau of Fisheries and Aquatic Resources and Department of the Interior and Local Government). (2001) *Philippine Coastal Management Guidebook Series.* Vols 1–8. Coastal Resource Management Project of the Department of Environment and Natural Resources. Cebu City, Philippines.

Ferrer, E.M., Polotan-De la Cruz, L. and Agoncillo-Domingo, M. (1996) *Seeds of Hope: A Collection of Case Studies on Community-based Coastal Resources Management in the Philippines.* College of Social Work and Community Development, University of the Philippines, Diliman, Quezon City.

Gomez, E.D., Aliño, P.M., Yap, H.T. and Licuanan, W.Y. (1994) A review of the status of Philippine reefs. *Marine Pollution Bulletin*, **29**: 62–8.

Green, S.J., Alexander, R.D., Gulayan, A.M., Migrio, C.C., III, Jarantilla-Paler, J. and Courtney, C.A. (2002) *Bohol Island: Its Coastal Environmental Profile*. Bohol Environmental Management Office and Coastal Resource Management Project, Cebu City, Philippines.

Green, S.J., White, A.T., Flores, J.O., Carreon, M.F., III, and Sia, A.E. (2003) *Philippine Fisheries in Crisis: a Framework for Management*. Coastal Resource Management Project of the Department of Environment and Natural Resources, Cebu City, Philippines.

Katon, B.M., Pomeroy, R.S., Salamanca, A. and Garces, L. (1999) Fisheries management of San Salvador Island: a shared responsibility. *Society and Natural Resources*, **12**: 777–95.

Kuperan, K., Abdullah, N.M.R., Pomeroy, R.S., Genio, E.L. and Salamanca, A.M. (1999) Measuring transaction costs of fisheries co-management in San Salvador Island, Philippines. *Naga, the ICLARM Quarterly*, **22**: 45–8.

Maypa, A.P., Russ, G.R., Alcala, A.C. and Calumpong, H.P. (2002) Long-term trends in yield and catch rates of the coral reef fishery at Apo Island, central Philippines. *Marine and Freshwater Research*, **53**: 207–13.

MCDP (Marine Conservation and Development Program) (1986) *Final Report and Evaluation for the Marine Conservation and Development Program of Silliman University.* The Asia Foundation and Silliman University, Dumaguete, Philippines.

Nanola, C.L., Hazel, A., Uychiaoco, A. and Alino, P.M. (2004) *Monitoring the Effects of Marine Protected Areas in CRMP Learning Areas (1997–2003).* Coastal Resource Management Project of the Department of Environment and Natural Resources and the University of the Philippines Marine Science Institute, Cebu City, Philippines.

NEDA (National Economic Development Authority), Region I. (1992) *The Lingayen Gulf Coastal Area Management Plan.* ICLARM Technical Report, 32.

NIPAP (National Integrated Protected Area Program) (1999) *Annual Report to the Public.* Department of Environment and Natural Resources, Manila, Philippines.

NSO (National Statistics Office) (2000) *Census Data for 2000.* National Statistics Office, Manila.

Pajaro, M., Olano, F., San Juan, B. and Nozawa, C.M. (1999) Inventory of marine protected areas in the Philippines. In: A.J. Uychiaoco, S. Schoppe, P. Alino and R. Hermes (Eds) *Proceedings of the Workshop on Marine Protected Areas in the Philippines.* Coral Reef Information Network of the Philippines.

Pauly, D., Christensen, V., Dalsgaard, J., Froese, R. and Torres, F.C., Jr (1998) Fishing down marine food webs. *Science*, **279**: 860–3.

Pauly, D., Christensen, V., Guenette, S., Pitcher, T.J., Sumaila, U.R., Walters, C.J., Watson, R. and Zeller, D. (2002) Towards sustainability in world fisheries. *Nature*, **418**: 689–95.

Pomeroy, R.S. and Carlos, M.B. (1997) Community-based coastal resource management in the Philippines: a review and evaluation of programs and projects, 1984–1994. *Marine Policy*, **21**: 445–64.

Russ, G.R. and Alcala, A.C. (1996) Do marine reserves export adult fish biomass? Evidence from Apo Island, Central Philippines. *Marine Ecology Progress Series*, **132**: 1–9.

Russ, G.R. and Alcala, A.C. (1999) Management histories of Sumilon and Apo marine reserves, Philippines, and their influence on national marine resource policy. *Coral Reefs*, **18**: 307–19.

Russ, G.R. and Alcala, A.C. (2003) Marine reserves: rates and patterns of recovery and decline of predatory fish, 1983–2000. *Ecological Applications*, **13**: 1553–65.

Russ, G.R., Alcala, A.C. and Maypa, A.P. (2003) Spillover from marine reserves: the case of *Naso vlamingii* at Apo Island, Philippines. *Marine Ecology Progress Series*, **264**: 15–20.

Russ, G.R., Alcala, A.C., Maypa, A.P., Calumpong, H.P. and White, A.T. (2004) Marine reserve benefits local fisheries. *Ecological Applications*, **14**: 597–606.

Sorensen, J.D., McCreary, S.T. and Hershman, M.J. (1990) *Institutional Arrangements for Management of Coastal Resources.* University of Rhode Island Coastal Resources Center, Narragansett, Rhode Island.

Sotto, F.B., Gatus, L.I., Ross, M.A., Portigo, Ma.L.Fe and Freire, F.M. (2001) *Coastal Environmental Profile of Olango Island, Cebu, Philippines.* Coastal Resource Management Project, Cebu City, Philippines.

Vogt, H. (1997) The economic benefits of tourism in the marine reserve of Apo Island, Philippines. *Proceedings of the 8th International Coral Reef Symposium, Panama*, **2**: 2101–4.

White, A.T. (1987) Philippine marine park pilot site: benefits and management conflicts. *Environmental Conservation*, **14**: 355–9.

White, A.T. (1988a) *Marine Parks and Reserves: Management for Coastal Environments in Southeast Asia.* ICLARM Education Series No. 2.

White, A.T. (1988b) The effect of community-managed marine reserves in the Philippines on their associated coral reef fish populations. *Asian Fisheries Science*, **1**: 27–42.

White, A.T. (1989) Two community-based marine reserves: lessons for coastal management. In: T.-E. Chua and D. Pauly (Eds) *Coastal Management in Southeast Asia: Policies, Management Strategies and Case Studies. ICLARM Conference Proceedings* 19.

White, A.T. (1996) Philippines: community management of coral reef resources. In: J. Clark (Ed.) *Coastal Zone Management Handbook.* CRC Lewis Publishers, Boca Raton.

White, A.T., and Cruz-Trinidad, A. (1998) *The Values of Philippine Coastal Resources: Why Protection and Management are Critical.* Coastal Resource Management Project, Cebu City, Philippines.

White, A.T. and De Leon, R.O. (2004) Mangrove resource decline in the Philippines: government and community look for new solutions. In: DA–BFAR (Department of Agriculture–Bureau of Fisheries and Aquatic Resources) *In Turbulent Seas: the Status of Philippine Marine Fisheries.* Coastal Resource Management Project, Cebu City, Philippines.

White, A.T. and Lopez, N. (1991) Coastal resources management planning and implementation for the Fishery Sector Program of the Philippines. *Proceedings of the 7th Symposium on Coastal and Ocean Management, Long Beach, California.*

White, A.T. and Vogt, H.P. (2000) Philippine coral reefs under threat: lessons learned after 25 years of community-based reef conservation. *Marine Pollution Bulletin*, **40**: 537–50.

White, A.T., Hale, L.Z., Renard, Y. and Cortesi, L. (Eds) (1994) *Collaborative and Community-based Management of Coral Reefs.* Kumarian Press, Hartford, Connecticut.

White, A.T., Salamanca, A. and Courtney, C.A. (2002) Experience with marine protected area planning and management in the Philippines. *Coastal Management*, **30**: 1–26.

White, A.T., Meneses, A.T. and Ovenden, M.F. (2004) Management rating system for marine protected areas: an important tool to improve management. In: DA–BFAR (Department of Agriculture–Bureau of Fisheries and Aquatic Resources) *In Turbulent Seas: the Status of Philippine Marine Fisheries.* Coastal Resource Management Project, Cebu City, Philippines.

White, A.T., Christie, P., d'Agnes, H., Lowry, K. and Milne, N. (2005a) Designing ICM projects for sustainability: lessons from the Philippines and Indonesia. *Ocean and Coastal Management*, **48**: 271–96.

White, A.T., Eisma-Osorio, R.L. and Green, S.J. (2005b) Integrated coastal management and marine protected areas: complementarity in the Philippines. *Ocean and Coastal Management*, **48**: 948–71.

Chapter 6

Redressing Access Inequities and Implementing Formal Management Systems for Marine and Estuarine Subsistence Fisheries in South Africa

Jean M. Harris, George M. Branch, Barry M. Clark, and Sandile C. Sibiya

Introduction

In many parts of the world, subsistence and artisanal fishers have been marginalized during coastal development and allocation of resources, often resulting in overharvesting and conflict with managers and other users (Berkes 1990; Fall 1990; Satia 1993). In South Africa, this phenomenon has been exacerbated by the country's unique political history. First, the racially-based 'homeland system' enforced in South Africa during the apartheid period (1948–1994) entrenched impoverished rural communities and perpetuated and concentrated their reliance on natural resources. Second, the policies then in existence resulted in grossly unequal access to natural resources. Subsistence activities nevertheless continued despite restrictions that deemed them illegal. This resulted in a large 'informal sector' of poachers. Towards the end of the apartheid era these fishers could theoretically gain legal access to resources by participating in recreational or commercial licensing systems. In effect, however, the quantities required by subsistence fishers did not conform to conditions for recreational permits, and subsistence fishers could simply not afford commercial license fees. In many ways the problems surrounding subsistence fisheries both reflect and illustrate other political maladies in South Africa, and the processes now underway to redress these problems have relevance in the broader social context (Harris et al. 2002a).

Following the first democratic elections in South Africa in April 1994, many of the policies and laws were revisited (Cochrane 1995; Hutton et al. 1997). In an attempt to achieve a new fisheries policy acceptable to all, a Fisheries Policy Development

Committee was set up by the Minister of the Department of Environmental Affairs and Tourism in April 1995 to represent the views of all interested parties (Cochrane and Payne 1998). The history and evaluation of the resulting fisheries policy is described in detail in Cochrane (1995), Hutton and Pitcher (1998), Martin and Nielsen (1998), and Hersorg and Holm (1999). The Fisheries Policy Development Committee appointed an Access Rights Technical Committee to advise on a system of fair and equitable access to fisheries stocks and their recommendations yielded the first step towards the formal recognition of subsistence fishers as a distinct sector (van der Elst et al. 1997).

The process of revising fisheries policy in South Africa culminated in the promulgation in 1998 of the Marine Living Resources Act No. 18 (DEAT 1998). This Act recognizes subsistence as a formal sector, and defines a 'subsistence fisher' as 'a natural person who regularly catches fish for personal consumption or for the consumption of his or her dependents, including one who engages from time to time in the local sale or barter of excess catch, but does not include a person who engages on a substantial scale in the sale of fish on a commercial basis'. This was a significant development for subsistence fishers who depend on marine resources for food. However, implementation of this Act with regard to subsistence fishing has presented major challenges. At the time the Act was promulgated, no management systems existed for these fisheries and little information about the sector was available to inform the process of formalizing and controlling their activities.

In this chapter we first provide a summary of the recent evolution of subsistence fisheries management in South Africa, particularly: (1) the initiation of a process at the end of 1998 by the national government to investigate the extent, nature, and needs of subsistence fisheries around the entire coast of South Africa; and (2) the resultant findings and recommendations for recognizing and defining subsistence fisheries and developing management systems. We then focus on a specific case study, the Sokhulu Intertidal Mussel Harvesting Project, which served as a model for co-management and informed the development of national policy.

Implementation of formal management systems for subsistence fisheries in South Africa is still underway. To exemplify, we explore the approach taken on the east coast in the province of KwaZulu-Natal, where the co-management model developed in the Sokhulu pilot project is now being implemented in 19 communities for 43 individual fishery groups. Finally, we evaluate challenges and successes, and highlight lessons learnt and limitations that still hinder the process of developing an equitable allocation of fishing rights in South Africa.

Testing the waters

In 1998, a national Subsistence Fisheries Task Group (SFTG) was appointed to conduct research, consult with fishers and managers, and provide recommendations for the management of subsistence fishers to the responsible (national) management agency: Marine and Coastal Management (MCM) of the Department of Environment Affairs and Tourism. The outcomes of their activities are documented in a series of papers outlining the process (Harris et al. 2002a), the survey of fishing communities (Clark et al. 2002; Branch et al. 2002a), perceptions of the fishing communities (Hauck et al. 2002), development of a definition of subsistence fishers (Branch et al. 2002b), a description of the resources harvested and their appropriateness for subsistence fisheries (Cockcroft et al. 2002), and recommendations for management (Harris et al. 2002b). Amongst the recommendations, a strong case was made for the establishment of local co-management structures that would allow joint responsibility by the fishers and the authorities for the management of subsistence fisheries.

Documenting a nationwide phenomenon

Prior to developing recommendations for the management of subsistence fishers, the SFTG needed an overview of the distribution and numbers of subsistence fishers in the country, the nature of their harvesting activities and the resources they harvested and their socio-economic status. This was undertaken through two nationwide surveys implemented by eight regionally-based enumerators. In the first survey, each enumerator was allocated a segment of coastline (areas A–H in Fig. 6.1), and was

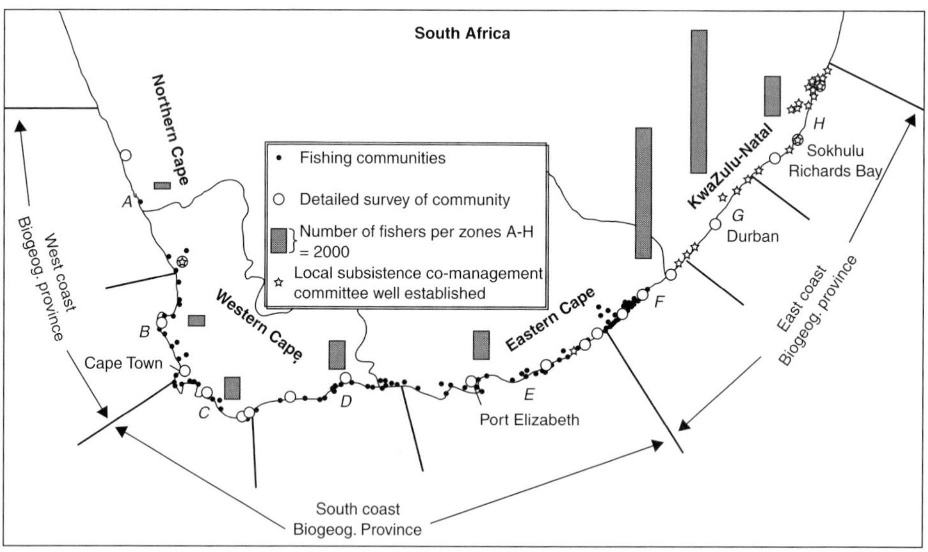

Figure 6.1 Map of South Africa showing the names of localities mentioned in the text, the positions of subsistence fishing communities, and identifying those where co-management has been initiated. Large white circles indicate communities that were surveyed in detail by Branch and colleagues (2002a). The bar graphs indicate the relative numbers of subsistence fishers in different sections of the coast.

instructed to target selected knowledgeable key informants. In total, 200 people were interviewed, resulting in the identification of 147 fishing communities and 29 233 people who could potentially be considered as subsistence fishers (Clark et al. 2002). The bulk of these fishers resided on the east coast, with more than 75% present in former homeland areas in southern KwaZulu-Natal and Eastern Cape (Fig. 6.1). Throughout the country, a wide variety of species was harvested from intertidal, shallow subtidal, and nearshore environments, with some important regional differences evident. Intertidal and shallow subtidal harvesting (neither of which require the use of a boat) accounted for most of the harvesting effort on the east and south coasts, while on the west coast, boat-based harvesting of nearshore resources predominated. Rocky intertidal invertebrates and fish were important in all areas, while estuarine invertebrates were significant only on the east and south coasts. 'High value resources', such as abalone and rock lobster, were mainly harvested on the west and south coast, particularly in urban areas. Most fishers lived close to their points of harvest (<10 km away) and tended to consume rather than sell a large portion of the resources harvested. Fishers in some areas were, however, reported to sell significant portions of their catch, and to harvest considerable distances from the places of residence, especially on the west coast. These fishers frequently made use of motorized vessels and also tended to be the ones harvesting higher value resources, and hence showed greater affinities to commercial rather than 'true' subsistence fishers.

Following the completion of the first broad-scale survey, a second more focused survey was initiated. In this survey, enumerators were instructed to focus on a subset of subsistence fishing communities, selected to span the full range of community types and harvesting strategies identified in the previous study. This survey elicited information on the socio-economic characteristics, more details about resource use and harvesting techniques employed by fishers in different parts of the country, and provided an insight into the fishers' needs and perceptions (Branch et al. 2002a; Hauck et al. 2002). Information was elicited directly from the fishers themselves through household surveys, focus group discussions, and key informant interviews. In total, 4549 people from 20 communities were surveyed. Approximately one-third of the people in each community listed themselves as fishers, where 84% considered themselves subsistence fishers, and the rest categorized themselves as commercial fishers. By far the majority of the fishers were male, with the exception of intertidal fishers in KwaZulu-Natal, where women were in the majority. Poverty was endemic in the surveyed communities, with mean adult equivalent income equating to less than US$70 per month. Mean adult equivalent income was substantially lower on the east coast (US$29–55 per month) than on the south or west coasts (US$64 and U$110), and was also lower in rural relative to small town or metro areas. Most of the fishers had no waged employment, with only half of the households surveyed having at least one occupant with waged employment. The level of education among the fishers was low with almost two-thirds of the sample aged 20 years and above having no education at all or only primary school education.

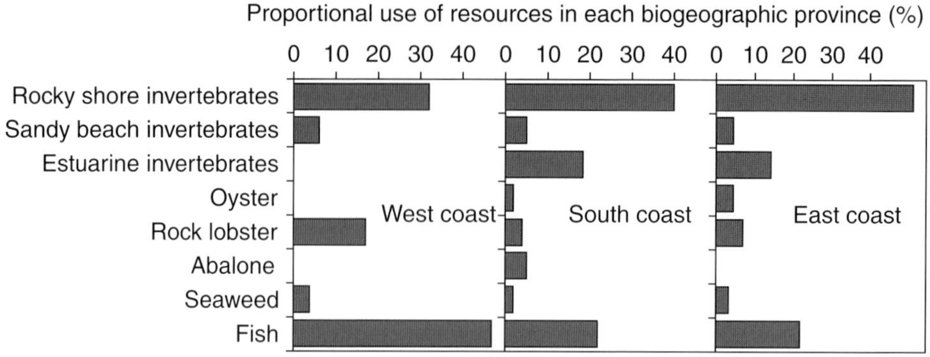

Figure 6.2 Relative frequencies of different resource groups harvested on the west, south, and east coasts (adapted from Branch et al. 2002a).

The relative amounts of different resources harvested depended both on the value placed on the species and their availability in the three biogeographic regions. Fish and rocky shore invertebrates were universally the most frequently harvested groups. Estuarine species were third in importance on both south and east coasts but were virtually untapped in the west coast. Rock lobsters were the third most important species on the west coast (Fig. 6.2).

Four main types of resources were harvested: commercially-lucrative species that are better suited to commercial than subsistence harvesting; species that are used primarily for bait; and intertidal invertebrates that are harvested either from rocky shores or estuaries and sandy beaches (Fig. 6.3a–d). In the case of the commercially-lucrative species, the amount sold tended to be greater than those eaten or used as bait, except for fish on the east coast, most of which were consumed (Fig. 6.3a). Bait species were used by the fishers themselves as bait, except for 'redbait', a solitary ascidian that is popular as bait in the south, but was eaten on the east coast. A small proportion of bait was sold (Fig. 6.3b). Fishers or their families consumed most of the rocky intertidal invertebrate species harvested in all three regions of the country (Fig. 6.3c). Oysters were an exception, due to the existence of a better market for this species than for other intertidal organisms. Estuarine and sandy-beach invertebrates, such as crab, were mostly eaten and sand mussels used as bait (Fig. 6.3d).

Overall, the patterns of resource use were dictated by two geographic trends. First, the west coast biogeographic province of South Africa is dominated by the slow northward-moving Benguela Current and is subject to wind-induced upwelling. As a result the waters are cold (9–14°C) and nutrient-rich and sustain a biota that is characterized by low diversity but high productivity and biomass. Most of the commercially-important fisheries are concentrated on the west coast. The east coast biogeographic province is a complete contrast, being fueled by the rapid southward-moving Agulhas Current that brings warm (20–24°C), nutrient-poor waters to the coast. The biota is highly diverse but less productive and there are few single-species fisheries of commercial importance. Between these two extremes lies the

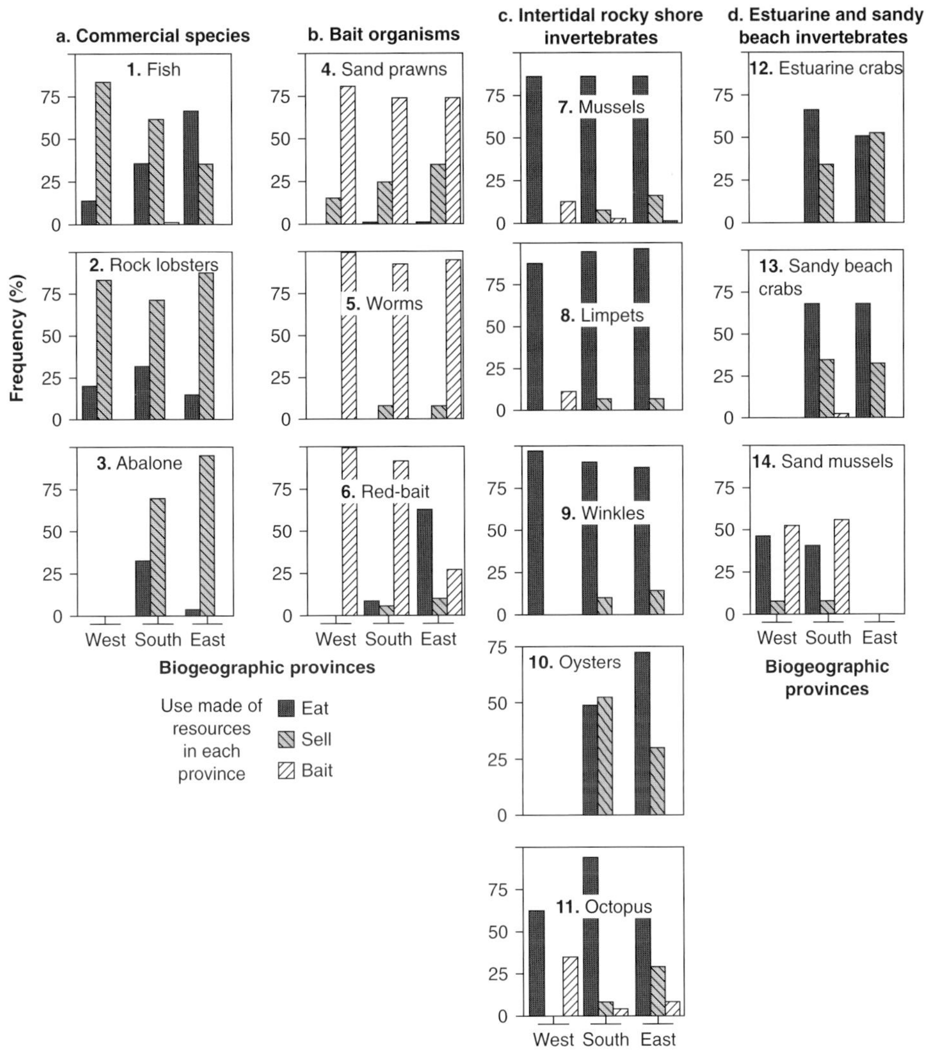

Figure 6.3 A summary of the types of resources harvested on the west, south, and east coasts of South Africa and how they are used (adapted from Branch et al. 2002a).

south-coast biogeographic province, where the Agulhas Current veers away from the coast and conditions are intermediate, being warm temperate (14–20°C) and with moderate nutrient concentrations. High levels of endemism are characteristic, and several important fisheries lie in this province.

Second, an important consequence of the cold water on the west coast is that the land is arid. This results in two other important gradients around the coast: estuaries are rare on the west coast but increase in number eastwards; and human occupation is concentrated on the east and south coasts and diminishes on the west coast.

Superimposed on this, past political influences dictated where people could live. 'Black' communities were largely limited to 'homelands' in northern KwaZulu-Natal and in Transkei and Ciskei on the southeast sections of the present Eastern Cape Province. The net effect of these factors was that subsistence communities were concentrated on the southeast and east coasts where productivity of the coast is intrinsically low, whereas the major commercial fisheries lie on the west and southwest coasts.

Defining subsistence fishers

A key function of the SFTG was to refine the definition of subsistence fishers provided by the new Act (DEAT 1998) so that acceptable criteria could be developed to distinguish who should qualify in this category. Features that stood out from the results of the surveys and during consultations with both fishers and managers were poverty, long-term reliance on the resources, simple fishing equipment and residency close to resources. This led to consensus about the following definition:

> 'Subsistence fishers are poor people who personally harvest marine resources as a source of food or to sell them to meet the basic needs of food security; they operate on or near to the shore or in estuaries, live in close proximity to the resource, consume or sell the resources locally, use low-technology gear (often as part of a long-standing community-based or cultural practice), and the kinds of resources they harvest generate only sufficient returns to meet the basic needs of food security.'
>
> – Branch et al. 2002b.

While developing this definition, it became clear that 'small-scale commercial fishers' or 'artisanal fishers' were distinct from both subsistence fishers and industrial-scale commercial fishers, and had not received adequate attention. They are distinguished from subsistence fishers through earning a living from fishing rather than being in survival mode, and differ from industrial fishers by living on or close to the coast, having a history of involvement with fishing, being personally involved in hands-on day-to-day running of their enterprises, operating with limited capital investment and low levels of technology, and employing small numbers of people. The special plight of these people was emphasized by the SFTG, with the recommendation that they receive separate attention, and that certain resources such as nearshore rock lobsters and linefish might best be preferentially allocated to them in view of their long-standing history of fishing and dependency on fishing.

Most fishers expressed dissatisfaction with the management procedures then in existence because they claimed to have been marginalized during allocation procedures and never to have been consulted when regulations were introduced.

Lack of communication and inadequate access to information were key complaints (Hauck et al. 2002).

A catalytic local case study

A number of projects aimed at developing appropriate management systems at a local level for subsistence fishers are long-standing in South Africa (see Hauck and Sowman 2003), and preceded the legal redress provided by the new post-apartheid fisheries legislation (DEAT 1998).

One of these, the Sokhulu project, was initiated in June 1995 to address problems of illegal and overharvesting of the intertidal brown mussel *Perna perna*, the lack of traditional access by local harvesters, and conflict between fishers and the provincial agency, Ezemvelo KwaZulu-Natal Wildlife (EKZN Wildlife) (Harris et al. 2003). It aimed to facilitate co-management of the mussel stocks by the subsistence gatherers and the management authority. At inception, an integral part of the project was participatory experimental harvesting to determine sustainable off-take and to increase the capability of members of the fishing community to participate in management decisions.

Specific measures of success of the Sokhulu project and challenges encountered are documented and analyzed in detail in Harris and colleagues (2003). Overall progress can perhaps best be assessed by some of its significant outcomes. First, there is now lively participation by both fishers and the authorities in the management of the fishery. Second, there has been a shift from an initial situation where external project staff would argue for sustainability, to one where fisher representatives themselves assumed an advocacy role to persuade other fishers of the need for sustainable management. Third, formal restoration of access rights has occurred.

A number of core elements integral to achievement of co-management of subsistence fisheries emerged from the Sokhulu project: (1) establishment of a local management structure, the co-management committee that is a democratically-elected body representative of fishers and the authorities and provides a structure to engage national government; (2) provision of training to both partners of co-management in basic fisheries management and resource sustainability, committee structure and functioning, and conflict resolution, which includes exchange visits between fisher communities, trains community members as trainers, and places special focus on children of fishers as subjects; (3) local joint decision-making; (4) participative processes where there is direct involvement of fishers in research to determine quotas and evaluate harvesting methods, and a community monitoring program is established; (5) information sharing in the form of report-backs on research results, access to information from government, and insights into indigenous knowledge; (6) restoration of legal access to the resources; and (7) trust-building activities (Fig. 6.4). Although the Sokhulu project focused on use of an intertidal resource at a single locality, the results have provided principles that are now being applied elsewhere.

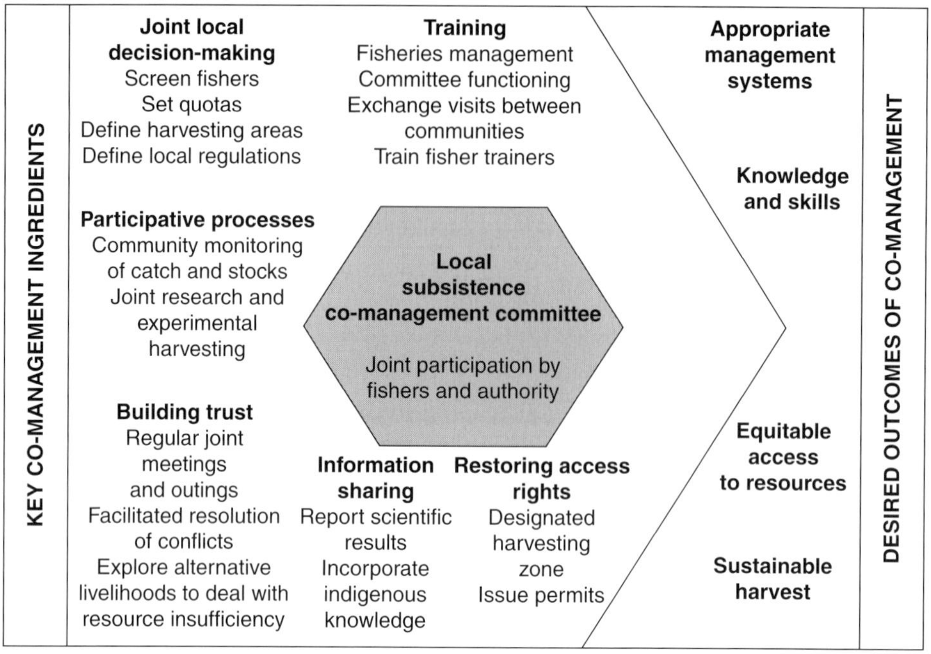

Figure 6.4 Key elements in the crafting of co-management at Sokhulu.

The Sokhulu project assumed importance because it both informed the process of developing a national policy as well as becoming a model approach that has since been exported to 19 other fishing communities. Two features enhanced the catalytic effect of this project. First, the timing of its execution was serendipitous with respect to the changing political and legal climate so that it provided a relevant example at a critical time in the evolution of fisheries management. Second, the project yielded key people with experience in implementing co-management arrangements at local scale who were deployed into positions with strategic influence on provincial and national processes. The Sokhulu project's executant was appointed as the chairperson of the national SFTG, which was instrumental in developing recommendations for management of subsistence fisheries in South Africa. The project also provided pivotal personnel for the wider implementation of subsistence fishery management systems. The project's Community Liaison Officer was appointed as the first provincial Subsistence Fisheries Implementation Unit Manager (in KwaZulu-Natal), and subsequently took up the position of national Subsistence Fisheries Programme Manager at MCM. Furthermore, two of the current provincial field extension workers who now facilitate the co-management arrangements for 43 fisher groups and train and supervise community monitors throughout the province of KwaZulu-Natal, were themselves community monitors trained within the Sokhulu project. This embedding of staff from the local case study into national and provincial government is likely to maximize the use of lessons learnt and sustain institutional application of the project's approach.

Providing access

National recommendations for management

A suite of recommendations was developed by the SFTG following their research and consultation and was submitted to the Chief Director of MCM in February 2000 (Harris et al. 2002b). The recommendations included: planning for implementation; definitions of subsistence fishers and other sectors; assessment and categorization of resources; determination of types of fishing activities; zonation; management systems; training; communication mechanisms; application and allocation procedures; compliance processes; research and monitoring; and development of training and institutional capacity. Revised definitions for subsistence, small-scale commercial and full commercial fishing sectors were provided in the recommendations, together with suggestions for the reclassification of all informal fisheries in terms of these definitions. Resources were also classified on the basis of their suitability for the different fishing sectors, and a total of 12 different categories of subsistence and small-scale fishing activities were identified together with a potential suite of resources to be made available to each sector (Table 6.1).

It was recommended that subsistence fishing rights be area-based and that where necessary, resources within these areas be reserved for the exclusive use of local subsistence fishers in order to reduce user conflict among communities and with other fishing sectors. A multi-tiered institutional management structure was recommended for the management of subsistence fishing activity, with a strong emphasis on co-management. Key agents within the management structures were identified, and included a dedicated national Subsistence Fishery Management Unit, provincial management agencies, regional fieldworkers, local co-management structures, community monitors, and an independent Subsistence Fisheries Advisory Group (see Fig. 6.5). A detailed implementation plan was also prepared by the SFTG that identified goals, summarized and prioritized actions to achieve each goal, assigned roles and responsibilities to identified agencies, and provided timelines for the achievement of each goal and action.

Provincial implementation

A large proportion of subsistence fishers identified in the processes described above, were found to be living on the east coast of South Africa, in the provinces of KwaZulu-Natal and Eastern Cape (Clarke et al. 2002). A void in the capacity (in both physical infrastructure and staff) of the national agency, MCM, exists in provinces other than the Western Cape, and this is a major impediment to implementing subsistence fisheries in remote places at local level. In KwaZulu-Natal, the nature conservation agency EKZN Wildlife was willing to take responsibility for implementing management systems for subsistence fishers in the province. In fact, this agency had already begun to engage with subsistence fishers to address problems surrounding traditional access and unsustainable harvesting. This

Table 6.1 Categories of subsistence and small-scale commercial fisheries identified by the Subsistence Fisheries Task Group (SFTG) (after Cockcroft et al. 2002).

Fishery categories	Target species
1. Rocky-shore intertidal benthic invertebrates	Limpets
	Polychaete worms
	Mussels
	Oysters
	Gastropod snails
	Octopus
	Red-bait
2. Sandy-beach invertebrates	Gastropods
	Bivalves
	Ghost crabs
	Mole crabs
3. Marine subtidal invertebrates and seaweeds	Spiny lobsters
	Abalone
	Kelps and seaweeds
4. Marine shore-based rod and hand line fishery	Fish
5. Marine non-motorized boats; rod and hand line	Fish
6. Marine motorized boats; rod and hand line	Fish
7. Marine gill/seine/net fishery	Fish
8. Estuarine benthic invertebrates	Mud/sand prawns
	Polychaete worms
	Bivalves
	Mangrove crabs
9. Estuarine seine and cast nets	Swimming prawns
	Crabs
	Fish
10. Estuarine gill nets	Fish
	Swimming crabs
11. Estuarine trap fishery	Fish
12. Estuarine shore-based rod and hand line fishery	Fish

included facilitation of experimental artisanal fisheries such as the St Lucia gill net fishery (Mann 1995), long-term programs (>15 years) to involve subsistence fishers in monitoring their estuarine and marine intertidal resource use in the Kosi area (Tomalin and Kyle 1998; Kyle 2003), and pilot projects such as the Sokhulu initiative that were aimed at involving local fishers in the management of fisheries (Harris et al. 2003; Mann 2003).

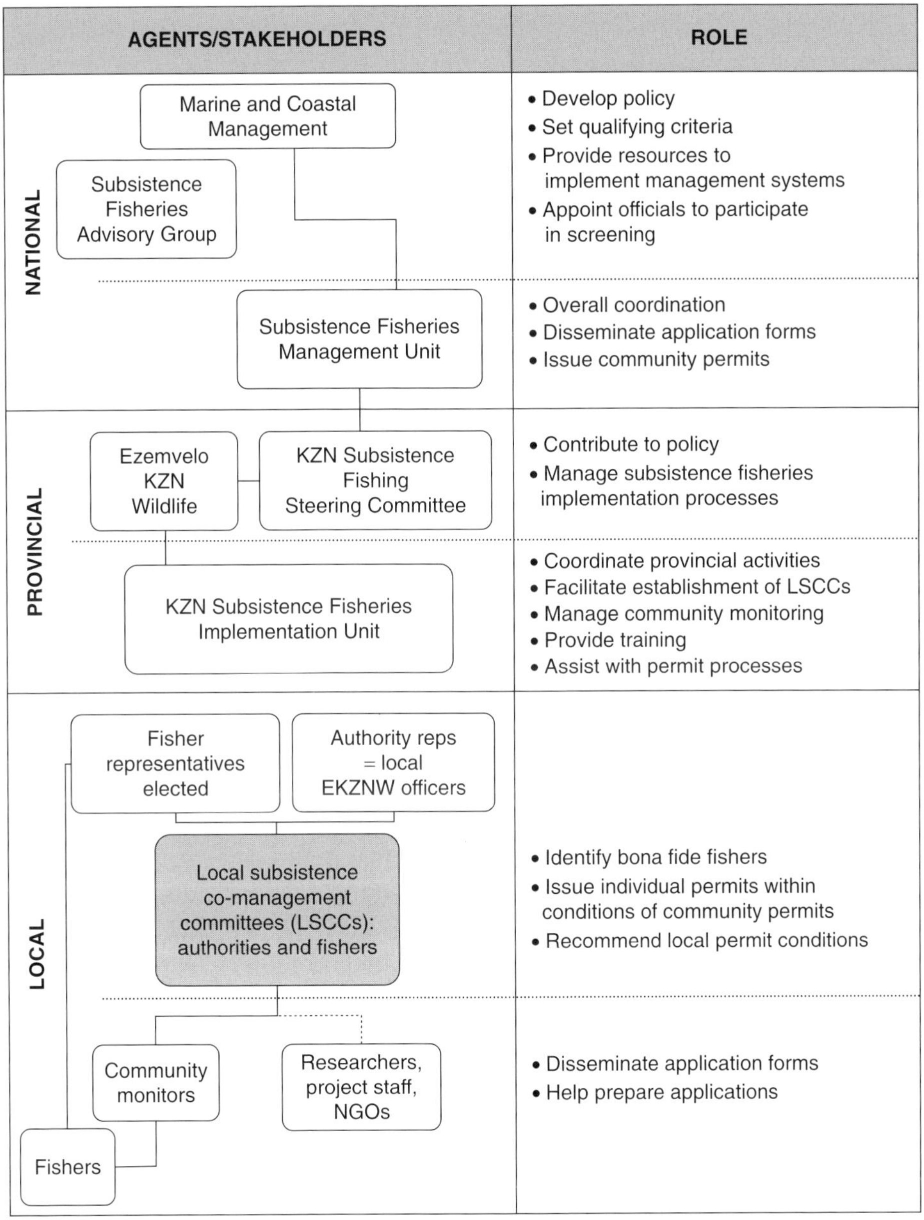

Figure 6.5 Institutional capacity and processes required for implementing subsistence fisheries effectively.

In 2000, the KwaZulu-Natal Subsistence Fisheries Management Implementation Programme was launched as a partnership between the provincial agency EKZN Wildlife and the national department MCM. It aimed to: (1) identify all subsistence and small-scale commercial fisheries; (2) develop management plans for each of these fisheries; (3) locate all subsistence-fishing communities; (4) establish local co-management structures in each of these communities; (5) develop and co-ordinate permit application procedures; and (6) implement resource monitoring and assess sustainability of resource use. These activities have been, and are, funded from three major sources: the Norwegian government; MCM has allocated funds from the Marine Living Resources Fund (MLRF); and The Green Trust, WWF–SA, which provided key support for the pilot projects. National government has shown commitment towards future support: MCM currently pays for community monitoring via the MLRF, and plans to absorb the costs of co-management arrangements when the Norwegian funding is phased out. The provincial government agency EKZN Wildlife contributes considerable staff time and administrative and logistic support.

Creating institutional context for management

The development of a multi-tiered institutional structure for management of subsistence fisheries in KwaZulu-Natal was considered critical to ensure consistent and fair management of the fisheries and to coordinate activities in support of local management structures (Fig. 6.5).

At a national level, MCM is responsible for managing subsistence fisheries. To accomplish this, a Subsistence Fisheries Management Unit was established in 2000. Scientific working groups and individual scientists contribute advice on a species-specific basis, but the advisory group, recommended to allow input of knowledge and advice from external participants, has not yet been formed. In the province of Kwazulu-Natal, MCM brokered a contractual agreement with EKZN Wildlife for implementation of management for subsistence fisheries in KwaZulu–Natal province that is now in its sixth year, and has included the establishment of a dedicated provincial Subsistence Fisheries Implementation Unit (KZN–SFIU) (Fig. 6.5).

The local co-management model developed by the Sokhulu case study, informed by lessons learnt through the other pilot projects (Mann 1995, 2003; Kyle 2003), was adopted and applied throughout the province of KwaZulu-Natal. By early 2005, a total of 43 Local Subsistence Co-management Committees (LSCCs) had been established in 19 fishing communities. The function of these local co-management committees is to involve fishers and the local authority in the joint management of the linefish and intertidal resources (Fig. 6.5).

Finding the fishers and fisheries

The national survey undertaken by the SFTG provided an initial 'snap-shot' of the nature, distribution, and abundance of informal fishers around the South African coast. The first task undertaken in the KwaZulu-Natal by the new KZN–SFIU was to verify and improve on this information, specifically to identify all communities in KwaZulu-Natal that harbor subsistence fishers, to estimate the numbers of fishers, and to identify the different fishing types. This was done by following up on information provided by EKZN Wildlife management staff involved with existing illegal or informal fisheries and through visits by field extension workers to potential subsistence communities.

Five subsistence fisheries have been recognized in KwaZulu-Natal (Table 6.2), namely rocky-shore invertebrate harvesting, sandy-shore invertebrate harvesting, mangrove crab harvesting, line fishing, and traditional estuarine spearing. Five artisanal or small-scale commercial fisheries were also identified: sand and mud prawn harvesting; shove net fishing; fish trap (kraal) fishery; cast netting; and oyster harvesting. A number of other informal fishing activities were considered unsustainable and these have been halted or are being phased out, including estuarine gill netting (Table 6.2).

Table 6.2 Five types of informal fisheries identified in KwaZulu-Natal and recognized as subsistence or small-scale commercial with lists of the most common species targeted.

Fisheries	Main target species
1. *Subsistence*	
Rocky-shore invertebrates	Mussels, red-bait, limpets, winkles, octopus
Sandy-shore invertebrates	Ghost crabs, mole crabs
Linefish	Marine and estuarine linefish species: most commonly shad, blacktail, stumpnose, karanteen, pompano and stonebream in the sea, and grunter, kob and perch in estuaries
Kosi lake spearing (graining)	Estuarine teleost fish
Estuarine (mangrove) crabs	Sesarmid crabs
2. *Artisanal (traditional small-scale)*	
Kosi lakes fish traps	Estuarine teleost fish
3. *Small-scale commercial*	
Sand and mud prawn (bait)	*Callianassa kraussi* and *Upogebia africana*
Shove nets	Fish and prawns
Shoreline cast netting	Fish
Oyster fishery	Edible oysters
4. *Undesirable – to be phased out*	
Estuarine gill net	Estuarine teleost fish and prawns
5. *Illegal*	
Shadecloth pull nets in estuaries	Estuarine fish and prawns
Cast netting in estuaries	Estuarine fish

In keeping with the definition developed by the SFTG (Branch et al. 2002b), individual fishers qualify for subsistence status in KwaZulu-Natal if they are engaged in a subsistence fishing activity (Table 6.2), belong to and are resident in a 'subsistence fishing community', and are poor. This latter criterion can be quantified, but generally only subjective socio-economic data are available. A pragmatic approach has been taken in that the community representatives on the LSCCs are given the responsibility of identifying (screening) which applicants are 'poor' and thus deserving of access to the fishery. This system depends heavily on establishment of democratic, representative LSCCs that have integrity and are not influenced by local politics or intimidation. It is also sensitive to resource abundance, since screening and access to the fishery becomes stricter as resources dwindle, particularly in the quota-based fisheries.

Nineteen subsistence fishing communities have been recognized along the KwaZulu-Natal coast (Fig. 6.1), these being communities with long-standing engagement in one or more subsistence (or small-scale commercial) fishing activities and proximity to a traditional harvesting area. All are rural or semi-rural, and can be delineated by traditional tribal boundaries.

The way in which the definition has been applied in KwaZulu-Natal is intended to formalize long-standing subsistence activities and legalize traditional rights, while excluding new entrants to the sector. It resists the use of subsistence fishing by government as a solution to unemployment and poverty. The reasoning behind this approach is that the natural resources available are already insufficient to meet the food security needs of existing traditional fishers and many stocks are already overexploited (Hockey et al. 1988; Lasiak and Field 1995; Siegfried et al.1995; Griffiths 2000), and additional use of these resources would threaten the food security of existing subsistence fishers. Being a subsistence fisher, i.e. the 'poorest of the poor', is inarguably undesirable and the goal is to develop and promote alternative options that will reduce reliance on direct use of natural resources and ultimately result in a shrinking of this sector over time. Unsurprisingly, the provision of alternatives for subsistence fishers remains a challenge.

Crafting local co-management

A step-by-step approach was followed to identify fisheries and fishers in each of the 19 subsistence communities, and to establish local co-management structures for each subsistence fishery in each community. Processes for formation of co-management arrangements were informed by the key elements identified by the Sokhulu case study (see Fig. 6.4).

Step 1. The first step towards co-management in each community was establishing contact with the fishers to inform them of the new legislation recognizing the subsistence sector and to initiate their involvement. This step is not trivial and proper protocols that take into account cultural etiquette are needed to ensure full participation and a democratic process. More than one meeting is necessary, and there is a need to ensure that local political interests and cultural issues do not compromise the process. Although the first meeting in each community usually involves the leadership representatives, the broader population, and fishers, subsequent communication and consultation is best done separately with fishers for each fishery in each community. Gender bias can influence progress and participation, and mixing fisher groups that are dominated by women with male-dominated fisher groups often results in the sidelining of women's issues.

Step 2. Facilitate establishment of local subsistence co-management committees (LSCCs) for each fishery in each community (Fig. 6.5). Here it is important to ensure the majority of fishers are involved, and that election of fisher and authority representatives and the preparation of a constitution are democratic and transparent, leading to a body that is genuinely representative and a constitution that is accepted.

Step 3. Provide training for management staff, extension staff, community monitors, and LSCCs in co-management principles, committee roles and functioning, legislative frameworks, basic fisheries management, stock assessment techniques and facilitation skills, as appropriate for the different role players. A key feature in this component has been the training of extension staff and community monitors as 'trainers' in co-management, conflict resolution and sustainability principles, so that they are able to transfer these concepts to fishers and particularly to children of fishers. Training 'modules' from the Sokhulu project have been applied as tools for use by the monitors. Exchange visits between communities have also been used as an effective form of training.

Step 4. Appoint community monitors for each fishery (Fig. 6.5). Community monitors have been trained to record fisher activities, off-take and utilization patterns, and assist with stock surveys. They also play an integral role in communication and as trainers. Monitors from the Sokhulu co-management pilot project have assisted in training all new monitors on basic marine biology and sustainable resource utilization. Presently there are 38 community monitors and four supervisors employed for the 19 subsistence-fishing communities. An integrated database has been designed for the data collected by the monitors, and a data technician captures and assists with analysis of the data.

Step 5. Participative research to determine stock availability, set quotas or bag limits and determine if limitations must be imposed on the methods of harvesting. Community monitors do the fieldwork of the stock surveys and the information is centrally collated and analyzed. This activity both draws from and informs the fishery management plans that must be developed for each of the subsistence and small-scale fisheries that have been identified.

Step 6. Local joint decision-making. Fishery management plans document the history of each fishery, status of the resources, and outline generic management rules and allocation procedures. The applications for community permits are prepared by the LSCCs, facilitated by the field extension workers. The LSCCs identify the location and extent of a desired subsistence fishery area and indicate the quantity of resources recommended, including annual quota or daily bag limits, while final approval is the responsibility of MCM and is subject to issues of access for other users and stock availability. At the local level, the LSCCs have the mandate to screen and determine who qualifies as a subsistence fisher, the number of fishers to be issued individual permits by the LSCC in terms of the community permit, and how the allocated resource will be shared amongst the participating fishers.

Step 7. Allocation of access rights. The intention is that community permits be issued to the LSCCs that then manage the allocation of the quota to the fishers through issuing individual permits. Unfortunately, the current legislation does not allow community permits to be issued to LSCCs, so rights awarded thus far to groups qualifying for subsistence permits have taken the form of exemptions issued to individual fishers. In order that the role of the LSCCs is not undermined, these exemptions are distributed via the LSCCs.

A related and closely-associated project funded by MCM has developed a database for information about the fisheries and communities, which is linked to a geographical information system for mapping fishing areas, so that these can be tracked over time.

Evaluating progress and success

Implementing systems for fisheries management

Commitment to developing systems and capacity for allocation of rights to subsistence fisheries and management of these fisheries has been demonstrated by the national agency (MCM) through the provision of staff and budget for a dedicated national Subsistence Fisheries Management Unit, and allocation of increasing amounts of funds to subsistence fisheries management in the provinces over the past six years. This is notable in KwaZulu-Natal where the community-monitoring program is now fully funded by MCM. However, the implementation program in this province is still dependent for a significant proportion of its budget on donor funding, and on EKZN Wildlife for free staff involvement and use of facilities, and therefore long-term sustainability of the program is not yet assured.

The progress of the subsistence fisheries program in South Africa has been uneven and non-systematic, with the four coastal provinces following strikingly different routes in developing management structures for subsistence fishers. In KwaZulu-Natal, MCM assigned responsibility to the provincial authority, EKZN Wildlife, which has developed a Subsistence Fisheries Management Unit. Drawing on the experiences of co-management pilot projects, Local Subsistence Co-management Committees have been created in all identified subsistence fishing communities and permits either have been, or in the process of being, issued to these communities. In the Eastern Cape, management of subsistence fisheries has been slower to take off due to lower provincial capacity, but co-management structures have been initiated in 15 fishing communities, of which Coffee Bay has been underway for several years. In the Northern Cape, virtually nothing has yet been accomplished, reflecting the scarcity of local capacity and the scarcity of the population there.

In the Western Cape, only one community, Ebenhaeser on the Olifants Estuary, has an established co-management committee, while a second is in the process of being created. Part of the reason for the relative neglect of the Western Cape is that many of the resources most heavily in demand, such as rock lobsters, abalone, and linefish, are not suited to subsistence fishing and are better reserved for small-scale commercial or artisanal operations. Compounding this, overfishing has depleted the linefish fishery (Griffiths 2000), and the rock lobsters and especially abalone are severely threatened by international poaching syndicates (Hauck and Sweijd 1999). Intense poaching at a level approaching local warfare inhibits cooperative co-management and has diminished the potential yields, thus discouraging legal fishing.

It is easy to criticize the slow implementation of subsistence fisheries management in three of the four provinces, but it does take time to roll out a country-wide initiative and it has been only six years since these communities were identified. KwaZulu-Natal had a head start in two important respects: it uniquely had its own autonomous and long-established conservation authority that assumed responsibility for fisheries management, and it benefited from pilot studies funded by an external agency. The latter kick-started local co-management and provided a model that quickly spread to other KwaZulu-Natal communities.

In strong contrast to the successful introduction of co-management for subsistence fishers in KwaZulu-Natal, most artisanal fishers throughout the country remain largely marginalized, ironically by the very process of transformation legislated to protect them (van Sittert et al. 2006). As part of their recommendations, the SFTG urged that attention be paid to 'small-scale commercial fishers', including 'limited commercial-rights holders' and 'artisanal fishers'. This coincided with government policy to promote micro-scale enterprises to reduce poverty and promote transformation from the apartheid situation, where whites dominated the fishing industry. From some perspectives, transformation has been successful, with previously disadvantaged individuals now holding a majority of the commercial rights and owning the majority of fishing companies in most commercial sectors. The number of rights-holders has increased substantially to a large number of relatively small companies rather than the previous monopolization by a few large white-owned companies (Anon 2002). The process has, however, tended to introduce new black entrepreneurs rather than securing access for existing fishing communities. This has occurred despite the pleas of the SFTG that preference be given to people historically connected with, actively participating in, and dependent on fishing (Branch and Clark 2006).

A key shortcoming in the implementation process since the SFTG provided its recommendations is that for various reasons MCM did not appoint an advisory group. That would have allowed the inexperienced management unit to draw on the knowledge of external field practitioners and provide continuity between the activities of the SFTG and the newly-created management unit. There has been a tendency for MCM to 'go it alone' in trying to solve complex problems, despite the obvious benefits and specific recommendations that such a group should be created (Harris et al. 2002b).

Despite the shortfalls, inconsistencies, and problems experienced, the various active attempts to engage informal fishers have, nevertheless, provided important lessons and examples to inform policy and practice. A positive recent development is that subsistence and small-scale commercial policies are now being prepared by MCM in consultation with stakeholders for submission to the Minister of Environment and Tourism for approval. It is important that these provide clear definitions and qualifying criteria for separation of these sectors, and that they pave the way to legal mechanisms for granting community (instead of individual) rights to traditional resources.

Co-management as a local fisheries management model

The likelihood that co-management will succeed depends on a range of factors. Ostrom (1990, 1992) pioneered thoughts about conditions that would favor successful co-management and a number of authors have suggested prerequisites for successful co-management (Pinkerton and Weinstein 1995; Baland and Platteau 1996; Berkes et al. 2001; Pomeroy et al. 2001). Pursuing this approach, Sowman et al. (2003) developed a list of conditions that they felt were important in the context of South African co-management, and developed measures of the success of co-management (Table 6.3).

Developing such lists does not, however, test whether these conditions are truly important or not, nor does it address whether co-management actually achieves what it sets out to do – improve the lots of fishers and managers and the sustainability of the resource. The KwaZulu-Natal implementation process provides a good opportunity to explore this, in that it has yielded 43 co-management arrangements that are at different stages of development and cover a number of distinct fisheries at different localities. Napier and colleagues (2005) attempted an analysis of 11 fishing communities in KwaZulu-Natal by quantifying measures of: (a) the extent that 'conditions' were achieved; and (b) whether co-management was regarded as a success or not. Success was measured as improvements in: (1) the circumstances of fishers; (2) management efficacy; and (c) the status of the resource. Both conditions and success indices were measured on a scale of 0–5, based on the independent views of the fishing communities and the authorities, and this allowed quantitative correlation between the attained conditions and success of co-management.

Several patterns emerged. First, the level of dependence of each community on natural marine resources was strongly correlated with an index of poverty ($r_s = 0.96$, $p < 0.01$), upholding the view that poverty is an underlying cause of subsistence reliance on natural resources and probably a major contributor to overexploitation.

Table 6.3 Measures of the success of co-management (adapted from Sowman et al. 2003).

Social indicators
1. Has communication improved between government and resource users?
2. Has trust between partners improved?
3a. Have appropriate institutional structures been put in place?
3b. Are these structures representative and accountable?
3c. Has cooperation and leadership in the community been improved?
4. Are there agreed objectives among project partners and have they been recorded?
5. Have resource users gained greater access to information?
6a. Has training enhanced the capability of resource users?
6b. Has training enhanced understanding and capability of managers?
6c. Has knowledge about the socio-economic circumstances of the users improved?
6d. Have resource users gained greater scientific knowledge about the resource?
7. Has indigenous knowledge been acknowledged by the authorities?
8. Have resource users gained decision-making power and management responsibility?
9a. Has there been buy-in (commitment) by authorities to co-management?
9b. Has there been buy-in (commitment) by resource users to co-management?
10. Has compliance improved amongst participating resource users?
Economic indicators
11a. Has access to resources been secured or improved?
11b. Has resource allocation been equitable?
12. Has there been sufficient funding for the process to be sustained?
13. Have attempts been made to improve alternative or supplemental livelihoods?
Resource management indicators
14. Is there broad agreement on the rules governing resource management?
15. Has resource monitoring been instituted?
16. Has there been an evaluation of the co-management process?
17a. Has there been an improvement in the scientific knowledge about the resources?
17b. Has this knowledge been applied by managers to improve management?
18a. Is use of resources now more sustainable?
18b. Has co-management reduced any adverse effects of human activities on the ecosystem?

Second, the overall level of attainment of the conditions was tightly correlated with the overall perceived success of co-management in each community ($r = 0.90$, $p < 0.0001$), supporting the intuitively obvious but previously untested conclusion that the more the conditions are fulfilled, the greater the likelihood of success. Third, the level of satisfaction about co-management was very high. In answer to questions such as 'Has co-management improved the life of the community?' or 'Are you happy with co-management?' fishers registered 87% satisfaction. Fourth, although the scores provided by the fishers and the authorities were significantly correlated in most cases, fishers were generally less enthusiastic than the authorities, providing scores that were almost invariably less than those of the authorities. This was most obvious in cases where communities had experienced unacceptably long delays in receiving subsistence permits: delays killed confidence in co-management. Fifth, the length of time individual co-management programs had been in operation was correlated with both the degree to which 'conditions' had been implemented ($r = 0.97$; $p < 0.0001$) and the perceived success of co-management ($r = 0.85$; $p < 0.001$) – co-management takes time to become established. The extent to which individual conditions were correlated with success varied widely, and conditions previously cherished as being important were not correlated with success, while others were strongly correlated (Fig. 6.6). Of the greatest significance is that three of the four conditions most strongly correlated with success received the lowest scores for the extent that they had been achieved: (1) 'Are the benefits of participating greater than the costs?' (2) 'Was training and empowerment supplied to the community?' (3) 'Has decentralization of authority taken place?' If co-management is to succeed in the long run, it is imperative that the conditions contributing most significantly to success are addressed as a priority.

There are both encouraging and 'red-flag' signals that emerge from this analysis of the programme in KwaZulu-Natal. On the plus side, following pioneering work in Kosi Bay (Kyle et al. 1997a, b; Kyle 2003) and the pilot study at Sokhulu (Harris et al. 2003), co-management has now been implemented for

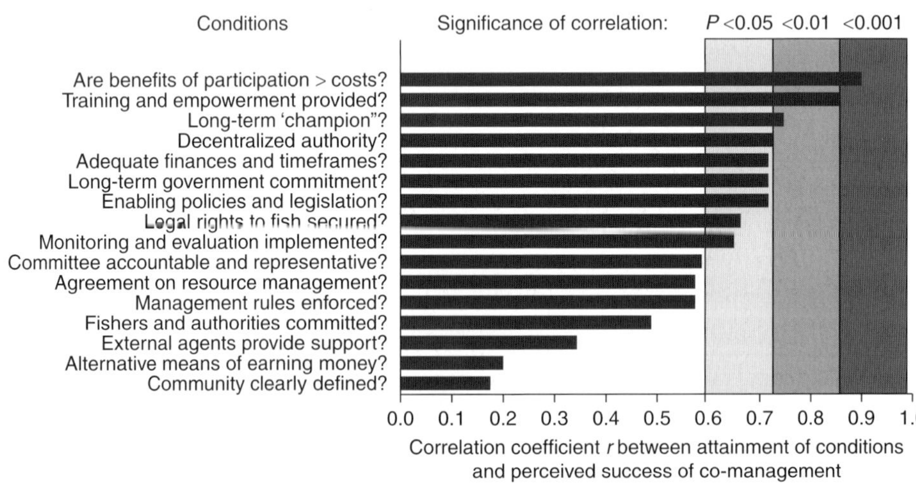

Figure 6.6 Coefficient of correlation (r) between the degree that conditions considered to be important for co-management have been implemented and the perceived success of co-management in 11 fisheries in KwaZulu-Natal. Shading on the right indicates the significance of the correlation coefficients (adapted from Napier et al. 2005).

43 fisheries in KwaZulu-Natal and has been widely perceived by both fishers and authorities as an improvement over top-down authoritarian imposition of regulations. In KwaZulu-Natal, the program has, to date, provided important information, established communication links with fisher communities, created provincial and local institutional structures to manage subsistence fisheries, and introduced a comprehensive monitoring system. This progress has exemplified what can be accomplished, and has provided lessons and case studies in establishment of co-management structures and the devolution of management that can be applied throughout the country.

On the negative side, the authorities were slow to act in granting permits and designating subsistence fishing areas, and attention still needs to be given to achieving conditions that are vital to success. More particularly, although co-management has been embraced in KwaZulu-Natal, the successes are being dulled by the slow rate that they are being extended to the entire country, and the failure to attend adequately to the needs of artisanal and small-scale fishers.

Five key points emerge. (1) Co-management is a viable and improved way of managing subsistence fisheries, but for it to succeed: (2) genuine sharing of responsibility; (3) devolution of authority; (4) provision of adequate personnel, funding, and time; and (5) training of participants, are imperative. The steps taken in South Africa are youthful; expansion of the successes to a national scale is an exhilarating but challenging vision (see Figs. 6.7 and 6.8).

Figure 6.7 In areas where mussel harvesting is co-managed, harvesters are limited to defined areas and restricted to controlled quantities, which may only be removed with non-destructive implements such as screwdrivers. Photograph by George Branch.

Figure 6.8 In co-managed areas, community monitors record the details of the catch on each collecting day and report to the local subsistence co-management committee. Photograph by Jean Harris.

Conclusions

The allocation of real access rights to subsistence fishers in South Africa has been a long and difficult process and is unfortunately far from finished. Subsistence fishers were formally recognized for the first time with the promulgation of new fisheries legislation in 1998, following the election of a democratic government in 1994. Prior to this, these fishers were treated mostly as poachers when they did not conform to the rules laid down for the formally recognized recreational or commercial fishing sectors. Understanding of the needs and ways that subsistence fishers operated was poor, being limited to a handful of case studies where researchers and managers had attempted to engage these fishers in some sort of quasi-formal management processes. Appointment of an SFTG in 1998 was the first sign that the government had real intentions of improving the plight of this group of vulnerable and marginalized fishers. Following a period of extensive consultation, research, and deliberation, a set of recommendations was produced by the SFTG that included a detailed implementation plan that identified goals, summarized and prioritized actions required to achieve each goal, assigned roles and responsibilities to identified agencies, and provided very strict timelines for implementation of these actions. These recommendations were accepted by government, and an implementation plan prepared, but regrettably upheavals and severe shortages in capacity within the key national and provincial government agencies mandated with the management of marine living resources in South Africa, resulted in significant delays in the full national implementation of these recommendations. This is very unfortunate as huge expectations were created amongst fishers through the SFTG process, and many fishers have since become even more disappointed and disillusioned by the apparent lack of action. Subsistence fishers continue to operate outside of the law in many areas and are frequently prosecuted and even imprisoned by management authorities. Others have taken advantage of the confusion surrounding the allocation of subsistence rights, using this opportunity to poach and plunder certain key resources.

On the positive side, however, significant progress towards formalizing management of subsistence fishers has been made in one of the four coastal provinces in this country (KwaZulu-Natal), due principally to the highly proactive role played by the management agency in this province. The KwaZulu-Natal Subsistence Fisheries Management Implementation Programme (a partnership between the provincial and national government agencies) was launched in the same year that the SFTG tabled their recommendations with the aim of identifying and locating all subsistence and small-scale commercial fishers and fisheries, developing management plans and establishing local co-management structures for each, and developing and coordinating permit application and resource monitoring procedures. This program has achieved the formalization of a large number of local co-management committees, management plans have been developed and approved for a number of these fisheries and permits have been issued to the fishers.

A number of very important lessons have been learned through this process, particularly in respect of those factors that affect success and failure in respect of co-management. Implementation of subsistence fisheries management structures to the rest of the country is now required as a matter of urgency.

References

Anonymous (2002) *Where Have All the Fish Gone? Measuring Transformation in the South African Fishing Industry*. Department of Environmental Affairs and Tourism, South Africa.

Baland, J.M. and Platteau, J.P. (1996) *Halting Degradation of Natural Resources: is there a Role for Rural Communities?* Clarendon Press, Oxford.

Berkes, F. (1990) Native subsistence fisheries: a synthesis of harvest studies in Canada. *Arctic*, **43**: 35–42.

Berkes, F., Mahon, R., McConney, P., Pollnac, R., and Pomeroy, R. (2001) *Managing Small-scale Fisheries: Alternative Directions and Methods*. International Development Research Centre, Ottawa, Canada.

Branch, G.M. and Clark, B.M. (2006) Fish stocks and their management: the changing face of fisheries in South Africa. *Marine Policy*, **30**: 3–17.

Branch, G.M., May, J., Roberts, B., Russell, E. and Clark, B. (2002a) Case studies on the socio-economic characteristics and lifestyles of subsistence and informal fishers in South Africa. *South African Journal of Marine Science*, **24**: 439–62.

Branch, G.M., Hauck, M., Siqwana-Ndulo, N. and Dye, A. (2002b) Defining fishers in the South African context: subsistence, artisanal and small-scale commercial sectors. *South African Journal of Marine Science*, **24**: 475–87.

Clark, B., Hauck, M., Harris, J., Salo, K. and Russell, E. (2002) Identification of subsistence fishers, fishing areas, resource use and activities. *South African Journal of Marine Science*, **24**: 425–37.

Cochrane, K.L. (1995) Anticipating impacts of recent political changes on fisheries management in South Africa. *Naga*, **18**: 4–8.

Cochrane, K.L. and Payne, A.I.L. (1998) People, purses and power: developing fisheries policy for the new South Africa. In: T.J. Pitcher, P.J.B. Hart and D. Pauly (Eds) *Reinventing Fisheries Management.* Kluwer Academic Publishers, London.

Cockcroft, A., Sauer, W., Branch, G.M., Clark, B., Dye, A.H. and Russell, E. (2002) Assessment of resource availability and suitability for subsistence fishers in South Africa with a review of management procedures. *South African Journal of Marine Science*, **24**: 489–501.

DEAT (Department of Environmental Affairs and Tourism) (1998) Marine Living Resources Act, No. 18 of 1998. *Republic of South Africa Government Gazette*, **395** (18930): 1–66.

Fall, J.A. (1990) The division of subsistence of the Alaska Department of Fish and Game: an overview of its research and program and findings: 1980–1990. *Arctic Anthropology*, **27**: 69–92.

Griffiths, M. (2000) Long-term trends in catch and effort of commercial linefish of South Africa's Cape Province: snap shots of the 20th century. *South African Journal of Marine Science*, **22**: 81–110.

Harris, J.M., Sowman, M., Branch, G.M., Clark, B.M., Cockcroft, A.C., Coetzee, C., Dye, A.H., Hauck, M., Johnston, A., Kati-Kati, L., Maseko, Z., Salo, K., Sauer, W.H.H., Siqwana-Ndulo, N. and Beaumont, J. (2002a) The process of developing a management system for subsistence fisheries in South Africa: recognizing and formalizing a marginalized fishing sector in post-apartheid South Africa. *South African Journal of Marine Science*, **24**: 405–24.

Harris, J.M., Branch, G.M., Clark, B.M., Cockcroft, A.C., Coetzee, C., Dye, A.H., Hauck, M., Johnson, A., Kati-Kati, L., Maseko, Z., Salo, K., Sauer, W.H.H., Siqwana-Ndulo, N. and Sowman, M. (2002b) Recommendations for the management of subsistence fisheries in South Africa. *South African Journal of Marine Science*, **24**: 503–23.

Harris, J., Branch, G., Sibiya, S. and Bill, C. (2003) The Sokhulu subsistence mussel harvesting project – a case study for fisheries co-management in South Africa. In: M. Hauck and M. Sowman (Eds) *Waves of Change: Coastal and Fisheries Co-Management in South Africa.* University of Cape Town Press, Cape Town.

Hauck, M. and Sowman, M. (Eds) (2003) *Waves of Change: Coastal and Fisheries Co-management in South Africa.* University of Cape Town Press, Cape Town.

Hauck, M. and Sweijd, N.A. (1999) A case study of abalone poaching in South Africa and its impact on fisheries management. *ICES Journal of Marine Science*, **56**: 1024–32.

Hauck, M., Sowman, M., Clark, B., Russell, E., Harris, J.M., Venter, A., Beaumont, J. and Maseko, Z. (2002) Perceptions of subsistence and informal fishers in South Africa about management of and access to living marine resources. *South African Journal of Marine Science*, **24**: 463–74.

Hersorg, B. and Holm, P. (1999) Change without redistribution: an institutional perspective on South Africa's new fisheries policy. *Marine Policy*, **24**: 221–31.

Hockey, P.A., Bosman, A.L. and Siegfried, R.W. (1988) Patterns and correlates of shellfish exploitation by coastal people in Transkei: an enigma of protein production. *Journal of Applied Ecology*, **25**: 353–64.

Hutton, T. and Pitcher, T.J. (1998) Current directions in fisheries management policy: a perspective on co-management and its application to South Africa. *South African Journal of Marine Science*, **19**: 471–86.

Hutton, T., Cochrane, K.L. and Pitcher T.J. (1997) Post-apartheid fisheries management policy in South Africa: the need for a change in management philosophy. In: D.A. Hancock, D.R. Smits, A. Grant and J.P. Beumer (Eds) *Developing and Sustaining World Fisheries Resources: the State of Science and Management*. CSIRO Publishers, Australia.

Kyle, R. (2003) Kosi Bay gillnetting: a community-based joint management fishery inside a nature reserve. In: M. Hauck and M. Sowman (Eds) *Waves of Change: Coastal and Fisheries Co-management in South Africa*. University of Cape Town Press, Cape Town.

Kyle, R., Pearson, B., Fielding, P.J. and Robertson, W.D. (1997a) Subsistence shell-fish harvesting in the Maputaland marine reserve in northern KwaZulu-Natal, South Africa: rocky shore organisms. *Biological Conservation*, **82**: 183–92.

Kyle, R., Robertson, W.D. and Birnie, S.L. (1997b) Subsistence shell-fish harvesting in northern KwaZulu-Natal, South Africa: sandy beach organisms. *Biological Conservation*, **82**: 173–82.

Lasiak, T. and Field, J.G. (1995) Community-level attributes of exploited and non-exploited rocky infratidal macrofaunal assemblages in Transkei. *Journal Experimental Marine Biology and Ecology*, **185**: 33–53.

Mann, B.Q. (1995) Quantification of illicit fish harvesting in the lake St Lucia Game Reserve, South Africa. *Biological Conservation*, **74**: 107–13.

Mann, B. (2003) The St Lucia subsistence gillnet fishery. In: M. Hauck and M. Sowman (Eds) *Waves of Change: Coastal and Fisheries Co-management in South Africa*. University of Cape Town Press, Cape Town.

Martin, R. and Nielsen, J. R. (1998) Creation of a new fisheries policy in South Africa: the development process and achievements. In: A.K. Normann, J.R. Nielsen and S. Sverdrup-Jensen (Eds) *Fisheries Co-management in Africa: Proceedings from a Regional Workshop on Fisheries Co-management Research (Mangochi, Malawi, 18–20 March 1997).* Institute for Fisheries Management and Coastal Community Development. Fisheries Co-management Research Project Research Report 12.

Napier, V.R., Branch, G.M. and Harris, J.M. (2005) Evaluating conditions for successful co-management of subsistence fisheries in KwaZulu-Natal, South Africa. *Environmental Conservation*, **32**: 165–77.

Ostrom, E. (1990) *Governing the Commons: Evolution of Institutions for Collective Action*. Cambridge University Press, Cambridge.

Ostrom, E. (1992) *Crafting Institutions for Self-governing Irrigation Systems*. Institute for Contemporary Studies, San Francisco.

Pinkerton, E. and Weinstein, M. (1995) *Fisheries That Work: Sustainability through Community-based Management.* The David Suzuki Foundation, Vancouver, Canada.

Pomeroy, R.S., Katon, B.M. and Harkes, I. (2001) Conditions affecting the success of fisheries co-management: lessons from Asia. *Marine Policy*, **25**: 197–208.

Satia, B.P. (1993) *Ten Years of Integrated Development of Artisanal Fisheries in West Africa: Origin, Evolution and Lessons Learned. Programme for Integrated Development of Artisanal Fisheries in West Africa*. IDAF Technical Report 50, FAO, Cotonou, Benin.

Siegfried, R.W., Hockey, P.A.R. and Crowe, A.A. (1985) Exploitation and conservation of brown mussel stocks by coastal people of Transkei. *Environmental Conservation*, **12**: 305–20.

Sowman, M., Hauck, M. and Branch, G.M. (2003) Lessons learned from nine coastal and fisheries co-management case studies. In: M. Hauck and M. Sowman (Eds) *Waves of Change: Coastal and Fisheries Co-management in South Africa*. University of Cape Town Press, Cape Town.

Tomalin, B.J. and Kyle, R. (1998) Subsistence and recreational mussel (*Perna perna*) collecting in KwaZulu-Natal, South Africa: fishing mortality and precautionary management. *South African Journal of Marine Science*, **33**: 12–22.

van Der Elst, R., Branch, G., Butterworth, D., Wickens, P. and Cochrane, K. (1997) How can fisheries resources be allocated . . . who owns the fish? In: D.A. Hancock, D.R. Smits, A. Grant and J.P. Beumer (Eds) *Developing and Sustaining World Fisheries Resources: the State of Science and Management*. CSIRO Publishing, Australia.

van Sittert, L., Branch, G.M., Hauck, M. and Sowman, M. (2006) Benchmarking the first decade of post-apartheid fisheries reform in South Africa. *Marine Policy*, **30**: 96–110.

Chapter 7

Collaborative Fisheries Management in Tanga, Northern Tanzania

Sue Wells, Melita Samoilys, Jim Anderson, Hassan Kalombo, and Solomon Makoloweka

Introduction

Tanga Region is the northernmost coastal region of Tanzania and encompasses three districts, namely Muheza District, Pangani District, and Tanga City that extend about 180 km south from the Kenyan border. The Tanga Coastal Zone Conservation and Development Programme (TCZCDP) was initiated in 1994 as a result of a marked decline in marine resources and deterioration of reef health that was leading to decreased fisheries incomes. Government statistics showed that the number of licensed fishers had more than doubled between 1970 and 1993 to over 4000 (Beckley et al. 1997), presumed to be due to several factors including the decline of the sisal trade in the 1970s and 1980s with a concomitant loss of jobs and greater unemployment, combined with natural population growth and immigration. Greater fishing intensity was compounded by the increase in labor-intensive but destructive fisheries techniques such as beach seine nets and purse seining (Bensted-Smith 1988).

Prior to the TCZCDP, little work had been carried out in the Western Indian Ocean on management of artisanal fisheries or on reef conservation, despite increasing recognition of the need to do so. The work that had been done focused on setting up marine protected areas with the primary aim of preserving biodiversity and to benefit the nascent tourist industry. The needs of fishers were frequently underemphasized. In contrast, the importance of involving local people in conservation and coastal management was already well understood and being taken into consideration in Southeast Asia and the Caribbean (White et al. 1994). It was, however, recognized that community management alone was often not sustainable and this had led to the development of the collaborative management approach, whereby the agency with responsibility for natural resources develops a partnership with resource users and other local residents that specifies and guarantees their respective functions, rights, and responsibilities (Borrini-Feyerabend 1996). With decentralization in progress in Tanzania, and a clear desire on the part of local communities to increase their role in the management of adjacent resources, the collaborative management approach provided an ideal mechanism to implement government and community management objectives. The aim was to build the capacity of local communities to identify the causes of marine resource degradation, find the solutions, and play a major role in implementing these and enforcing any necessary regulations.

The TCZCDP was, therefore, one of the first coastal management programs in the region to make livelihood improvement a key objective, and to start with a community-based approach. As with many 'integrated conservation and development projects' of the time (McShane and Wells 2004), a very broad and ambitious approach was taken, incorporating attempts to trial and implement new livelihood activities, develop and implement mangrove management plans, establish and mainstream new institutional arrangements for coastal management, and build capacity for management through a major training and environmental education program. It was soon realized that the number and extent of the objectives of TCZCDP were unrealistic and the rationale of the program was changed after four years to focus more on developing a mechanism for sustainable management of the artisanal fishery, while still retaining the overall original approach.

The TCZCDP was initially carried out through the Tanga Regional Government, but when decentralization came into effect in the late 1990s under the Local Government Reform Programme (LGRP), responsibility for natural resource management was shifted to the three individual districts, with the Tanga Regional Government providing an advisory role. This required major changes in implementation and administrative arrangements to the TCZCDP, which drew financial and technical input away from immediate fisheries issues. The Government of Ireland provides financial support and the Eastern Africa Regional Office of IUCN – the World Conservation Union provided technical assistance until 2005.

Biophysical characteristics

In Tanga Region, the continental shelf to 200 m depth varies in width, from 5 to 10 km between Tanga and Pangani, to over 40 km near the Kenya border and south of Pangani (Bensted-Smith 1988) (Fig. 7.1). There is one large river, the Pangani, which is one of the largest in Tanzania, and several smaller permanent rivers and seasonal streams. The region has diverse coral reefs, mangroves, and seagrass beds, and is of regional and possibly global importance, on account of its migratory water birds, a possible remnant population of dugong, and the recent discovery of coelacanths. The reefs vary both from north to south and from inshore to offshore. Those in the north are more extensive and interspersed with sandy areas, whereas those in the south are smaller and more precisely defined with higher coral cover. Fringing reefs occur along the coast, shallow patch reefs are found in the inshore waters, and offshore there are deeper reefs near the drop-off. A total of 47

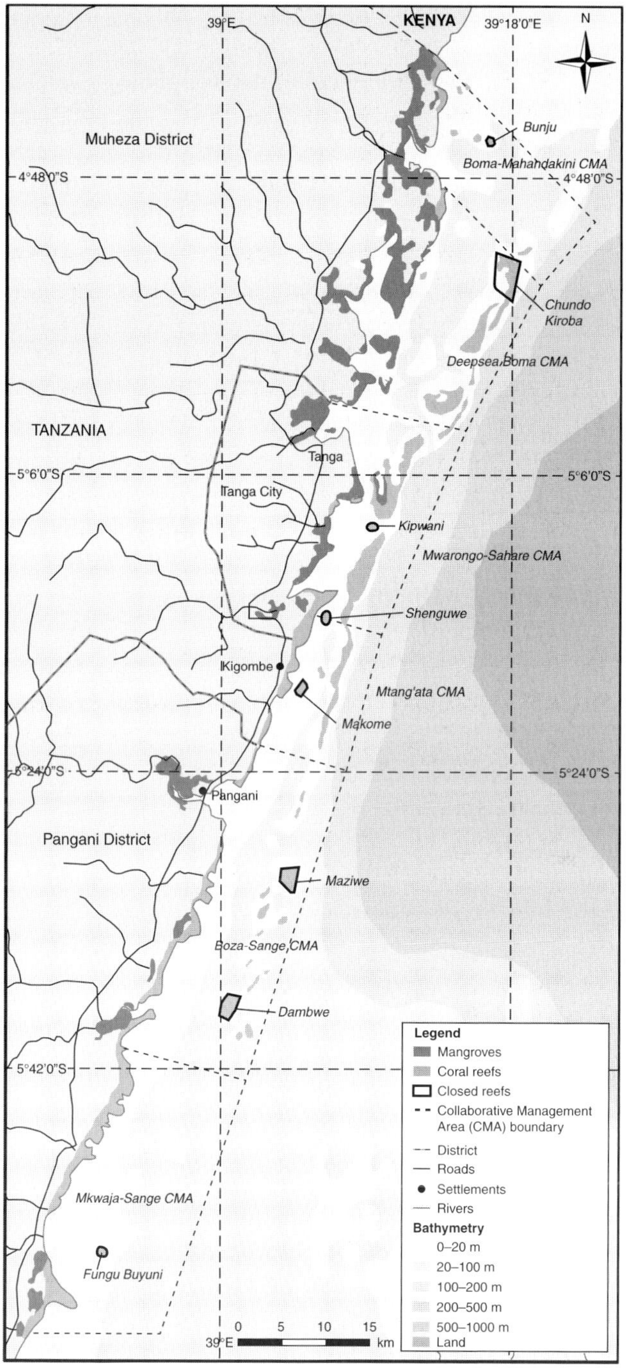

Figure 7.1 Map of collaborative management areas in Tanga Region.

coral genera have been recorded with diversity increasing from 20 genera on the inshore fringing reefs, to 24 on the inner patch reefs, and 28 on the outer patch reefs (Horrill et al. 2000). The differences in geomorphology and reef structure are also reflected in the fish community structure, which is significantly different on the more southern reefs than on those to the north, and on some of the offshore reefs compared to reefs closer to shore (Othina and Samoilys 2005).

In the late 1960s, reefs adjacent to Tanga town were described as being 'among the best along Tanzania's coastline' (Ray 1968). By the late 1980s, however, the situation had drastically changed, with coral cover of less than 20% in most areas (Bensted-Smith 1988; UNEP 1989), a value generally considered to represent coral reefs in poor condition (Wilkinson 2004). By 1995, an estimated 12% of reefs were destroyed, 64% were in poor or moderate condition, and only 24% were in good condition. Abundance of commercially-important fish families, such as snappers (Lutjanidae), emperors (Lethrinidae), and rabbitfish (Siganidae), was low on 90% of the reefs, as were populations of commercially important lobster, sea cucumbers, and molluscs (Horrill et al. 2000). Much of the decline in reef health was attributed to dynamite fishing, although it can be assumed that the large increase in fishers also played a role, particularly in the low fish abundance.

Socio-economic characteristics – the artisanal fishery

According to national census data, the region had a coastal population of under 400 000 in 1994 at the beginning of the TCZCDP, but over 500 000 by 2005. There are two towns: Tanga which accounts for at least half of the total population, and Pangani with about 6000 people. In addition there are just under 50 coastal villages. These depend on a wide range of activities for their livelihoods, including farming, petty trade, and salt production, but artisanal fishing is the most important economic activity. Most people are below or close to the poverty line, reflecting the national situation where an estimated 85% of people living on the coast survive on less than US$1 a day (Ruitenbeek et al. 2004). A few of the more rural villages have up to 70–80% of the adult male population involved in fishing, but this percentage drops to 25–30% in the semi-urban areas of Tanga City (Gorman 1995). Men do most of the fishing but women fish in shallow water for shrimp, and a larger number process fish and collect octopus and molluscs at low tide.

The coastal fishery in Tanga Region is multigear and multispecies, as seen in many tropical parts of the world (Munro and Williams 1985; Wright and Hill 1993; Munro 1996), and the main gears are hand lines and fish traps, followed by gill nets and shark nets (Table 7.1). The species composition varies according to gear type (Table 7.2), and with the season, habitat, and fishing ground. An analysis of catches landed at Kigombe village from 1995–1999 showed that 40% was inshore fish, 32% rays, 21% offshore fish, 6% sharks, and 1% other species (Horrill 1999) and this has changed little since (Anderson 2004).

Table 7.1 Trends in numbers of fishing gears used in Tanga Region (data source: Government frame surveys with figures for long lines in 2001 amended by Samoilys 2004).

Gear type	1995	1998	2001
Hand line	2898	2654	1883
Fish trap	1294	1258	2162
Gill net	1182	631	953
Shark net	814	608	404
Ring net/purse seine	79	128	102
Scoop net	66	255	106
Beach seine	57	84	0
Long line	38	34	56

There are a variety of gears and vessels (Tables 7.1 and 7.3), and villages have particular preferences for gear types (Bensted-Smith 1988; Gorman 1995; Horrill et al. 2001; Anderson 2004). Hand lines and traps are traditional, the least expensive, and the most common. Two types of gill nets, distinguished by their mesh size, were introduced from the 1960s onwards through a number of fishery development programs (Bensted-Smith 1988). Beach seine, or pull seine nets, reportedly in use since the 1960s, were legal when the TCZCDP started but were banned on mainland Tanzania in November 1998 because of their destructive effects on bottom habitat, and this explains the zero recorded for the number of these gears in 2001 (Table 7.1). However, patrol data from TCZCDP show that they are still used (TCZCDP 2004), particularly by certain villages and by visiting fishers from Zanzibar, where their use was banned before the TCZCDP started (Bensted-Smith 1988). Large beach seines, as well as ring nets that are legal, require crews of 10–20 men, and are expensive and therefore owned by wealthier individuals or larger commercial interests. Spear guns are also still widely used, as they are inexpensive to make although they have been illegal since 1970. Various other nets and traps are also used, as well as illegal dynamite fishing.

Table 7.2 Predominant species groups caught with different gears from outrigger canoes at Kigombe village (compiled from Anderson 2004).

Gear	Species group	Proportion of catch
Hand line	Snappers and emperors	39%
	Rabbitfish	19%
Fish trap	Rabbitfish	81%
Shark net	Rays	72%

Table 7.3 Numbers of landing sites, fishers, and vessels in Tanga Region (source: Government frame survey).

			Vessel categories					
Year	Landing sites	Fishers	Outrigger canoes	Dugout canoes	Sailing boats	Dhows	Motor boats	Total vessels
1995	48	4202	394	261	150	50	41	896
1998	46	4380	452	262	84	137	34	969
2001	52	4361	502	237	100	84	21	944

During the southeast monsoon (April–October), which has stronger winds and rough seas, fishing activities are confined to inshore waters but, in the calmer period of the northeast monsoon (November–March), boats fish further offshore. Distances travelled are also limited by the lack of ice and storage facilities and by the size of the fishing vessels. Catches are landed mainly at about 50 government-designated landing sites, although it is thought that other sites are also used to evade payment of the local fisheries sales tax.

There is growing commercial fishing for prawns, octopus, lobster, sea cucumbers, and shells for both export and local sale. One commercial company has been operating since 1993, and a second since 2002. At present, export of marine finfish is prohibited in Tanzania but this ban is being reconsidered and in 2004 the Government offered a limited number of finfish export licences on a trial basis. Some export of finfish is now occurring from Tanga City as part of the trial (Allard unpublished data, 2005). Other targeted marine resources include corals for building, lime production, and curios, and the red seaweed *Eucheuma* spp., which is farmed (Gorman 1995). Other livelihoods that the coastal fishing communities are involved in include agriculture, petty trade, fish trading, boat building, salt boiling, solar salt production, charcoal making, mangrove pole cutting, and tourism.

There are about 4300 licensed fishers in the three districts; numbers have stayed relatively constant between 1995 and 2001 (Table 7.3) and are presumed to have remained stable since then. Muheza has the greatest number of fishers (1578), followed by Tanga (1371) and Pangani (301). Tanga City, with a short coastline but large population, has the greatest density of fishers per km of coastline (16 per km) (Horrill 1999). Accurate estimates of fisher numbers have proved difficult, partly because of the presence of migrant fishers. Migrant fishers come from Pemba and Unguja (Zanzibar), other areas on the mainland coast and within Tanga Region itself, and set up camp for 2–3 months at certain locations, depending on seasonal variations in the productivity of fishing grounds. Camp fishing, or *dago,* is also a 'style of life' for many fishers, allowing them to save money by living away from home.

Fishing vessel numbers have also remained relatively constant since the early 1990s, at just over 900 (Table 7.3). Most are small, sail-powered, and operated by their owners. The most common are dugout canoes, but there are also a number of larger sailing vessels and a few boats with engines. The relatively constant number of fishers and vessels from 1995 to 2001 suggests that fishing effort has remained fairly steady despite human population increases. This is encouraging and also useful when analyzing changes in catch rates.

The TCZCDP strategy

The focus of the TCZCDP has been the establishment of six collaborative management areas (CMAs), covering the entire coast of Tanga Region, a total area of 1604 km^2. Each CMA comprises the 'home' fishing grounds shared by a group of villages. There are no traditional management systems in Tanzania and fishers are allowed to access all fishing areas. However, they tend to use their closest grounds, apart from migrant fishers. Fishers in Tanga Region have named about 500 specific fishing grounds and these may be whole reefs, parts of reefs, or non-reefal areas. Some are used more than others, such as those close to Tanga City where the main markets are located (Anderson 2004).

In setting up the CMAs, the TCZCDP took a learning approach, based on a model developed for the World Bank by Picciotto and Weaving (1994) that recognizes the complexity, time required, and assistance needed for development projects. The model has four stages:

(1) Listening – to ensure that the priority issues and actions are those of the beneficiaries themselves.
(2) Piloting – to test how well proposed actions work and to try alternatives in order to avoid mistakes on a larger scale.
(3) Demonstrating – to fine-tune and adapt processes and actions to a wider range of cases.
(4) Mainstreaming – to adopt processes, actions, and methods as normal practice.

The 'listening' stage consisted of participatory socio-economic and resource assessments with three villages in each district, followed by workshops to identify priority issues and potential solutions. Seven priority issues were identified, ranging from overfishing and use of destructive fishing methods, to poor agricultural production and lack of basic sanitation (Makoloweka and Shurcliff 1997). The piloting stage involved selection, in each district, of one pilot village that developed an action plan to address two of the seven priority issues. The development of the action plans involved analysis by the communities and districts of the cause and effects of the priority problems, followed by identification of the activities needed to solve the problems. Such 'action planning' became a core strategy for all components of the program. During the demonstration phase, the village action plans were combined into collaborative management area plans (CMAPs), each involving several villages. Ultimately all coastal villages in the three districts were included. 'Mainstreaming' began around 2001 and is aimed at continuing efforts to build capacity for implementation of the CMAPs, with a sound financial basis and professionally-managed government departments that are held accountable for implementing government policies and procedures.

Process for developing the CMAPs

The process for preparing CMAPs evolved over a series of stages. Initially, the three pilot villages prepared action plans that addressed specific issues such as dynamite fishing and use of small mesh nets. Subsequently, the process was adapted to cover issues broader than fishing, such as beach erosion and vermin control, and neighbouring villages were consulted. The prevailing understanding at the time was that communities often had exclusive rights to certain resources and areas. However, in the case of one pilot village in the TCZCDP, disagreements arose because two adjacent villages used the same fishing grounds as the pilot village and did not feel that they were being adequately involved. This made it clear that all villages using an area needed to be actively involved from the outset. The process was therefore adapted and a Central Coordinating Committee (CCC) established to ensure coordination between the villages. The final process that is now being recommended is as follows:

(1) A survey is carried out in the area of the proposed plan to determine the villages that use it and the stakeholders.
(2) A chart or map of the area is prepared to show key features and natural resource use.
(3) The CMA area is negotiated, delineated, and the boundaries demarcated.
(4) Village environmental management committees (VEMCs) are elected if not already in place; these are statutory bodies, responsible for representing individual village interests and implementing village-based activities related to environmental issues.
(5) Village action plans are prepared by the VEMCs with assistance from district staff, and presented to the village government for comment and approval.
(6) CCC elections are held. The CCC comprises representatives of each VEMC. It has no legal mandate but plays a key role in resolving inter-village disputes.
(7) Village action plans are integrated into a single CMAP by the CCC, which is then sent back to the villages for approval.
(8) The CMAP is approved at village and district levels.
(9) The CMAP is approved nationally by the Fisheries Division.

The CMAs vary in size from about 100 to 400 km^2 (Table 7.4) and differ in bathymetry, cover of coral reef, mangroves and seagrass beds, and numbers of villages, landing sites, fishers, and boats. As agreed by the communities, the landward boundaries were defined approximately by the mean high water mark and the seaward boundaries by a line connecting the reef base of the seaward side of the

Table 7.4 Key features of the CMAs (listed from north to south).

Name	District	Villages	Size (km^2)	Established (approved)
Boma-Mahandakini	Muheza	8, with 5 landing sites	145	2000 (2001)
Deepsea Boma	Tanga, Muheza	14 with 13 landing sites	377	Sept 2000 (2001)
Mwarongo-Sahare	Tanga	12 with 20 landing sites	196	July 1999 (2000)
Mtang'ata	Muheza, Tanga	4 with 3 landing sites	97	July 1996 (1999)
Boza-Sange	Pangani	8 with 8 landing sites	400	Sept 1996 (1999/2000)
Mkwaja-Sange	Pangani	3 with 3 landing sites	389	Sept 2000 (2001)

outermost series of reefs. Boundaries between the CMAs were determined through a process of negotiation and represent a line where there was the smallest amount of overlap in resource use by the villages of each CMA. Exceptions to this are the northernmost CMA boundary that is defined by the border with Kenya and the southernmost that is the border between Tanga and Pwani Region (Fig. 7.1).

Content of the CMAPs

Each CMAP comprises several sections and the general arrangement is as follows.

(1) A brief introduction explaining the plan.
(2) The geographical coordinates of the area to be managed and a list of the stakeholders.
(3) The principles to be used in guiding management decisions, such that all stakeholders should participate in management activities and decisions, and that gears and fishing techniques prohibited by national legislation shall not be used in the CMA.
(4) The responsibilities of each stakeholder group.
(5) A brief description of how the plan was formulated, including the villages involved, meetings held, and other details.
(6) A summary in matrix form of the overall purpose of the plan, expected results, indicators and assumptions. For most CMAs, the overall objective is to 'increase income through the use of marine and coastal resources'. At the time the plans were first produced, the TCZCDP still had a very broad approach and each CMAP therefore has four to eight expected results. These include not only fisheries-related results such as reduction in illegal fishing, but also other environmental issues such as seaward production, mangrove management, and turtle conservation.
(7) The actions to be taken to achieve the desired results, such as who will do what and when it should be done, the output of these actions, and the specified assumptions.

(8) Arrangements for monitoring the indicators identified in (6), such as the frequency of monitoring, and the groups or individuals responsible.
(9) The regulations, which include requirements that all relevant national legislation for fisheries and mangroves must be enforced, that specified reefs are closed to fishing but remain open for tourism or other non-extractive activities, and that beach sand, corals, and seaweed farms are protected.

Closed reefs

Villagers and district officers drew up criteria for closure of reefs at the beginning of the TCZCDP with technical assistance from IUCN. These criteria require that closed reefs should be:

- Less favored as fishing grounds, in order to ease implementation
- Further from shore, but not too distant for patrols
- Have average reef health conditions

Reefs are closed under village by-laws but must subsequently be approved by the government minister responsible for fisheries. The current closed areas cover over 28 km^2 in total (1.7% of the total CMA area) and vary in number (from 1–3) and in size from 1.5 to 10.0 km^2, averaging 4 km^2 (Table 7.5). Initially closures were planned to be for just one year, with the exception of Maziwe, but all of the current closures have now been in place for more than four years and several are now considered permanent.

Mtanga'ata CMA previously had two different closed reefs. The first two, Kitanga and Upangu, were reopened after only a year, following a disagreement between the villages when the CMAP was being reviewed and politics prior to local elections. However, overall catches in the CMA, as recorded by the TCZCDP monitoring program, showed an increase since the reefs were closed and appeared to decline to previous levels two months after the reefs were reopened (Horrill et al. 2001). As a result, in the next review of the CMAP, villagers agreed that two other reefs (Makome and Shenguwe) should be closed. The period of closure is five years, unless the two-yearly reviews recommend otherwise or in cases of severe famine or bad weather.

One of the closed reefs in Boza Sange is Maziwe Marine Reserve which had been legally gazetted in 1975 and thus, already theoretically closed to fishing. The marine reserve designation was, however, done without user consultation and there was previously much illegal fishing. The Marine Parks and Reserve Unit that is responsible for the Maziwe reserve, with Pangani District Council on behalf of the CMA, now jointly manage the area. A similar collaborative arrangement has been developed for the closed reef in Mkwaja CMA that is also the marine extension of Saadani National Park, gazetted in 2005 and under the central authority of the Tanzanian National Parks Authority (TANAPA).

Table 7.5 Main features of closed reefs (as of 2005). Italics = reefs that are no longer closed.

Collaborative management area name	Closure area name	Area km^2	Percentage of total CMA	Date closed	Date opened	Length of closure	Planned length of closure
Boma-Mahandakini	Bunju	2.0	1.4%	2001		4 y	8 y
Deepsea Boma	Chundo/ Kiroba	10.0	2.7%	2000		5 y	5 y, but reviewed every 2 y
Mwarongo-Sahare	Kipwani	1.5	0.8%	2000		5 y	Reviewed every 2 y
Mtang'ata	*Kitanga*	*1.0*		*1997*	*1998*	1 yr	
	Upangu	*4.25*		*1997*	*1998*	1 yr	
	Makome (Kigombe)	2.5	4.7%	2001		4 y	5 y but reviewed every 2 y
	Shenguwe	2.0		2001		4 y	5 y but reviewed every 2 y
Boza-Sange	Dambwe	5.5	2.5%	1998		7 y	Permanent, but reviewed at intervals
	Maziwe	4.5		1975		~20 y	Permanent
Mkwaja	Fungu Buyuni*			2005			Permanent

*Details of size not available at time of going to press, because of the recent designation.

Implementation

Although the CMAs have a broad overall objective and results that address a wide range of issues, at the time the plans were established, the donor-funded aspect of the TCZCDP had been oriented specifically to marine resource issues. Implementation of the CMAPs has, therefore, focused largely on the fisheries components as described in this section. The national fisheries legislation that regulates use of gear types, requirements for boat and fisher licenses, and allows for the establishment of closed areas, provides the legal backing for the CMAPs. Details of the closed reefs and regulations concerning activities not covered in national legislation are specified through by-laws made by the district and village councils.

Compliance and enforcement mechanisms

From the start, all those involved in the TCZCDP believed it was important to involve the villages in enforcement. Initially, a patrol unit was set up in each pilot village, run by the VEMC and supported by the TCZCDP with boats and equipment. This was successful in reducing illegal fishing by fishers from the pilot villages, but not illegal fishing by fishers of neighboring villages. At a later stage, each CMA had a patrol unit, based in one of the villages and equipped by the TCZCDP with boat and radio equipment. The composition of the patrols underwent various changes, including contributions from the marine police, trained village militias, and the navy, which proved the most effective.

In 2005, three new and larger patrol boats were provided, one for each district, to service all six CMAs. Each district has retained one of the old small boats in case the larger boat breaks down. The VEMC and village government, of the village where the boat is kept, manage it and the equipment, plan the patrols, and in some cases pay for security, with the district providing fuel. Each patrol comprises one or more community representatives from villages in the relevant CMA, district personnel and the marine police. Patrols should be carried out twice a week but lack of funds means this is often reduced to once a week and in response to reported incidences. If there is an outbreak of illegal activities, patrols may go out daily for a period. Incidents are logged, along with the action taken and its outcome.

The TCZCDP has also been helping to improve the arrest procedures and speed up the time taken to bring cases to court. The prosecution process tended to be time consuming and complicated, with cases often poorly prepared. Penalties were often poorly implemented and fines were lower than required under the regulations. The TCZCDP, therefore, organized two-day marine ecology courses for magistrates, prosecutors, police, and decision-makers in regional and district government, and district councillors to raise awareness of the effect of illegal fishing and use of dynamite.

Resource monitoring

Although monitoring activities are laid out in the CMAPs, lack of expertise at village and district level meant that an overall monitoring program was designed with technical assistance from IUCN and is overseen by the TCZCDP, essentially as a regional activity. Participatory monitoring methods were introduced from the beginning, and the overall program covers monitoring of mangrove and coral reef ecosystems, fish catches, seaweed production, participation of women in CMA activities, and implementation of quarterly and annual work plans. The fisheries monitoring component is described here.

The national government's Fisheries Division has, for many years, operated a creel-survey at six landing sites and monitors fish catch and effort. The TCZCDP operates an independent system that was initiated in 1995. It was originally focused on the fish auction sites in pilot villages and was limited to boat type, gear type, and number of fishers and trips, excluding the parameter of time spent fishing. The protocol was revised in 2001 to include hours spent fishing, which allowed for analysis of CPUE. At the same time, data collection was moved to the landing sites and this allowed the collection of a more representative range of catch and effort data, including zero-catch data and small catch-weights, and more villages were involved. Sampling stratification was further improved in 2003, such that data could be reported by district and month.

Efforts were made to ensure that the results of the monitoring were analyzed simply and quickly and presented to villagers, since the findings are used in the regular review of the CMAPs. The data storage and analysis procedures were also refined over time. This has culminated in a monitoring and assessment database and guide (Pabari et al. 2005). The system incorporates standardized analyses for district officers to be able to run monthly or half-yearly trends in catch rates for combinations of selected fisheries, vessels, gears, CMAs, and reefs.

Review and re-negotiation of CMAPs

Plans are reviewed approximately every two years, but more frequently if a new issue needs resolution. The review process involves analysis and evaluation of data from the monitoring and enforcement activities, and discussions with all villages, relevant district officers and program staff to update information relevant to the CMAP. The effectiveness of the plan is assessed and recommendations are made, such as opening or closing of reefs. Efforts are made to resolve disagreements between stakeholders. If a resolution cannot be found, village assemblies vote on the issue, recommendations from each village are presented to the CCC, and democratic decisions are made based on the votes from all villages in the CMA. Reviewed plans are then presented for approval to the relevant higher authorities as described in the process above.

Effect of the CMAs

Improving compliance

Compliance with the regulations of the CMAPs has proved to be difficult. Some enforcement capacity has been built in the form of the participatory patrols but, despite considerable efforts, the use of beach seines and dynamite fishing has not been eliminated. The original prevailing perception was that dynamite was used because of a lack of alternative fishing gears, but the TCZCDP has demonstrated that other factors are involved. Incidences of dynamite fishing were common during the 1980s and early 1990s, with an estimated 30–75 blasts occurring each month based on fishers' perceptions. When blast monitoring started, there were over 180 blasts in the single month of December 1995. By the late 1990s, when the TCZCDP was well underway, reported incidents had dropped to an average of less than one a month, with a few exceptions (Horrill et al. 2001). This trend continued through to 2003 (TCZCDP 2004). Awareness of the long-term negative effects of dynamite fishing by communities resulted in more requests for enforcement against the few persistent dynamite fishers. This led to the perception that TCZCDP had brought this practice to a near halt (Verheij et al. 2004).

There was evidence early on that the number of incidents of dynamite fishing was linked to the effectiveness of enforcement. Incidents increased during periods of withdrawal of navy support, for example around Christmas. Dynamite fishing also increased in early 2003, associated with nearby bridge construction, because dynamite was more easily available. At the end of 2004, the navy unexpectedly withdrew its support, and this led to a rapid increase in blasting. Large numbers of blasts were heard on a regular basis in 2005, including nearly 50 in one month, resembling rates seen in the early 1990s. This was despite an intensive patrolling schedule in conjunction with the marine police and increased funding from the Fisheries Division. It is important to note that this increase is not unique to Tanga Region, dynamite fishing having resurfaced in many other parts of mainland Tanzania in 2004 and 2005.

Enforcement of the ban on beach seine nets has also been difficult. These were fully banned nationally in 1998 through improved legislation developed partly through the efforts of the TCZCDP. In 1999, the TCZCDP exchanged beach seines for less damaging gear for those fishers willing to be involved. The District Fisheries office, with TCZCDP assistance, drafted agreements with the beach seine groups on how to use the new gear, which was mainly gill nets and, in a few cases, hand lines and traps. Although the frame survey for 2001 showed that beach seines dropped to zero and gill nets and some other gears increased, particularly traps (Table 7.3), the ban is not being fully enforced. Based on apprehensions by the patrols, there are about 100 incidents of illegal beach seining a year in Tanga, less than 20 in Muheza, and none in Pangani (TCZCDP 2004).

The number of incidents of illegal fishing in closed areas has declined from around 50–60 in Tanga and Muheza in 2001 to less than 20 in 2003. In Pangani the incidence was low at around two a year (TCZCDP 2004). No management restrictions are in force for visiting fishers although this is one of the issues specified in the CMAPs for attention.

Effects on fisheries

Given that the fundamental goal of the TCZCDP has been to improve the livelihoods of coastal communities dependent on marine resources, attempts have been made to determine whether the CMAs have brought about any increases in total catches, catch rates, and revenue. Government statistics show total catches for Tanga Region of between 3400 and 7600 metric tons (MT) up to 1993 (Fig. 7.2). Estimations for the years 2002–2004, using TCZCDP monitoring data, suggest there has been little change with an annual average of about 7000 MT for the region (Table 7.6).

Tanga City records the largest annual catch because it is the main market and best distribution and selling point.The second largest catch is Pangani, although it is often reported as the poorest district, and Muheza records the lowest annual catch (Table 7.6). Much of the catch from the CMAs located on the Muheza coast may be

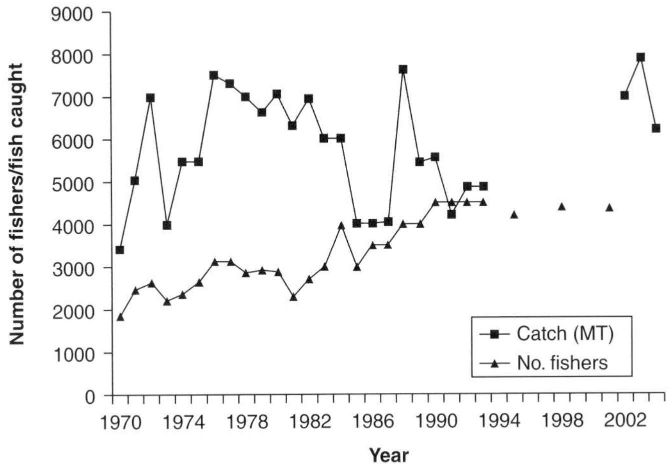

Figure 7.2 Trends in fish catch and fisher numbers, 1970–2004, Tanga Region.

landed in Tanga City and, thus may not be recorded in the Muheza District fishery statistics. Given the major differences between districts, it can be expected that catches within the different CMAs also vary significantly, reflecting their different ecological and socio-economic characteristics. The monitoring data available did not allow comparison between CMAs until amended in 2003 (Anderson 2004), and

Table 7.6 Annual estimates of catch and value by district, 2002–2004 (source: Anderson 2004).

District	Year	Annual estimate of catch (MT)	Annual estimate of value (Tsh)	Annual estimate of value (US$*)
Muheza	2002	1016	527 192 239	483 663
Muheza	2003	639	377 365 104	346 207
Muheza	2004	676	481 237 536	441 502
Pangani	2002	1760	579 271 330	531 442
Pangani	2003	1329	426 303 171	391 104
Pangani	2004	941	344 601 172	316 148
Tanga	2002	4190	2 365 675 317	2 170 344
Tanga	2003	5912	2 514 534 532	2 306 912
Tanga	2004	4571	1 767 872 393	1 612 901
Total	2002	6966	3 472 138 886	3 185 448
Total	2003	7880	3 318 202 807	3 044 222
Total	2004	6188	2 593 711 101	2 379 551

*Tsh 1090 = US$ 1.

a full analysis of the location and use of different fishing grounds in relation to the CMA boundaries is not given here, but will be possible by 2006.

Before the TCZCDP started, annual catch per fisher in Tanga Region had declined from 2.4 MT in 1970 to 1.0 MT in 1993 (Beckley et al. 1997). Using the average annual catch for 2002–2004 (7011 MT, Table 7.6) and the average number of fishers for 1995–2001 (4361 in 2001, Table 7.3), the estimated current annual catch per fisher is about 1.6 MT. This indicates an increase on 1993, which is encouraging but this result should be treated with some caution because of the approximation of the catch data (Anderson 2004).

Results from the TCZCDP monitoring program are also not conclusive since effort data were not collected during the earlier years of the TCZCDP, sample sizes were highly variable and long-term monitoring was limited to only one of the pilot villages, Kigombe in Mtang'ata CMA. Three vessel and gear combinations, namely hand lines, traps, and gill nets deployed from outrigger canoes, dominated catches in the Kigombe data set, and have made more detailed analyses possible (Anderson 2004). In order to maximize the data set available, while minimizing bias caused by within-year seasonal variations in catch rates, a 12 month period of March–February, rather than a calendar year, was used in the analysis.

There is no significant increase in the CPUE (kg/gear/trip) for fishers using hand lines between the years 1996 and 2002, but the CPUE for 2003/04 was statistically higher than previous years at nearly 4 kg/line/trip for all species aggregated (Fig. 7.3a), which may be a fluctuation or the beginning of a trend. Looking at snappers and emperors only, which comprise about 40% of the catch weight of this gear and are the largest single contributors to the catch, the upward trend in mean annual CPUE is stronger, with a catch rate of about 3 kg/line/trip (Fig. 7.3b). Although there was an apparent decline in CPUE in 2002 this decline was not significant, suggesting that catches of snappers and emperors caught by hand lines are at least stable, and increasing recently (Anderson 2004).

Between 1995 and 2001, TCZCDP fishery data collection recorded all snappers and emperors as a single group and so trends in catch rates of different species were masked. In a study of a similar small-scale coastal fishery in neighbouring Kenya, which recorded individual species, *Lethrinus mahsena* (~28%) and *L. lentjan* (~27%) were the dominant snappers in hand line and large fish trap catches. *L. harak* was the most common snapper (~13%) in gill net catches but was less significant (~2–3%) in fish trap catches (McClanahan and Mangi 2004). If these species also dominate the fisheries in Tanga, then the recent increases in catch rates can probably be attributed to them. Since 2001, *L. harak* and *L. lentjan* are being recorded specifically in the TCZCDP monitoring program.

Mean annual catch rates for fish traps in the Kigombe fishery show a significant decrease from 1996 through 2001, with a statistically insignificant increase to 2003 (Fig. 7.3c). The CPUE trend for Siganidae (rabbitfish) alone, which typically comprises about 80% of the catch weight of this gear, is similar with a decline from 1997 and indications of an increase in 2003 (Fig 7.3d). The decline could have been

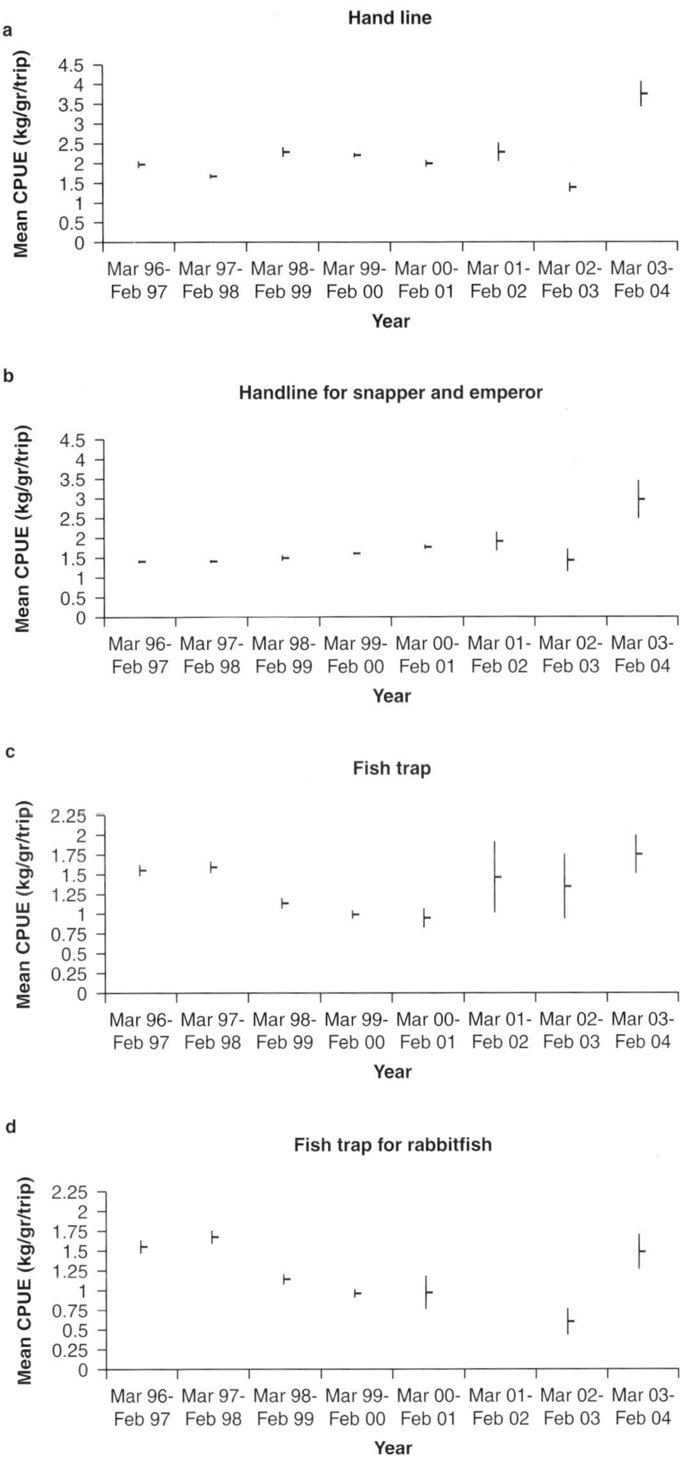

Figure 7.3 Mean annual CPUE (kg/gear/trip, with 90% CI), landed at Kigombe village from outrigger canoes during 1996–2004, for: (a) all species caught with hand lines; (b) snappers and emperors caught with hand lines; (c) all species caught in traps; (d) rabbitfish caught in traps.

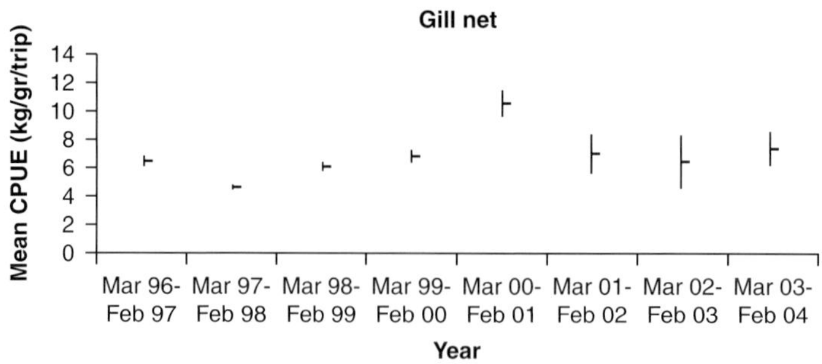

Figure 7.4 Mean annual CPUE (kg/gear/trip, with 90% CI) for gill nets using outrigger canoes that target rays (72% of the catch), landed at Kigombe village during 1996–2004.

due to the increase in the number of traps in 2001 (Table 7.1), which may have been linked to the gear exchange program.

The ray fishery can be monitored by combining data for gill nets and shark nets. Rays generally occur in soft bottom habitats between reefs, though some species are associated with reefs. Since the primary interventions of the TCZCDP have been to eliminate dynamite fishing and beach seining, neither of which target rays, the gill net fishery is not likely to have been affected directly by the TCZCDP. Mean annual catch rates for this fishery show a decrease in CPUE between 1996 and 1997, with a recovery in 1998 that continued through to a peak in 2000, but catch rates then dropped again (Fig. 7.4) to levels that do not differ significantly from the earlier years. It is difficult to ascribe cause and effect for this pattern, but the decline may be a simple case of recruitment overfishing that has been reversed perhaps due to fishers moving out of this fishery and thus reducing effort. To our knowledge there has been no research on the ray fishery in East Africa, a gap that requires attention.

Determining whether the trends observed through the monitoring program are due to management interventions or to population fluctuations resulting from natural causes requires careful analysis. Snappers and emperors are species that are primarily associated with coral reefs, and the increase in their catch could be related to the reduction in dynamite fishing that occurred over the period analyzed here. Equally, the reef closures in Mtang'ata CMA, namely Kitanga and Upangu from 1997 to 1998, and Makome and Shenguwe from 2001 to the present, may have had an effect. Many of the species involved in the fishery have juveniles that live in seagrass beds before settling on reefs, and the reduction in beach seining could have had a positive effect on their survival. However, it is also important to note that many of the reefs experienced substantial bleaching of corals in 1998 as a result of the El Niño event. It is known that such changes can have a significant effect on fish populations, increasing herbivores for example (Lindahl et al. 2001) and thus potentially on subsequent catches. Thus, although the results of the analysis may indicate the beginning of positive trends, they could also reflect short-term population fluctuations.

It is also useful to compare these results with a research study carried out in collaboration with TCZCDP that looked at the underwater fish communities in

Mtang'ata CMA in 1996 and again seven years later in 2004 (McClanahan et al. 2006). This involved three reefs open to fishing (Chanjale, Taa, and Funguni) and one of the reefs (Makome) that was closed in 2001. The study found that overall fish biomass on the Tanga reefs increased from 260 kg/ha in 1996 to 457 kg/ha in 2004. The increase was most noticeable in the herbivorous group of species, which included parrotfish and rabbitfish, but there was a significant decline in the carnivorous group comprising snappers, emperors, and grunts. These results do not fully correspond with the fish catch data obtained by the TCZCDP but, because the relationship between fish biomass and catch is unimodal and not linear, it is difficult to relate the two factors. McClanahan and colleagues (2006) also showed that fish biomass was low on the Tanga reefs compared with that on reefs that are totally protected in nearby southern Kenya. The study looked at reefs in Kisite Marine National Park (MNP), where the total wet weight estimate of fish was about 2.8 times greater than on the Tanga reefs. The biomass of the carnivorous group was particularly low on the Tanga reefs, with 8 kg/ha compared with 250–300 kg/ha on the unfished reefs in Kisite MNP.

Despite some indication of increasing catch rates in Tanga region, revenue from catches has decreased over the period of the TCZCDP. Horrill and colleagues (2001) calculated a 29% decrease in income per trip due to a fall in fish prices between 1996 and 2000. The estimated annual values of more recent catches also show a decline, from 584 Tanzanian shillings/kg in 2002 to 421 Tsh/kg in 2004 (Table 7.6).

Discussion

The main achievement of the TCZCDP has been the development of a collaborative system for preparing fisheries management plans, using a process that is broadly satisfactory to both communities and the government. Management plans are key tools for sustainable fisheries, recommended by the FAO Code for Responsible Fisheries (FAO 1995) and are described in Tanzania's national 2003 Fisheries Act as 'management agreements'. To date, the three districts in Tanga region are the only coastal areas of the country outside of marine parks that have such plans in place, although steps towards collaborative fisheries management planning have been taken along the shore of Lake Victoria. As in many countries, most efforts to manage artisanal fisheries have been based around the establishment of marine protected areas. For example, the two mainland marine parks, Mafia Island and Mnazi Bay–Ruvuma Estuary, have a major focus on fishery management but their primary objective is marine biodiversity conservation. A central government agency, the Marine Parks and Reserves Unit, has overall responsibility for management, although district fisheries officers also have a role in management decisions. The TCZCDP is the only program to have adopted the broader approach of managing the fisheries of an entire marine area.

The TCZCDP's work has also done much to explore structures for collaborative management within Tanzania's political and institutional framework. From a situation where there was no collective natural resource management system within the communities, the TCZCDP has introduced a strong sense of ownership of resources in the face of what has been largely open access. This is unlike situations where artisanal fishery management programs have been able to build on established traditional tenure systems (Anderson et al. 1999; Aswani and Hamilton 2004; Kuemlangan 2004).

Having started with pilot villages, an approach that was widely used at the time, the TCZCDP rapidly adapted the process to take account of the more complex relationships between adjacent villages in terms of their use of fishing grounds. Applying the collaborative approach to management needs careful attention to the individual characteristics of the stakeholder groups, as there are likely to be many differences that will affect their willingness and ability to participate, as well as the extent to which they will benefit. A study of community-based fishery management programs in Zanzibar similarly revealed how coastal communities that are superficially similar may differ markedly on closer analysis (Levine 2004). The success of the TCZCDP's approach is demonstrated by the fact that it was extended to mangrove management in the region (Nurse and Kabamba 2000) with the development of collaborative mangrove management plans in two districts. Other districts in the country have also taken a great interest in the approach and study visits have been made to the TCZCDP to obtain advice.

There is no specific legal framework for the CMAPs as established at present. The Fisheries Division approvals of the most recent plans have included the recommendation that once the by-laws have been approved, they should be considered as operational beach management units (BMUs) and this will give them legal backing. BMUs can be established under the 2003 Fisheries Act but, unlike the CMAPs, are based around individual landing sites and operate at village level (Ogwang et al. 2004). This expansion of the concept of the BMU, from single landing sites to several villages, will help to harmonize the fisheries planning approaches and promote a more ecosystem-based approach to management, focusing on areas of resource distribution and use, rather than administrative boundaries.

The fact that reef closures are included in all the CMAs, that these are being established for increasingly long periods, and that most communities see them as an acceptable management tool, is a second major achievement. There is substantial scientific evidence that permanent reef closures are a vital tool in the suite of measures required to manage coral reef fish populations sustainably (Gell and Roberts 2003; Halpern 2003, Russ 2004; McClanahan and Graham 2005), but this is not always easy for communities to understand. Scheinman and Mabrook (1996) found that communities in Tanga were already familiar with the concept of closed reefs, having heard of the marine parks in Kenya that are closed to fishing. This may explain why reef closures were fairly readily adopted compared with Mafia Island Marine Park and Mnazi Bay–Ruvuma Estuary Marine Park where no-take zones have been, or are being introduced, and Mnemba Island Marine Conservation Area in Zanzibar which is a closed area, and where there has been some opposition (Wells and Samoilys, personal observations).

Unfortunately, the evidence that the CMAPs are having a direct positive effect on fishing livelihoods is weak. This may be because of the complex and incomplete data set, the incomplete understanding of the fishery when the CMAPs were developed, the avoidance of landing sites and thus data collection points by the fishers, poor enforcement, and impacts on the marine environment unrelated to the management interventions. For example, the reefs have been affected by disease and coral bleaching, although there are some indications that management activities have improved reef health (Verheij et al. 2004). In addition, CMA management has focused on reefs, but rays, which are found largely in non-reefal habitats, are one of the main catch components and octopus fishing is also an important livelihood activity. The TCZCDP did not directly address either of these. Furthermore, the area closed to fishing is only a small proportion of the total stock area within the CMAs, much less than the 20–30% recommended in the literature (Halpern and Warner 2002; Roberts et al. 2003), and little or no areas of other important fisheries habitats are closed.

It should be noted that the TCZCDP is not alone in finding it difficult to demonstrate increased catches as a result of closed areas; there are still very few studies globally that clearly show a greater fisheries yield on a long-term basis in areas adjacent to no-take zones (Sale et al. 2005). This does not of course mean that closed reefs are not important, but rather that they need further study, particularly design components including the ratio of closed to open areas and their size (Chapter 8). The next steps for the TCZCDP, therefore, should include encouraging villages to increase the amount of reef that is closed to fishing and improving enforcement and management of these, as well as monitoring and evaluation.

Although the development of the CMAPs has been successful, implementation has proved far more difficult. There are a number of reasons for this. The TCZCDP covers 1600 km^2 and deals with over 200 000 relatively poor people in 49 communities. This is a vast undertaking given that the donor funding has averaged the relatively small sum of about US$500 000 a year. The area covered is at least twice the size of each of Tanzania's marine parks and involves many more people. Mafia Marine Park is 822 km^2, involves 19 000 people, and has an annual budget similar to that of TCZCDP; Mnazi Bay Marine Park is 650 km^2, involves 30 000 people and has an annual budget of around US$500 000. Even though the magnitude of the aims of the TCZCDP was recognized in the sense that it was decided to take a pilot village approach to start with, the complexity of the issues was probably not fully understood, and indeed may still not be fully comprehended by all stakeholders.

Despite intensive awareness campaigns by the TCZCDP and regular patrolling, dynamite fishing has not been eliminated. Although there was an initial reduction, a small number of law-breakers remain in the communities and when enforcement relaxes or dynamite becomes more accessible they increase their activities. The navy presence provided a strong deterrent, but enforcement alone is rarely a long-term sustainable option. At Danajon Bank, Bohol, in the Philippines, a Fishers Alliance numbering over 700 fishers has been set up to help stop dynamite fishing and to protect the coral reefs and other fishing grounds (Blanco and Vincent, personal communications). The Alliance operates across a distance of 145 km, and provides intense peer pressure within villages, as well as sending messages to the police on their cell phones with names of offending fishers, times, and locations of blasts. This approach has empowered fishing communities to take initiative and responsibility for their marine resources, and to work with the authorities. Such a structure exists in each CMA, through the VEMCs, the CCCs and a recently-established Tanga Dynamite Fishing Monitoring Network formed by concerned tourism operators and expatriate residents who report blasting incidents to the relevant authorities. Further lessons can perhaps be learned from the Philippines, in terms of the need to develop appropriate relationships with the enforcement agencies, to ensure that the necessary action is taken when they are alerted of problems.

Mechanisms needed are currently being assessed by the TCZCDP and will probably require input from the national level given that dynamite fishing has become a problem in many parts of the country. A previous outbreak of dynamite fishing, in the mid-1990s, was only halted when country-wide and international media publicity triggered support at ministerial level. Greater involvement of all the stakeholders in the general management of the TCZCDP, particularly the increasingly-important private sector, is also required. Again the structure for this exists, in the form of the Tanga Coastal Consultative Forum that was established part way through the TCZCDP to provide a more permanent mechanism for participation of all groups in decision-making and provision of advice.

The participatory monitoring program has contributed to a large extent to the sense of ownership, and has been an example for several other programs in Eastern Africa (Obura et al. 2002). As shown, it has not, however, been without problems. Data collection and analysis procedures have recently been streamlined and improved, so that all data sets are in one database allowing analysis across them: for example, patrolling versus benthic cover and fisheries versus reef health. The various data collection protocols have been adjusted to meet international standards and indicators established based on analysis of the longer data sets (Anderson 2004; Samoilys 2004; Othina and Samoilys 2005; Pabari et al. 2005). This should help to ensure the collection of more appropriate data and its regular analysis by district officers, who now have the facility to generate status reports and to assess the effects of management. A more detailed understanding of trends in fisher numbers and other relevant characteristics of their lifestyle, particularly those who migrate and those involved in dynamite fishing, is needed if current destructive practices are to be halted.

Despite the lack of evidence of a clear increase in fish catches, and the decline in their value, many fishers have reported that they perceived an improvement in their livelihoods over the course of the TCZCDP. Although this may in part be due to the TCZCDP, there have been many other changes since it was initiated, including a general improvement in the Tanzanian economy as well as a number of other donor-funded development projects in the three districts that may have had an effect. Nevertheless, there are several reports, for example in TCZCDP evaluations, of fishers stating that they now have higher catch rates. Arguably fishers' perception of catch rate is likely to be a sensitive indicator of change, since these fishers are highly dependent on fishing for their livelihoods. Monitoring fishers' perceptions is increasingly recognized as a useful method by which to evaluate the fishery effects of management interventions (Meeuwig et al. 2003; Malleret-King et al. 2006).

The TCZCDP has also had other components addressing income-generating activities, some of them have been unsuccessful, but others, such as promotion of vermin control to increase agricultural productivity and assistance with seaweed farming enterprises, have contributed to general improvements in coastal livelihoods. Monitoring of the acceptance and success of these activities has generally been poor but seaweed production has greatly increased since the TCZCDP started. This was due in large measure to the overall growth in the industry nationally and promotion by the private sector, but the TCZCDP's input, in terms of helping with zoning and reducing conflict between fishers and seaweed farmers, has probably also contributed. Although the adjustment of the TCZCDP to focus more on fisheries activities after a few years was justifiable in terms of available resources, the subsequent difficulties encountered demonstrate that improving the livelihoods of poor coastal communities ultimately requires that attention is paid to all livelihood strategies in use, a point increasingly being made in discussions on sustainable development (Allison and Ellis 2001; Ireland et al. 2004; Ruitenbeek et al. 2004).

Maintaining program activities in the long term will be a major challenge as larger annual budgets will be necessary once the activities are fully mainstreamed into district natural resource management plans. At present the districts still receive funding from the Government of Ireland, but external technical assistance has been substantially reduced, and is now provided on a consultancy base when the districts request it. The aim is that the TCZCDP will fund itself by 2007, but this will be difficult to achieve. A variety of options have been considered, such as improving collection of the fisheries levy and ensuring that this is retained for management activities, but greater political will is required to fully test and implement those measures with potential for success.

The need to build on previous work and for strong government commitment to implementation of the CMAPs is now more urgent than ever. The artisanal fishery continues to be a vitally important sector for the livelihoods of a large percentage of the population of these three districts. The negative effect of poorly-managed artisanal fisheries on marine resources and ecosystems has been clearly demonstrated in Tanzania and many parts of the world (McClanahan et al. 1999; Hawkins and Roberts 2004). There are increasing pressures on the fishery, including the growing human population and potential new uses of marine resources. Among these is the export of finfish that, if it were to become economically important, could have a major effect on Tanga's fisheries if appropriate management is not in place.

There is little doubt that fishers and coastal communities now have a much greater involvement in, and understanding of, fisheries management and a concomitant greater sense of ownership. This has been brought about through the process developed for preparing CMAPs that involves village and inter-village consultation, as well as the technique of 'action planning', where problems, causes, and solutions are clearly identified. The TCZCDP has thus laid a sound basis for future fisheries management in three districts that is potentially applicable in other districts. It has also demonstrated the difficulties of implementation and gathering the sound scientific data that are needed for successful adaptive management, and the urgency with which these problems need to be addressed.

Conclusions

The TCZCDP was initiated in 1994 as a result of declining marine resources and deteriorating reef health, which were leading to decreased incomes. Set up with the broad objective of improving the livelihood of coastal communities, the TCZCDP involved a wide range of activities including the development and implementation of collaborative fisheries and mangrove management plans, trialling and implementing new economic activities, building capacity and establishing institutional arrangements for coastal management, and also environmental education.

The primary achievement has been the establishment of the six CMAs, using a process that is agreed by both communities and the government, and each containing reefs that are closed to fishing. Participatory monitoring of the effects of the CMAs was accorded a high priority, but the TCZCDP demonstrated the difficulty of maintaining such activities and of measuring the success of management interventions where baseline surveys are inadequate and monitoring protocols were inappropriate.

Enforcement was also conceived as a participatory activity, and the patrols have always included community representatives as well as enforcement officers. Effective enforcement has, however, proved to be one of the most difficult aspects of the program. The use of dynamite and beach seines has not been eliminated completely and illegal fishing within the closed reefs still occurs occasionally. This may partially account for difficulty in detecting change over the life of the program and in the size of the total catch, despite the regulatory measures. There are probably a number of other contributing factors including natural events such as coral bleaching and disease that have affected habitat quality. Despite this, fishers generally perceive the TCZCDP as successful, possibly because they now have a much greater control over their marine resources.

References

Allison, E.H. and Ellis, F. (2001) The livelihoods approach and management of small-scale fisheries. *Marine Policy*, **25**: 377–88.

Anderson, J.D. (2004) *Analysis of Reef Fisheries under Co-management in Tanga*. IUCN–Eastern Africa Regional Programme and Tanga Coastal Zone Conservation and Development Programme, IUCN–EARO, Nairobi.

Anderson, J.D., Mees, C.C. and Cowan, V.J. (1999) The performance of customary marine tenure in the management of community fishery resources in Melanesia. *Guidelines Towards Effective Co-management of Coral Reef Fisheries in the Pacific Region*. Vol. 5, FMSP Project R6436, MRAG Ltd. http://www.fmsp.org.uk/

Aswani, S. and Hamilton, R.J. (2004) Integrating indigenous ecological knowledge and customary sea tenure with marine and social science for conservation of bumphead parrotfish *Bolbometopon muricatum* in the Roviana Lagoon, Solomon Islands. *Environmental Conservation*, **31**: 1–15.

Beckley, L.E., Govender, A., Mann, B.Q. and van der Elst, R.P. (1997) *Artisanal Fisheries Management*. Report to TCZCDP. Oceanographic Research Institute, Durban, South Africa.

Bensted-Smith, R. (Ed.) (1988) *The Coastal Resources of Tanga Region, Tanzania*. Report of a Preliminary Study in October 1987. Regional Natural Resources Office, Tanga Region and IUCN–EARO.

Borrini-Feyerabend, G. (1996) *Collaborative Management of Protected Areas: Tailoring the Approach to the Context*. Issues in Social Policy, IUCN, Gland, Switzerland.

FAO (1995) *Code of Conduct for Responsible Fisheries*. Food and Agriculture Organization, Rome.

Gell, F. and Roberts, C. (2003) Benefits beyond boundaries: the fishery effects of marine reserves. *Trends in Ecology and Evolution*, **18**: 448–55.

Gorman, M. (1995) *Report on Socio-economic Study/Participatory Rural Appraisal*. Tanga Coastal Zone Conservation and Development Programme, Tanga, Tanzania.

Halpern, B. (2003) The impact of marine reserves: do reserves work and does reserve size matter? *Ecological Applications*, **13**: S117–37.

Halpern, B.S. and Warner, R. (2003) Matching marine reserve design to reserve objectives. *Proceedings of Royal Society B*, **270**: 1871–8.

Hawkins, J.R. and Roberts, C.B. (2004) Effects of artisanal fishing on Caribbean coral reefs. *Conservation Biology*, **18**: 215–26.

Horrill, J.C. (1999) *Gillnets for Beach Seines in Tanga: Sustainable Fisheries Management?* Tanga Coastal Zone Conservation and Development Programme, Tanga, Tanzania.

Horrill, J.C., Kamukuru, A.T., Mgaya, Y.D. and Risk, M. (2000) Northern Tanzania, Zanzibar and Pemba. In: T.R. McClanahan, C.R.C. Sheppard and D.O. Obura (Eds) *Coral Reefs of the Indian Ocean: their Ecology and Conservation*. Oxford University Press, New York.

Horrill, J.C., Kalombo, H. and Makoloweka, S. (2001) *Collaborative Reef and Reef Fisheries Management in Tanga, Tanzania*. IUCN Eastern Africa Programme.

Ireland, C., Malleret, D. and Baker, L. (2004) *Alternative Sustainable Livelihoods for Coastal Communities. A Review of Experience and Guide to Best Practice*. IUCN Eastern Africa Regional Programme, Nairobi.

Kuemlangan, B. (2004) Creating legal space for community-based fisheries and customary marine tenure in the Pacific: issues and opportunities. *FishCode Review*, No. 7. FAO, Rome.

Levine, A. (2004) Local responses to marine conservation in Zanzibar, Tanzania. *Journal of International Wildlife Law and Policy*, **7**: 14.

Lindahl, U., Ohman, C.O. and Schelten, C.K. (2001) The 1997/1998 mass mortality of corals: effects on fish communities on a Tanzanian coral reef. *Marine Pollution Bulletin*, **42**: 127–31.

Makoloweka, S. and Shurcliff, K. (1997) Coastal management in Tanga, Tanzania: a decentralized, community-based approach. *Ocean and Coastal Management*, **37**: 349–57.

Malleret-King, D., Glass, A., Bunce, L. and Pomeroy, B. (2006) *Socioeconomic Monitoring Guidelines for Coastal Managers of the Western Indian Ocean.* Coral Reef Degradation in the Indian Ocean (CORDIO)/Global Coral Reef Monitoring Network (GCRMN) (in press).

McClanahan, T.R. and Graham, N.A.J. (2005) Recovery trajectories of coral reef fish assemblages within Kenyan marine protected areas. *Marine Ecology Progress Series*, **294**: 241–8.

McClanahan, T.R. and Mangi, S. (2004) Gear-based management of a tropical artisanal fishery based on species selectivity and capture size. *Fisheries Management and Ecology*, **11**: 51–60.

McClanahan, T.R., Muthiga, N.A., Kamukuru, A.T., Machano, H. and Kiambo, R.W. (1999) The effects of marine parks and fishing on coral reefs of northern Tanzania. *Biological Conservation*, **89**: 161–82.

McClanahan, T.R., Verheij, E. and Maina, J. (2006) Comparing the management effectiveness of a marine park and a multiple use collaborative management area in East Africa. *Aquatic Conservation: Marine and Freshwater Ecosystem*, **16**: 147–65.

McShane, T.O. and Wells, M.P. (2004) *Getting Biodiversity Projects to Work: Towards More Effective Conservation and Development.* Columbia University Press, New York.

Meeuwig, J.J., Samoilys, M.A., Erediano, J. and Hall, H. (2003) Fishers' perceptions of the seahorse fishery in central Philippines: interactive approaches and an evaluation of results. In: N. Haggan, C. Brignall and L. Wood (Eds) *Putting Fishers' Knowledge to Work: Conference Proceedings.* Fisheries Center, University of British Columbia.

Munro, J.L. (1996) The scope of tropical reef fisheries and their management. In: N. Polunin and C. Roberts (Eds) *Reef Fisheries*. Chapman and Hall, London.

Munro, J.L. and Williams, D.M. (1985) Assessment and management of coral reef fisheries: biological, environmental and socio-economic aspects. In: C. Gabrie and B. Salvat (Eds) *Proceedings of the 5th International Coral Reef Congress, Tahiti.* Vol. 4.

Nurse, M. and Kabamba, J. (2000) *Defining Institutions for Collaborative Mangrove Management: a Case Study from Tanga, Tanzania.* IUCN Eastern African Regional Office, Nairobi.

Obura, D.O., Wells, S., Church, J. and Horrill, C. (2002) Monitoring of fish and fish catches by local fishermen in Kenya and Tanzania. *Marine and Freshwater Research*, **53**: 215–22.

Ogwang, V., Medard, M. and Nyeko, J.I. (2004) *Harmonised Beach Management Unit (BMU) Operational Guidelines for Fishing Communities of East African States.* Implementation of the Fisheries Management Plan for Lake Victoria, Lake Victoria Fisheries Organisation.

Othina, A. and Samoilys, M. (2005) *Impacts of Management on Coral Reefs in Tanga, Tanzania*. IUCN–EARO, Nairobi.

Pabari, M., Samoilys, M., Muniu, H., Othina, A., Thande, G., Mijifha, P. and Matiru, V. (2005) *Using Monitoring and Assessment for Adaptive Management: a Guide to the TCZCDP Information Management System*. IUCN–EARO, Nairobi.

Picciotto, R. and Weaving, R. (1994) *A New Project Cycle for the World Bank?* Finance and Development, World Bank, December 1994.

Ray, G.C. (1968) *Marine Parks of Tanzania*. Conservation Foundation, Washington.

Roberts, C.M., Andelman, S., Branch, G., Bustamente, R.H., Castilla, J.C., Dugan, J., Halpern, B.S., Lafferty, K.D., Leslie, H., Lubchenco, J., McArdle, S., Possingham, H.P., Ruckelshaus, M. and Warner, R.E. (2003) Ecological criteria for evaluating candidate sites for marine reserves. *Ecological Applications*, **13**: S199–214.

Ruitenbeek, J., Hewawasam, I. and Ngoile, M. (2004) *Blueprint 2050: Sustaining the Marine Environment in Mainland Tanzania and Zanzibar*. The World Bank, Washington.

Russ, G.R. (2004) Marine reserves: long-term protection is required for full recovery of predatory fish population. *Oecologia*, **138**: 622–7.

Sale, P.F., Cowen, R.K., Danilowicz, B.S., Jones, G.P., Kritzer, J.P., Lindeman, K.C., Planes, S., Polunin, N.V.C., Russ, G.R., Sadovy, Y.J. and Steneck, R.S. (2005) Critical science gaps impede use of no-take fishery reserves. *Trends in Ecology and Evolution*, **20**: 74–80.

Samoilys, M. (2004) *Review of the Village Monitoring Team's Coral Reef Monitoring Programme in Tanga*. IUCN–EARO, Nairobi.

Scheinman, D. and Mabrook, A. (1996) *The Traditional Management of Coastal Resources*. A Consultancy Report for Tanga Coastal Zone Conservation and Development Programme, Tanzania.

TCZCDP (2004) *End of Phase III Report. Muheza, Pangani, Tanga Districts*. Support Unit, Tanga Coastal Zone Conservation and Development Programme.

UNEP (1989) *Coastal and Marine Environmental Problems of the United Republic of Tanzania*. UNEP Regional Seas Reports and Studies, No. 106. UNEP, Nairobi.

Verheij, E., Makoloweka, S. and Kalombo, H. (2004) Collaborative coastal management improves coral reefs and fisheries in Tanga, Tanzania. *Ocean and Coastal Management*, **47**: 309–20.

White, A.T., Hale, L.Z., Renard, Y. and Cortesi, L. (Eds) (1994) *Collaborative and Community Based Management of Coral Reefs. Lessons from Experience*. Kumarian Press, West Hartford.

Wilkinson, C. (2004) *Status of Coral Reefs of the World: 2004*. Australian Institute of Marine Science, Townsville, Australia.

Wright, A. and Hill, L. (1993) *Nearshore Marine Resources of the South Pacific*, XVI. Forum Fisheries Agency, Honiara, Solomon Islands.

Chapter 8
Management of Area and Gear in Kenyan Coral Reefs

Tim R. McClanahan

Introduction

Kenyan marine fisheries are exploited well beyond the maximum sustainable yields (MSYs) of most species, and many of the fished ecosystems are heavily degraded in terms of biodiversity and ecological functions (McClanahan 2000a). Consequently, it would be unfair to promote the marine fisheries in this country as a success. Nonetheless, like other countries that have reached this level of overexploitation, there is a growing awareness that practices need to change and associated initiatives are attempting to direct a way out of the current dilemma. The trial and error process of these initiatives are teaching important lessons that may eventually help to rectify the unregulated resource use that has created the current situation. This chapter will briefly present the biological and cultural environment, the recent history that has led to the current problem, and will focus on a few of the initiatives that have the potential to improve the current situation. The purpose of this chapter is, therefore, to describe how the current crisis in small-scale artisanal fisheries was created and the ongoing efforts to alleviate it. This situation described here is, perhaps, typical of many poor tropical countries, and many of the lessons are likely to be applicable to other countries.

The biophysical and economic environment

The majority of fishing in Kenya is focused on the nearshore habitats of creeks, reef lagoons, and shallow reef and seagrass environments. The shoreline of most of southern Kenya is fronted by a fringing reef that lies between a few hundred meters to a few kilometers offshore. Fishers travel to fishing grounds on foot, by swimming, or in boats, including small dugout canoes or outrigger canoes with small sails. The reefs are close to shore and the physical environment is highly seasonal, with a strong southeast monsoon from May to September. The factors of reef geography, strong seasonality, small boats, and reduced productivity offshore create conditions that promote heavy use of the nearshore environment. The calmer northeast monsoon allows easier access to areas further offshore, and these areas are used for less than half of the year, depending on the local geography and boat availability. Kenya also has a high tidal range of 4 m and many fishers follow the tidal cycle and use this tidal power to transport their small boats.

Some believe that poverty and its associated low and renewable technology alone restricts fishing effort close to shore. There are cases where offshore fishing is profitable in areas that were previously lightly fished (Young 2004) but, in some areas where offshore fishing is more common, such as off of Dar es Salaam, Tanzania, low net profits are common, despite moderately high yields, because of higher capitalization, running, and maintenance costs of offshore fishing (Kamukuru 1992). There are efforts to promote fishing offshore as a means to reduce onshore fishing effort. This is seen by many as the most obvious way to alleviate the overuse of nearshore fishing grounds, but there are considerable cultural, conservation, and economic hurdles that need to be overcome before this option is implemented in a sustainable manner.

Ensuring that offshore fishing actually reduces onshore fishing effort rather than simply adding an additional pressure to fisheries is critical to the promotion of sustainable offshore fishing. It may be that the offshore benthic environment provides a refuge for many species that allow the nearshore fishery to be exploited beyond a MSY without collapsing (McClanahan and Mangi 2000; Rodwell et al. 2002). If the offshore system were to be overexploited then the potential for a large-scale collapse of the national marine fishery will increase considerably.

The effect is a heavy use of ecosystems close to shore that have a high potential production that is likely to exceed 5 tons/km^2/year (McClanahan and Mangi 2000), while areas offshore have underutilized resources but lower potential production. Offshore areas have lower sustainable potential yields and, although underexploited resources may produce net profits (Young 2004), many do not currently have a net economic return. The one exception is offshore fishing for shrimp, which can produce net economic profits because of the high export price (McClanahan and Obura 1996). Similarly, game fishing is popular and profitable to game fish operators that take advantage of tourism and because game fishers are not concerned about net profits based on operating costs and the market price of their catch.

The potential for collapse of the national marine fishery by uncontrolled expansion of fishing into offshore areas indicates that it is prudent to develop controls and effective management on the nearshore fishery prior to promoting an expansion. Consequently, the discussion that follows will evaluate two of the more likely ways to promote sustainability in the nearshore fishery: closed areas and gear management. These management systems are currently active and developing and the discussion below will highlight the current understanding and potential for successful management.

The social-cultural and legal environment

History of fisheries and gears

The Kenyan coast is historically composed of many small area-restricted ethnic groups that are now overlain by a more vagile people and a post-colonial national government with various cultures and national institutions. National regulatory institutions associated with resource use frequently change with government ministries. Local and national government cultures can differ in their views of resource use and lead to potential conflicts. This occurred in the post-independence period when the high population growth rate and development ethic of the national government created a policy of unrestricted access to fishing grounds in an effort to promote resource extraction and wealth. For example, the Kenyan Fisheries Act (Cap 378; Kenya Gazette Supplement 1991, 2000) did not originally place any restrictions on use in the marine environment except for a required permit. Cultural traditions did, however, recognize area, gear, time, and access restrictions and this led to conflicts and reorganization of resource use as local and national leadership and institutions of enforcement conflicted (McClanahan et al. 1997). In most cases, traditions were altered or abandoned as there were no effective legal mechanisms to enforce them when they conflicted with the national government enforcement authorities.

It has taken some time for the national authorities to recognize that resources are limited and overexploited and that this unrestricted policy with its associated conflicts is not in the interest of maintaining high and sustainable yields. In 2001, a number of restrictions were put into law (Kenya Gazette Notice No. 7565) and these included seasonal restrictions on trawling, the need for an approved turtle excluder device on trawlers, a ban on the use of monofilament nets, seine nets, harpoons, and spear guns. The Fisheries Act allows the Director of Fisheries to make additional regulations and, in an effort to protect the deepwater resources, Legal Notice 214 of 2003 prohibited the use of scuba gear and spear guns for fishing lobster and sea cucumber. These laws have been weakly enforced with the exception of beach seine use in some of the nearshore areas as will be discussed below. Consequently, the current social environment would appear to be in a stage of reorganizing around the need for restrictions.

History of marine protected areas

A parallel area-based or national park management system that covers ~9% of the coastal shelf developed largely independently of the fisheries laws and regulations. The national park system was frequently in a different government ministry and was intended to serve a different purpose to fisheries. The national park system first developed in the late 1950s but with the recognition that wildlife would not survive hunting and that other human pressures and international tourism were likely to be a major source of income for this developing country. Consequently, marine national parks were legally established in 1968 in Malindi-Watamu and in 1973 in Kisite. In addition, in 1979, a national reserve was established in Kiunga, in 1986 a reserve and park was established in Mombasa, and in 1995, a reserve was established in Diani (Sheppard and Wells 1988). These were composed of two types of management systems: smaller fully-closed park areas with no resource extraction that compose 54 km^2 and larger limited-use reserves of 780 km^2. The smaller closed areas are functioning well, but not the reserves.

Regulations in the parks and reserves were originally described in the Kenyan Government Sessional Paper No. 3 of 1975 and formally in The Wildlife Act Cap 376 in 1989. This Act allows the minister to have discretionary powers to regulate areas in reserves, but this power is infrequently used. Therefore, fisheries restrictions are not explicit and regulations have variously been interpreted to permit only traditional or non-destructive gear, but the lack of clarity and various interpretations results in few effective restrictions. Furthermore, tourists pay entrance fees into the fully-protected but not reserve areas and this has resulted in most of the management effort being focused on the approximately ~55 km^2 areas of the four parks. Although many conservation scientists currently see closed areas such as these parks as a form of fisheries management (Lubchenco et al. 2003), the intention of these parks was largely to preserve unexploited marine ecosystems and encourage tourism. Nonetheless, as described below they have proved to be very useful for understanding the effects of resource extraction and the potential for fisheries management (McClanahan and Kaunda-Arara 1996).

Closed areas

Perceptions towards closed areas

Successful management is ultimately measured by the state of the resource, but deciding which of the many management options will work effectively should depend on the perceptions and compliance of the resource users and managers (Sutinen and Kuperan 1999). Fisheries can be managed in a variety of ways but most management of wild-caught species requires a combination of limits on area, gear, size, species, time, and effort (Chapter 1; McManus 1996). The acceptance of these restrictions can vary for a variety of cultural, economic, legal, and technological reasons but, when there is coincidence in the perceptions of users and managers, management is likely to be easier and less expensive to enforce and maintain. Consequently, as part of the effort to develop fisheries management with the highest potential for success and self-regulation, where enforcement resources are few, interviews with Kenyan resource managers and users were undertaken to determine what they perceived as the best management options (McClanahan et al. 2005).

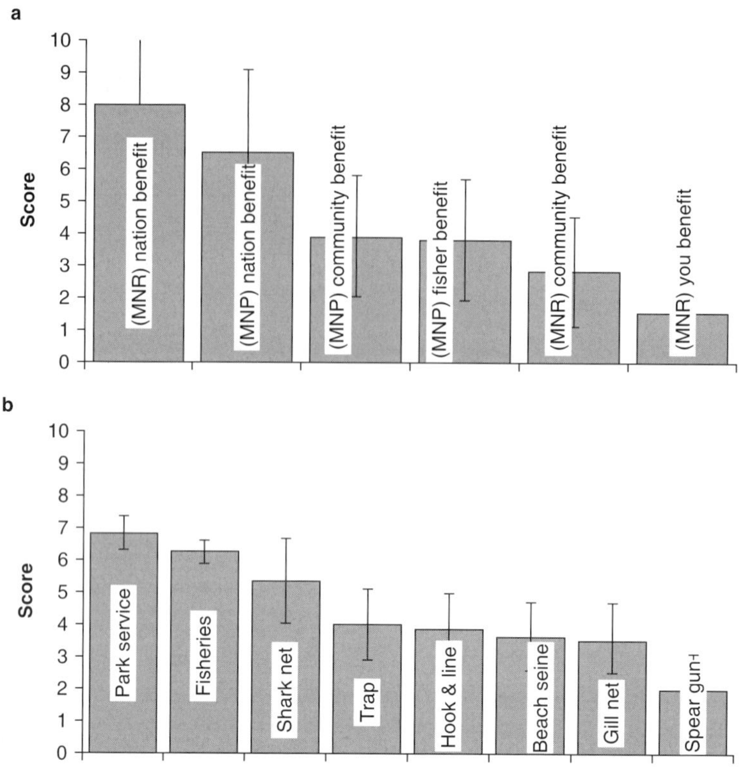

Figure 8.1 Scores of the perceived benefits of (a) the two types of area management in Kenya (marine national parks – MNP and marine national reserves – MNR) by all respondents combined and (b) for each of the respondents by their profession or gear use. The maximum benefit score is ten and is given for three types of benefits: to the nation, the community, and individual fisher.

Because the parks were started at different times and under different social circumstances, this allowed a test of the effects of the age of a park and the histories of community participation; the oldest parks having little community participation compared to the newest parks where community participation was considered a prerequisite to area management. Both resource users and managers agreed that closed and gear-restricted areas benefited the nation most, but park personnel saw communities and fishers benefiting from closed and gear-restricted areas while fishers did not share this view (Fig. 8.1). Overall, the most positive perceptions towards area-based management was found among the government employees, being significantly different than all other groups of fishers, except for shark-net fishers. Increasing wealth or community participation was not associated with more positive perceptions towards closed areas, but fishers working near the oldest marine protected area (MPA) had more positive perceptions than those living near the newest MPA.

Gear-restricted areas were viewed more positively by fishers, and increased with education and the age of the protected area. Individuals with a secondary education were more common in the park service than fishing and these individuals had a more positive perception of the value of gear-restricted areas towards fishers and their communities than less-educated individuals, but did not have a significantly more positive perception of the value of fully-closed areas. Despite the conflict between users and managers, there has been good compliance with closed and poorer compliance to gear-restricted areas. Consequently, there is not the expected close relationship between perceptions and compliance.

The results suggest that resource users' perceptions towards area management improves with time and that fishers and many government personnel are not aware of the indirect effects of MPAs. Increased opportunities for formal and informal education are likely to improve appreciation for indirect effects and reduce differences in the perceptions between these two groups. Unexpectedly, community programs prior to the park establishments were not a quick fix for improving perceptions. Nonetheless, it may be difficult to entirely eliminate the dichotomy as permanent and fully-closed national parks may, in fact, have the greatest benefit to the broader citizens of the nation. Managers are left with the need to reduce the negative effects or compensate those people that are compliant and reliant on resources near fully-closed areas.

Direct effects of marine protected areas

The long-term effects of the closed-area management on resource abundance and diversity can be determined due to the existence of multiple closed areas differing in the time since their closure. In particular, one of the protected areas, the Mombasa MNP, was studied from four years before to 13 years after it was protected from resource extraction. Using space-for-time substitutions, it is possible to evaluate ecological changes inside these MPAs over a 35 year time span or chronosequence. Fish are the main focus of resource extraction, and therefore the focus in what follows.

Fish abundance and diversity

The chronosequence indicates that fish population densities and biomass both produce convex polynomial relationships (Fig. 8.2a, b). During the first twenty years after closure both the fish densities and biomass increase but density begins to decline at 20 years while there is a smaller decline in biomass once the maximum biomass of 1200 kg/ha is reached at around 22 years. Analysis of the size of fish indicates that there is a general increase in larger fish over time, but there is considerable variation in sizes over time such that no equilibrium in size classes was found (McClanahan and Graham 2005). The value of 1200 kg/ha is a useful number for setting fishing targets for fish biomass in fished reefs as it is what fisheries biologists call the 'pristine biomass' of this fishery. Maximum sustainable fisheries yields of fish should be achieved at around half the pristine biomass (Clark 1985), and therefore fished reefs should maintain a fish biomass of around 600 kg/ha to

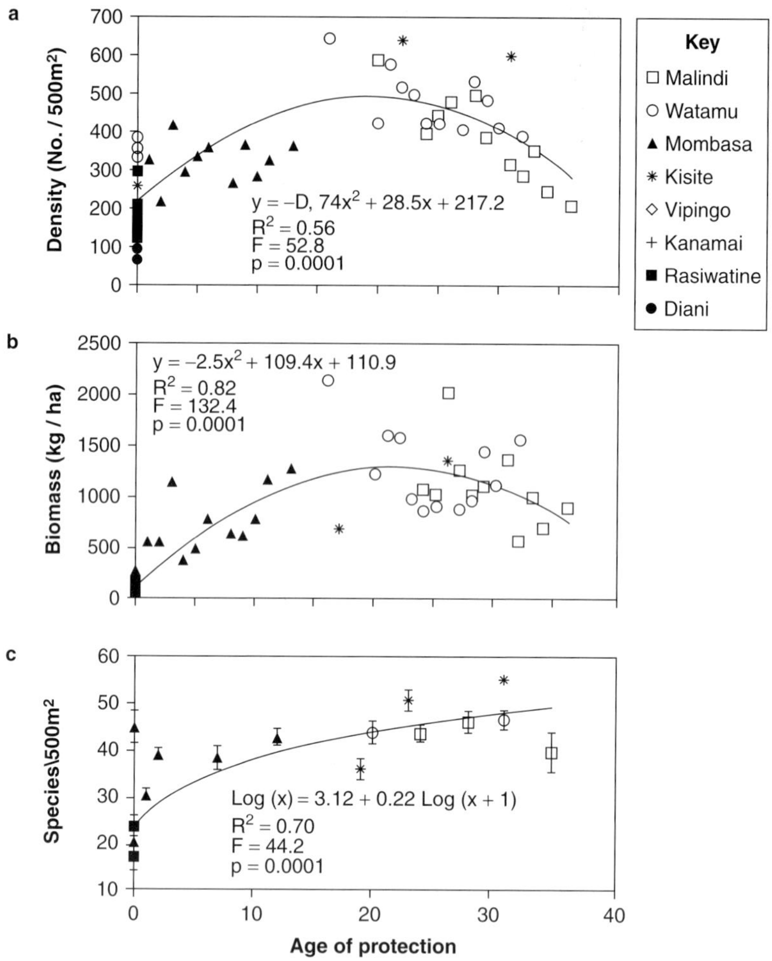

Figure 8.2 Scatter plots of (a) density, (b) wet weight, and (c) numbers of species as a function of the age of the protected area. Unprotected areas are presented for an age of 0 on the x-axis.

achieve a MSY, which is considerably higher than the values found for fished reefs at around 110 kg/ha (Fig. 8.2b).

The numbers of fish species appears to rise quickly in the first few years after closure and then level with no indication of a decline in the older parks (Fig. 8.2c). Studies on other taxonomic groups such as erect algae, corals, sea urchins, and snails indicate that closed areas have a considerably higher diversity of fish and corals than fished reefs but that there are small increases in the numbers of species of snails, sea urchins, and algae in the fished reefs (McClanahan and Obura 1996; McClanahan et al. 2006). The overall diversity of all studied groups combined is higher in the closed areas, but there appears to be some release from predation in the more susceptible prey groups of fishes that elevates their local diversity, as described below.

Ecological effects

Changes in fish abundance are associated with indirect or cascading effects on other unutilized resources in the coral reefs (Fig. 8.3). For example, predation on sea urchins increase with time since protection, and this is associated with a decline in sea urchin biomass such that sea urchins become relatively uncommon after 15 years of protection. There are also changes in the abundance of the algal groups, with erect fleshy algae becoming less and encrusting corallines more abundant over time. The decline in sea urchin abundance is due to an increase in the predators, most notably the redlined triggerfish (*Balistapus undulatus*), which is very sensitive and slow to recover from fishing (McClanahan 2000b). Large wrasses, such as the triple lobed wrasse (*Cheilinus trilobatus*), appear to be more tolerant of fishing

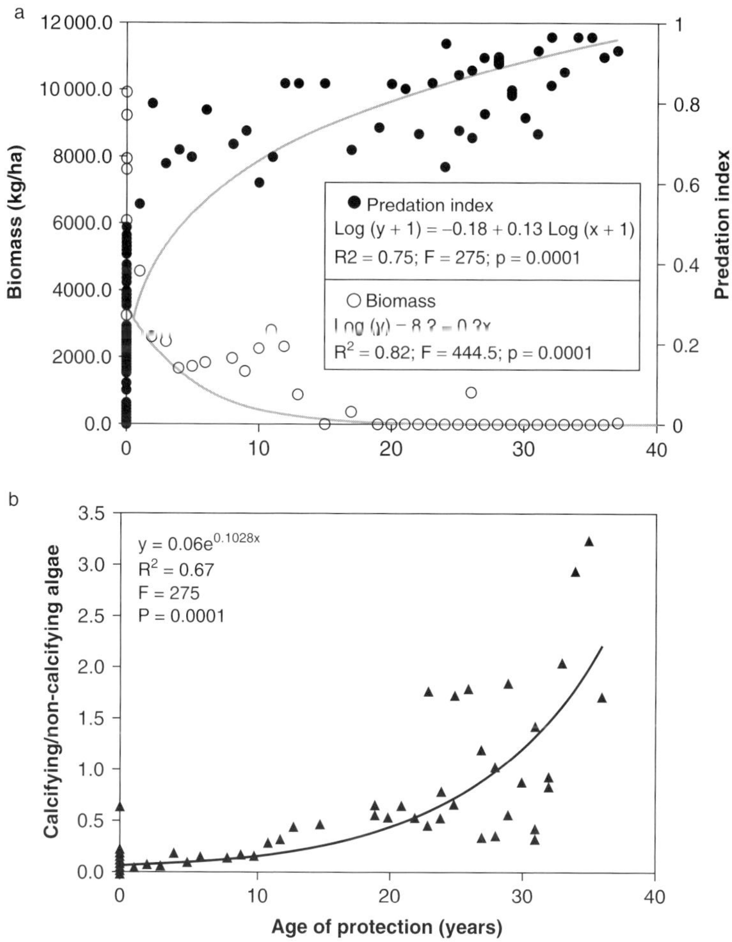

Figure 8.3 Chronosequences of (a) sea urchin abundance and predation rates on sea urchins and (b) turf and calcifying red and green algae as a function of the age of the marine protected area.

and recover more quickly after reduced fishing, but they also do not appear to be as effective as predators and are unable to maintain sea urchin abundance at low levels.

A combination of an increase in fish and a decline in sea urchin herbivores results in a change in the algal groups from turf to encrusting and calcifying algae after protection (Fig. 8.3b). Experiments on algal succession find that sea urchin grazing promotes a turf algal community that colonizes in less than one month and remains stable for many years, while high fish grazing produces a slower succession towards calcifying red and green algae that takes more than six months to complete (McClanahan 1997). The park chronosequence also suggests that there is a long-term increase in the abundance of these calcifying algae and consequent declines in turf algae. Turf algae are more productive than calcifying algae (Steneck and Dethier 1994) and this may explain the reduced biomass of fish after 20 years, but this should also be associated with higher calcification and reef growth.

It has been suggested from field data and modeling experiments that the reduction in sea urchin predators results in a rise in sea urchin numbers that competitively excludes herbivorous fishes (McClanahan 1992). The effect of fishing of herbivores and competition with sea urchins is, however, often interactive and difficult to measure. One field test separated the effect experimentally by reducing sea urchins in and out of protected areas and found evidence for competition between herbivorous sea urchins and fishes (McClanahan et al. 1996). Consequently, one of the roles of protected areas is as a refuge for those fish species that are sensitive to fishing but play an important keystone function in the reef ecosystem. Unfortunately, the redlined triggerfish appears to be so sensitive to fishing (Watson 1996; McClanahan 2000b; Kaunda-Arara and Rose 2004a) that it cannot persist and perform its ecological function without closed areas or perhaps a ban on the killing of this species in fishing grounds.

Indirect effects of marine protected areas

Indirect effects are often more difficult to measure and study and are often based on theoretical considerations that have only seldom been measured. For example, larger fishes in closed areas are expected to produce more eggs and to seed adjacent fished reefs, a process known as larval spillover. The distance larvae travel and the effect of big fish in enhancing recruitment are difficult to measure but, nonetheless, have been evaluated through models (Rodwell et al. 2002). Those aspects of spillover that can more easily be measured, such as the movement of fish along park borders and the catch of fish along borders, have been studied in some detail (Chapter 5).

Spillover of fish

Three studies on movement of fish from park borders have found that their movement is influenced by the species of fish, reef morphology, and fishing. In the isolated island reef of Kisite MNP, there was little evidence for movement of fish out of the park (Watson 1996). Given the moderate daily movements of many coral reef fish and the isolation of this reef, these results are not surprising. Tagging studies in the Malindi-Watamu MNP found that 88% of the tagged species did not move outside of parks, but a rabbitfish and two emperors of commercial importance did move and there was more movement of fish along a continuous reef than an isolated patch reef (Kaunda-Arara and Rose 1994a, b).

Studies along the continuous fringing reef of the Mombasa MNP found that rabbitfish, emperors, and surgeonfish were the main groups to move outside the park and their movement was greater on the managed side with less use of seine nets (McClanahan and Mangi 2000). These three groups of fish are only a small portion of the total fish diversity on these coral reefs but they compose about 60% of the fish catch, which means that spillover can be a significant contribution to fish catch. Numbers of species, mean sizes, and biomass of fish all declined away from the park and fishers responded to this effect by increasing the density of traps along the park boundary, which reduced the total catch per trap. The overall effect of the closed area on the fish catch was to double the catch per person but, because of the reduced area for fishing, the total catch declined during this initial period of protection (Fig. 8.4, McClanahan and Kaunda-Arara 1996; McClanahan and Mangi 2000).

The findings in Mombasa can be used to calibrate models to determine the effect of the closed areas on total fish yields. Most available models require that fish biomass equilibrates, which as shown above could take around 20 years. The three groups above have instantaneous immigration rates of around 0.50, which if used in a model where fish movement is the critical parameter indicates that closing around 15% of the fishing area will result in increased fish yields at equilibrium (DeMartini 1993).

A second model with assumptions about adult and larval dispersal and retention predicts the optimal closed area size for different levels of fisheries exploitation and finds that for a reasonable annual exploitation rate of 40%, the optimal reserve size is 10% of the area and the required closed area increases with the exploitation rate (Rodwell et al. 2002). This model was further modified to examine the effects of habitat quality in the reserve, where improved habitat quality could increase fish biomass in the park but might also reduce the movement of fish out of the park (Rodwell et al. 2003). If increased biomass is the main effect, than fish yields could be increased by as much as 40% but, if retention of fish in the closed area is the main effect, than catches could decline by around 20%.

The actual data from next to the Mombasa MNP indicates that there is a period of a significant loss in fish catch with lost fishing area but, with time and an increase in the park's fish biomass, there is some increase and compensation (Fig. 8.4). Other factors such as changes in the boundary and changes in fishing gear appear to have significant effects on the fish catch and it may require more than the 13 years of study presented here to determine the final equilibrium outcome on the fisheries

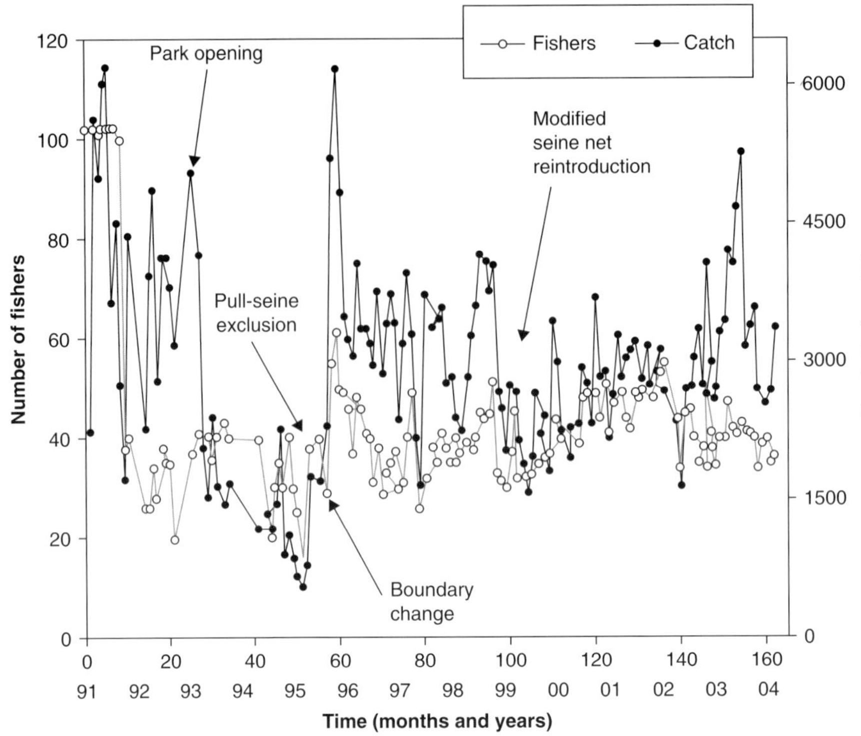

Figure 8.4 Fish catch adjacent to the Mombasa MNP from the inception of the park in 1991 to 2003. Also presented are important historical events in the regulation of the park size and fishery in the adjacent reserve.

catches. It does, however, suggest that there is a considerable time period in the early stages of closure where total fisheries catches can be low, and this period is likely to result in conflicts between fishers and park managers. In this case, many fishers left the fishery (McClanahan and Kaunda-Arara 1996) but, even with this reduction in effort, there was still continued tension and attempts by fishers to degazette the park.

In mid-1995, an agreement was reached to reduce the size of the park to appease fishers, and to eliminate beach seine nets to appease the park service. This resulted in a short-term increase in catch followed by a new and higher catch equilibrium. By early 1999, however, a modified seine net with a larger mesh size returned to the fishery and this appears to have lowered the total catch again, probably by reducing the mean size of fish caught (McClanahan and Mangi 2001). Nonetheless, there appears to be a slow rise in the total catch even after this change, which may be associated with the increasing biomass in the park (Fig. 8.2b). Consequently, fish catches in this site are a complex interaction between the area of closure, the time since closure, and the types of gear or size of the fish caught in the fishery. Managers will need to learn to manipulate these variables over the development of closed area management in order to optimize fish catch.

Economic effects

Closing areas has the potential to increase or decrease economic benefits dependent on the conditions of the local economy. It has been common for environmentalists to focus on the economic benefits of closed areas (Balmford et al. 2004) and the use and non-use value of the ecosystem (Cesar 2000). Those interested in alleviating the poverty of fishing communities will, however, not forget the lost opportunity costs of closed areas on fishers (Emerton and Tessema 2001). One can produce contradictory conclusions concerning economic value depending on the state of the resource, economy, estimates and recognition of values, and the scale of observation. For example, the spillover of fish described above will only benefit a fishery when the fishery is exploited beyond the MSY and only after near-equilibrium conditions have been reached in the recovery of fish biomass in the parks (Rodwell et al. 2002). In the case of Kenya, both conditions are true for the older parks and the areas around them and there should, therefore, be some net benefits to the fishery.

It has been argued that governments will not invest in protecting closed areas unless the incomes from tourism will equal or exceed the value of fishing, and this will depend heavily on the number of visitors to parks (McClanahan 1999). In the case of Kenya, the balance between the two is often close. For example, in Kisite during the late 1990s, 30 000 visitors generated about US$130 000 in fees at an operating cost to the government of US$19 000 per year (Emerton and Tessema 2001). If fishing were allowed and maintained at an estimated MSY of 5 tons/km^2 it is expected that the catch in this 28 km^2 area would generate a gross revenue of US$140 000 at US$1 per kg at a cost of US$10 000 to the fishers (McClanahan et al. 2002). This suggests a slight economic benefit to fishing if MSY can be maintained. Tour operators, however, generated US$1 600 000 per year from this park and therefore gained the largest economic benefit. Consequently, it is in the interest of the national economy to maintain these parks as long as they provide a better tourism experience than the non-park areas, which as shown above is the case. This might be seen to ignore the plight of poor fishing communities, who lose out to the more profitable interests of tour operators. Nonetheless, a study of the food security of people in this area found that people living closer to the park had higher food security than those living further away (Malleret-King 2000). Clearly, efforts to reduce this disparity in profits and losses of fishing versus tourist communities will reduce potential political conflicts and stability of the park system.

Gear management

Fishing effort is often the most frequently-used management option in wealthy countries but it has largely been unsuccessful in poor countries where many people fish because fishing is better than other poorly-paid economic alternatives. Attempts by governments to regulate fishing effort are often unsuccessful because it requires compliance by resource users, and considerable personnel and money to develop enforcement, which is uncommon in poor tropical countries. It is also often politically difficult to justify because the enforcement is focused on poor people with few alternatives. For example, the only fishery law in Kenya prior to 2001 was the need to register and obtain a fishing license from the Department of Fisheries. This law, common to many countries, gave the government the potential to regulate effort by restricting the number of licenses. Nevertheless, there was almost no compliance with this law by artisanal fishers with only game, export, and ornamental fishers following the regulation. Consequently, the government was not able to control the majority of fishing effort in the most common sector, artisanal fisheries.

Given the lack of compliance and difficulty of regulating effort, there is a need to utilize other management tools for artisanal fishers. Regulation of gear remains one of the better options as gear determines the species and size of the resource used, is an obvious object that can be seen and controlled if needed, and is often rooted in traditions of fishers and, therefore, is amenable to local opinions and informal regulation. The manager must, therefore, identify a mix of gears that do not overlap in resource selectivity, are not destructive to habitat and fisheries sustainability, and are also accepted by fishers (McClanahan and Mangi 2004). A recent effort to do this is described below.

Perceptions towards gear

The first step in the process was to determine the perceptions of resource users and managers towards the gear options (McClanahan et al. 2005). In this case it started with informal discussions, but this was eventually followed by quantitative questionnaires that allowed for a statistical analysis of opinions. Resource users and managers were asked to list the regulated gear and then to scale each gear by their personal belief in the ability of the gear to sustain fish catch, and also to scale the gear according to how they believed government and traditional leaders perceived the sustainability. When combining the scaling options, beach seines and spear guns were most commonly scaled as unsustainable gears (Fig. 8.5a). There were similarities in the perceptions among the users, with even those fishers that used beach seines and spear guns sharing similar perceptions (Fig. 8.5b). There were some minor differences between the park service and fisheries department personnel, with the park service more frequently listing mesh size as illegal compared to the fisheries department's more frequent listing of beach seines. Nonetheless, government regulations after 2001 supported traditions of gear use but despite this concordance there was still poor adherence to these laws as nearly two-thirds of the fishers used the prohibited gear.

Gear selectivity and competition

The second step was to evaluate the gears in term of their catch rates and the species selectivity, body sizes, and mean trophic levels to determine if findings from the catch coincided with perceptions of sustainability. A more thorough evaluation

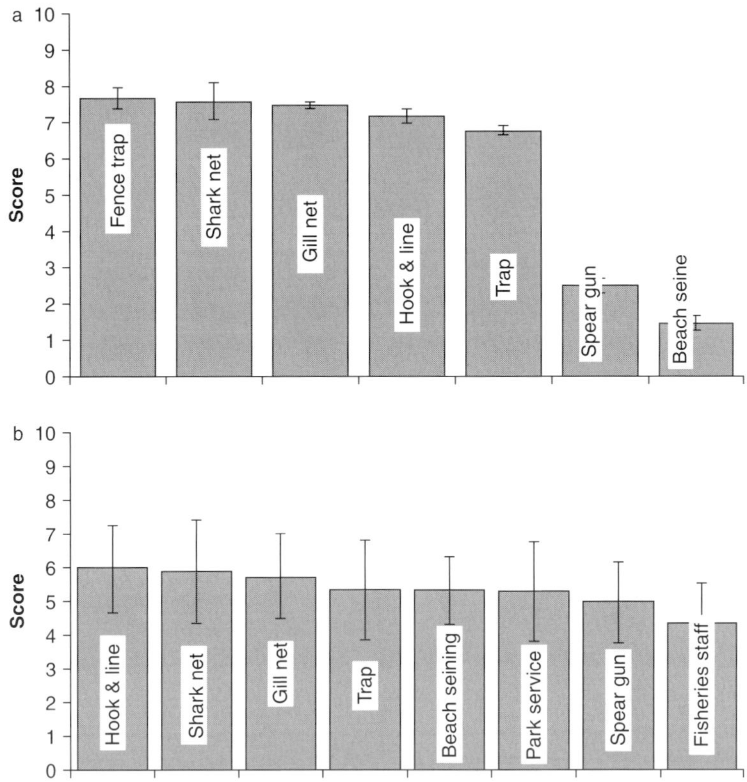

Figure 8.5 Scores of the sustainability, with a maximum of 10, for the various fishing gears based on (a) all respondents combined and (b) each respondent by their profession or gear use.

should also evaluate the effect of the various gears on habitat, but this has only been partially done (Cros and McClanahan 2003). Based on an evaluation of 1486 fisher days and 11 402 fish caught over a two year period it was found that the fishery was composed of 163 species but that four species accounted for about 60% of the catch. There were differences in the selectivity of the fish by gear with lines catching mostly emperors, but there was selectivity overlap between spear guns and small traps, which caught mostly the speckled parrotfish (*Leptoscarus vaigensis*), and overlap in the catch of seine nets, big traps, and gill nets (Fig. 8.6). Small traps caught smaller fish than spear guns and seine nets caught the highest number of species and the smallest individuals. Consequently, this evaluation suggests that seine nets and small traps were the two gears that were least sustainable, because they overlapped in species selectivity and caught smaller individuals than the other gear. There is also evidence that the catch per unit area in fishing grounds dominated by beach seines is lower than in areas without their frequent use (McClanahan et al. 1997; McClanahan and Mangi 2001). Small traps had not originally been distinguished from large traps in the questionnaire and are not that commonly used. Consequently, the perception that beach seines were unsustainable was upheld by this analysis but not spear guns. A commonly-held belief of fishers was that spear guns produced

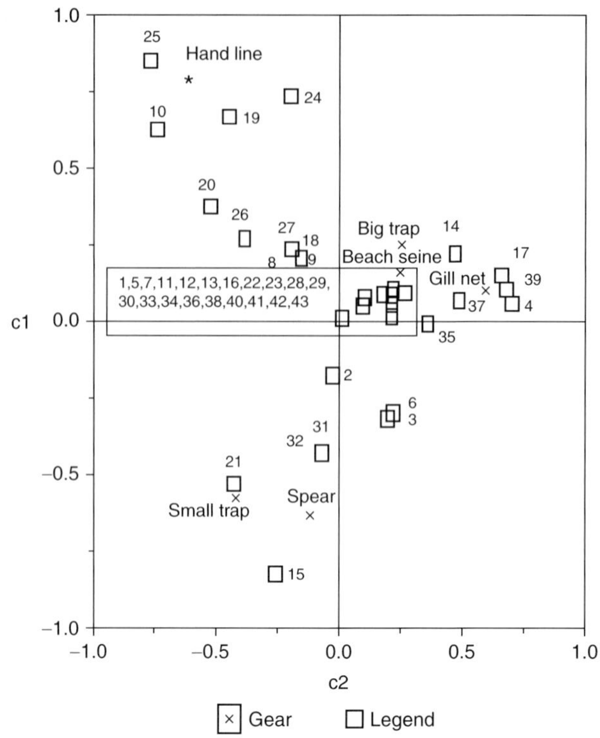

1 *Abudefduf sexsatalis*
2 *Abudefduf vaigiensis*
3 *Acanthurus nigrofuscus*
4 *Acanthurus triostegus*
5 *Anampses caeruleopunctatus*
6 *Calotomus carolinus*
7 *Calotomus spinidens*
8 *Cheilinus chlorourus*
9 *Cheilinus trilobatus*
10 *Cheilio inermis*
11 *Geres oyena*
12 *Helichores scapularis*
13 *Hempiramphus far*
14 *Leiognathus equula*
15 *Leptoscarus vaigiensis*
16 *Lethrinus elongatus*
17 *Lethrinus harak*
18 *Lethrinus lentjan*
19 *Lethrinus mahsena*
20 *Lethrinus mahsenoides*
21 *Lethrinus nebulosus*
22 *Lethrinus ramak*
23 *Lethrinus rubrioperculatus*
24 *Lethrinus sanguineus*
25 *Lethrinus xanthochilus*
26 *Lutjanus fulviflamma*
27 *Lutjanus gibbus*
28 *Mulloides flavolineatus*
29 *Parupeneus barberinus*
30 *Parupeneus macronema*
31 *Pempheris vanicolensis*
32 *Plectorhinchus flavomaculatus*
33 *Pomadasys spp.*
34 *Scarus ghobban*
35 *Scarus psittacus*
36 *Scarus sordidus*
37 *Scolopsis ghanam*
38 *Secutor insidiator*
39 *Siganus sutor*
40 *Sphyraena flavicauda*
41 *Spratoloides robustus*
42 *Carangoides sp*
43 *Sardinella sp*

Figure 8.6 Multivariate analysis (detrended correspondence analysis) of the catch composition of the major fishing gears used in Kenya. Gears that are least differentiated from the others are at the center of the graph.

blood that scared fish away, but this has not been experimentally tested and is unlikely to be a significant factor in the large scale of the national fishery.

Effect of seine-net removal on fish catches

Co-management and control

The final step in this process was to test the hypothesis that seine nets were unsustainable by removing them from some of the fish landing sites and evaluating the effects of this removal on the fish catch. This had not been done by the fisheries department since the change in the law and required an agreement between fishers and government personnel to eliminate seine nets from portions of the fishery. It was expected that until there was a joint agreement between the government and community leaders, this process would not be successful. Consequently, the questionnaire and field data were presented to groups of fishers and fisheries personnel and, after many discussions, seine nets were identified as the top priority for elimination. This led to a joint effort between the fisheries community leaders and fisheries department personnel to remove the seine nets in Diani, where most of the participants in this project were located. Fisheries leaders and government personnel travelled among the landing sites and confiscated the gear or asked that it not be used. This led to two periods in 2001 when the gear were eliminated from the area, and then a reorganization of the use of the gear such that it was only used at some of the landing sites (Fig. 8.7). The initial effect was to reverse a decline in catch that had been measured since 1995.

After two attempts to eliminate the gear it has remained at some sites and the current situation of spatial partitioning of gear by landing sites will probably be stable until the government and community instigate further actions. There is also an indication of a levelling to slight decline in fish catch after the initial increase in 2002. The reason that the total elimination of this gear did not occur is probably that this regulation did not receive support at all landing sites. Community fisheries leaders rarely have support outside of their own landing site and leaders will not always agree on what is expected by tradition or sustainability (McClanahan et al. 1997). Consequently, it is common for community leaders to have different opinions and for fishers to move among landing sites where their gear is encouraged or tolerated. Unless the government is actively involved in maintaining restrictions across all landing sites, this type of spatial organization and local community control of individual landing sites is likely to persist. Fishers almost always leave from and return to the same landing site to sell their catch, but it is possible to fish in the grounds of an adjacent landing site. This is often where the tension between landing sites occurs and leads to efforts to influence decisions in adjacent landing sites. It is likely that the short-term enforcement in 2001 led to a reorganization such that landing site communities that were most opposed to beach seines were sufficiently far away from those that were neutral or support seine nets, and this led to a new but patchy spatial equilibrium.

The study does suggest that seine nets are not maximizing the CPUE, but fishers using this gear are unlikely to stop by choice, as the gear catches smaller fish and is therefore more competitive than alternatives. The owners of this gear are often diversified investors that only deal with the marketing of the catch and many of the fishers that use this gear are either not full-time fishers or are foreigners (McClanahan et al. 1997). This creates a situation where there is little long-term interest in sustainability and the selfish rational choice is to continue to use this gear even in the face of evidence and agreement that it is not the best long-term option

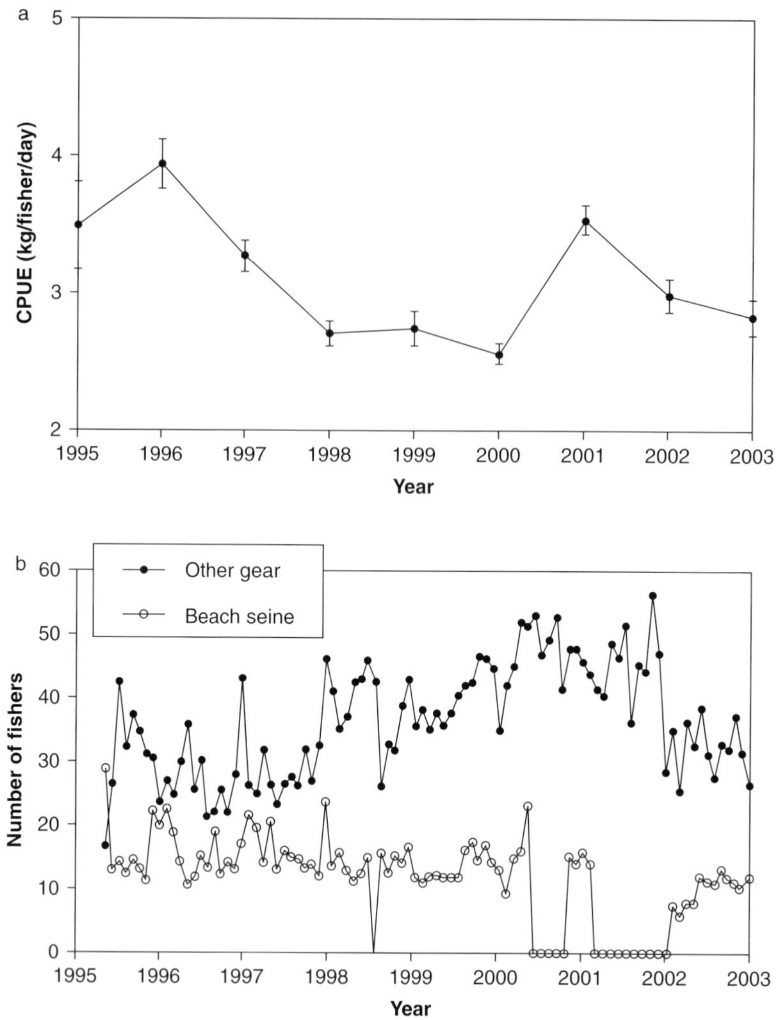

Figure 8.7 Plot of (a) the catch per unit effort and (b) gear use in Diani before and after the temporary beach-seine elimination in 2001.

for the fishery as a whole. Where there is disagreement between local communities and a system where decision makers benefit from the unsustainable gear, then the chances for its removal are slim without persistent government control.

Summary

The findings presented here reflect the complexity of managing species-rich fisheries in poor tropical countries. It shows that there are no simple solutions, and that there are consequences for both the fisheries and the ecosystems that support them.

It also indicates that, while perceptions are important to achieving actions, shared perceptions alone are insufficient to achieve high compliance and sustainability. There are times when control is required to change the habits of individuals or communities that act in their own but not the regional or national interests. Control at the community level is difficult because there are differences in opinion and subsequent behavior that can result in patchy spatial distribution of regulations. In order to achieve sustainability at the national level, active understanding of the consequences of management decisions is needed as well as enabling and enforcement of regulations by government managers that are in the interest of the fishery as a whole.

Closed-area management is not well liked by resource users but they, like government employees, recognize that they benefit the nation. Closed areas are a central part of national conservation and a fisheries management plan as shown in this and other studies. They preserve biomass, biodiversity, and ecological functions that are not maintained even in well-managed fisheries. Consequently, a combination of discouraging offshore fishing and closing nearshore fisheries is needed to maintain the states and functions of unexploited ecosystems and to generate tourist income. The total area closed to fishing should be around 15% of the nearshore area but given the pressure on resources and the lag time between establishment and benefits this number will become increasingly harder to achieve. This suggests the need to move some of the unregulated fisheries into gear-managed areas and then into a mix of gear and closed-area management as the benefits of gear management reduces poverty and negative perceptions towards management (Chapter 7). This study highlights a few limited successes, which if expanded to a larger scale will help to achieve the difficult goal of fisheries sustainability.

References

Balmford, A., Gravestock, P., Hockley, P., McClean, C.J. and Roberts, C.M. (2004) The worldwide costs of marine protected areas. *Proceedings of the National Academy of Science*, **101**: 9694–7.

Cesar, H. (2000) Coral reefs: their functions, threats and economic value. In: H. Cesar (Ed.) *Collected Essays on the Economics of Coral Reefs*. CORDIO, Sweden.

Clark, C.W. (1985) *Bioeconomic Modelling and Fisheries Management*. Wiley – Interscience Publication, New York.

Cros, A. and McClanahan, T.R. (2003) Coral transplant damage under various management conditions in the Mombasa Marine National Park, Kenya. *Western Indian Ocean Journal of Marine Science*, **2**: 127–36.

DeMartini, E.E. (1993) Modelling the potential of fishery reserves for managing Pacific coral reef fishes. *Fishery Bulletin*, **91**: 414–27.

Emerton, L. and Tessema, Y. (2001) *Economic Constraints to the Management of Marine Protected Areas: the Case of Kisite Marine National Park and Mpunguti Marine National Reserve, Kenya*. IUCN Eastern Africa Regional Office, Nairobi.

Kamukuru, A.T. (1992) Costs and Earnings of Basket Trap and Handline Fishery in the Dar-es-Salaam Region of Tanzania. MSc thesis, Department of Applied Zoology, University of Kuopio, Finland.

Kaunda-Arara, B. and Rose, G.A. (2004a) Out-migration of tagged fishes from marine reef National Parks to fisheries in coastal Kenya. *Environmental Biology of Fishes*, **70**: 363–72.

Kaunda-Arara, B. and Rose, G.A. (2004b) Effects of marine reef National Parks on fishery CPUE in coastal Kenya. *Biological Conservation*, **118**: 1–13.

Lubchenco, J., Palumbi, S., Gaines, S. and Andeman, S. (2003) Plugging a hole in the ocean: the emerging science of marine reserves. *Ecological Applications*, **13**: S3–7.

Malleret-King, D. (2000) A Food Security Approach to Marine Protected Area Impacts on Surrounding Fishing Communities: the Case of Kisite Marine National Park in Kenya. PhD thesis, University of Warwick.

McClanahan, T.R. (1992) Resource utilization, competition and predation: a model and example from coral reef grazers. *Ecological Modelling*, **61**: 195–215.

McClanahan, T.R. (1997) Primary succession of coral-reef algae: differing patterns on fished versus unfished reefs. *Journal of Experimental Marine Biology and Ecology*, **218**: 77–102.

McClanahan, T.R. (1999) Is there a future for coral reef parks in poor tropical countries? *Coral Reefs*, **18**: 321–5.

McClanahan, T.R. (2000a) Coral reef use and conservation. In: T.R. McClanahan, C.R.C. Sheppard and D.O. Obura (Eds) *Coral Reefs of the Indian Ocean: their Ecology and Conservation*. Oxford University Press, New York.

McClanahan, T.R. (2000b) Recovery of the coral reef keystone predator, *Balistapus undulatus*, in East African marine parks. *Biological Conservation*, **94**: 191–8.

McClanahan, T.R. and Graham, N.A.J. (2005) Recovery trajectories of coral reef fish assemblages within Kenyan marine protected areas. *Marine Ecology Progress Series*, **294**: 241–8.

McClanahan, T.R. and Kaunda-Arara, B. (1996) Fishery recovery in a coral-reef marine park and its effect on the adjacent fishery. *Conservation Biology*, **10**: 1187–99.

McClanahan, T.R. and Mangi, S. (2000) Spillover of exploitable fishes from a marine park and its effect on the adjacent fishery. *Ecological Applications*, **10**: 1792–805.

McClanahan, T.R. and Mangi, S. (2001) The effect of closed area and beach seine exclusion on coral reef fish catches. *Fisheries Management and Ecology*, **8**: 107–121.

McClanahan, T.R. and Mangi, S. (2004) Gear-based management of a tropical artisanal fishery based on species selectivity and capture size. *Fisheries Management and Ecology*, **11**: 51–60.

McClanahan, T.R. and Obura, D.O. (1996) Coral reefs and nearshore fisheries. In: T.R. McClanahan and T.P. Young (Eds) *East African Ecosystems: their Ecology and Management*. Oxford University Press, New York.

McClanahan, T.R., Kamukuru, A.T., Muthiga, N.A., Gilagabher Yebio, M. and Obura, D. (1996) Effect of sea urchin reductions on algae, coral and fish populations. *Conservation Biology*, **10**: 136–54.

McClanahan, T.R., Glaesel, H., Rubens, J. and Kiambo, R. (1997) The effects of traditional fisheries management on fisheries yields and the coral-reef ecosystems of southern Kenya. *Environmental Conservation*, **24**: 105–20.

McClanahan, T.R., Maina, J. and Pet-Soede, L. (2002) Effects of the 1998 coral mortality event on Kenyan coral reefs and fisheries. *Ambio*, **31**: 543–50.

McClanahan, T.R., Maina, J. and Davies, J. (2005) Perceptions of resource users and managers towards fisheries management options in Kenyan coral reefs. *Fisheries Management and Ecology*, **12**: 105–12.

McClanahan, T.R., Verheij, E. and Maina, J. (2006) Comparing management effectiveness of a marine park and a multiple-use collaborative management area in East Africa. *Aquatic Conservation: Marine and Freshwater Ecosystems*, **16**: 147–65.

McManus, J.W. (1996) Social and economic aspects of reef fisheries and their management. In: N.V.C. Polunin and C.M. Roberts (Eds) *Reef Fisheries*. Chapman and Hall, London.

Sheppard, C. and Wells, S.M. (Eds) (1988) *Coral Reefs of the World*. UNEP/IUCN, Gland, Switzerland.

Rodwell, L., Barbier, E.B., Robert, C.M. and McClanahan, T.R. (2002) A model of tropical marine reserve-fishery linkages. *Natural Resource Modeling*, **15**: 453–86.

Rodwell, L.D., Barbier, E.B., Roberts, C.M. and McClanahan, T.R. (2003) The importance of habitat quality for marine reserve–fishery linkages. *Canadian Journal of Fisheries and Aquatic Science,* **60**: 171–81.

Steneck, R.S. and Dethier M.N. (1994) A functional group approach to the structure of algal-dominated communities. *Oikos*, **69**: 476–98.

Sutinen, J.G. and Kuperan, K. (1999) A socio-economic theory of regulatory compliance. *International Journal of Social Economics*, **26**: 174–93.

Watson, M. (1996) The Role of Protected Areas in the Management of Kenyan Reef Fish Stocks. DPhil thesis, University of York.

Young, S.E. (2004) Artisanal Fishers' Perspective on a 'Rich Mans' Fishery: the Potential for Ring Net Development in Kenya. MSc thesis, University of Newcastle upon Tyne.

Chapter 9
Progress Towards the Cooperative Management of Marine Hunting in the Great Barrier Reef

Helene Marsh

Introduction

Agenda 21, a comprehensive global action plan for the environment, was adopted by178 governments at the United Nations Conference on Environment and Development in 1992 and was endorsed at the World Summit on Sustainable Development in 2002. Chapter 6 of Agenda 21 (Anon 1992) explicitly recognizes: 'The need for developing institutional arrangements for integrated coastal zone management which involve indigenous peoples to ensure the ecologically sustainable use of marine and coastal resources through establishing a process which empowers indigenous people, strengthens their participation in policy formulation and involvement in natural resource management'.

There are inevitable challenges in developing such arrangements. These challenges are particularly great for threatened species of considerable cultural and dietary value to indigenous people. In developed countries in particular, the wider community often perceives indigenous hunting as a major threat to wildlife, particularly charismatic megafauna, even when there is little scientific basis for this perception (Bomford and Caughley 1996; Freeman et al. 1998). There is resultant tension between indigenous peoples, policy makers, day-to-day managers, conservationists, and researchers. Indigenous people perceive that they are competing with conservationists for wildlife and are concerned about their future access to prey species and the survival of important components of their culture. Conservationists and researchers are concerned about overexploitation, potentially leading to local extinctions. Policy-makers and day-to-day managers have statutory responsibilities to conserve threatened species.

Two species of marine megafauna, the green turtle, *Chelonia mydas*, and the dugong, *Dugong dugon*, are of considerable cultural value to coastal Aborigines and Torres Strait Islanders in northern Australia (Johannes and MacFarlane 1991), including the Great Barrier Reef (GBR) region. The meat of both species is ranked highest among traditional foods. Hunting practices and prowess represent important aspects of Aboriginal and Islander traditions. The species are often hunted together using similar equipment.

Both species also have high conservation values and are threatened with extinction in many parts of the world. The green turtle is listed as ‘endangered’ by the World Conservation Union, the dugong as ‘vulnerable’ (Hilton-Taylor 2000). The green turtle is listed as ‘vulnerable’ and the dugong categorized as a ‘listed marine species’ under Australian legislation (Environment Protection and Biodiversity Conservation Act 1999). Both species are listed as vulnerable under Queensland state legislation (Qld Nature Conservation Act 1992). The Great Barrier Reef Marine Park Authority (GBRMPA) has statutory obligations to protect these threatened species, especially as the importance of the GBR region as habitat for dugongs and sea turtles was one of the reasons for its World Heritage listing (GBRMPA 1981). Australia also has obligations to protect these species under several other international conventions including the Convention on Biological Diversity, the Convention on the Conservation of Migratory Species, and the Convention on the International Trade of Endangered Species of Wild Fauna and Flora (CITES) (Stokes and Dobbs 2002).

As a signatory to the Convention on Biological Diversity, Australia also has obligations to ‘protect and encourage’ marine hunting by indigenous communities as ‘customary use of biological resources in accordance with traditional cultural practices that are compatible with conservation or sustainable use requirements’ and ‘to respect, preserve and maintain the knowledge, innovations, and practices of indigenous and local communities embodying traditional lifestyles relevant for the conservation and sustainable use of biological diversity’. Thus there is a tension between the need to protect threatened species and traditional cultures, lifestyles, and values.

The wider Australian community does not, in general, support special hunting rights for indigenous peoples. This situation creates problems for indigenous communities, placing them on the defensive when it comes to hunting matters. For example, Ponte et al. (1994) found that only 61% of the 400 people surveyed in north Queensland coastal cities approved of indigenous hunting in national parks, and that 75% of those that approved, only did so if pre-European weapons were used. The major reason for this disapproval of traditional hunting was an objection to Aboriginal peoples having special privileges. A much lower proportion of individuals were concerned about the sustainability of traditional hunting.

This review describes two decades of the journey towards the cooperative management of indigenous marine hunting by Hope Vale Aboriginal community, a major dugong and turtle hunting community in the remote Cape York region (Fig. 9.1), and the policy-makers and day-to-day managers who are responsible for the management of the adjacent Great Barrier Reef (GBR) World Heritage Area.

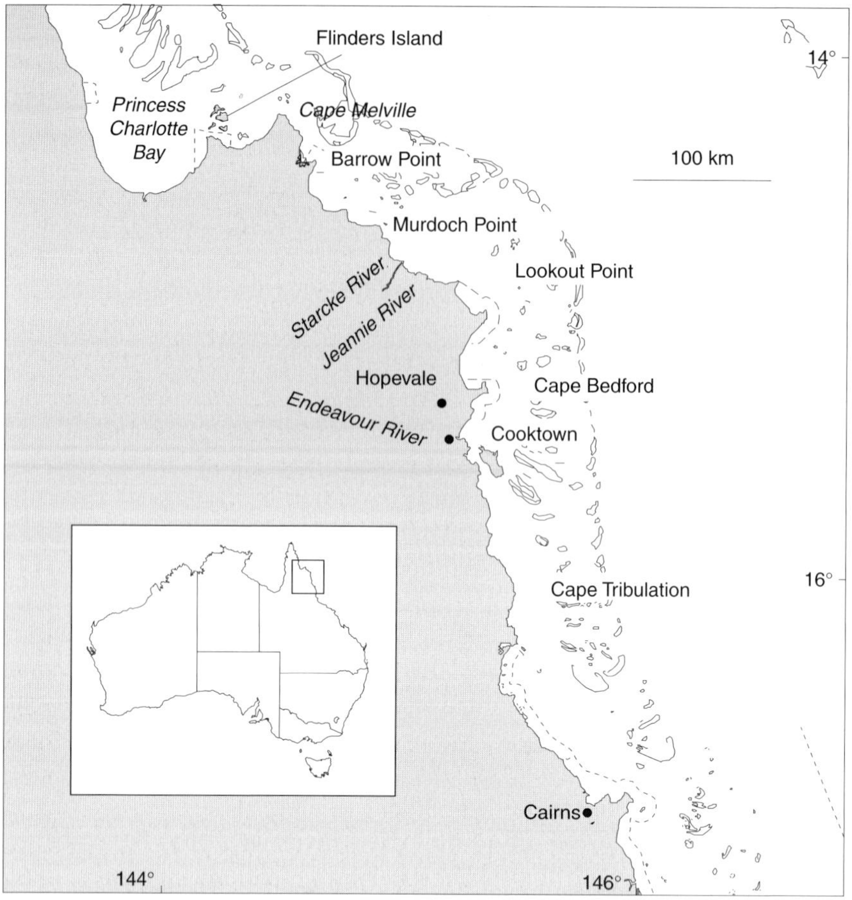

Figure 9.1 Map of the Hope Vale area illustrating the place names mentioned in the text.

Site profile

The Cape York region of Queensland in Australia is large (150 000 km^2), supports a very small human population (~15 000 including 10 000 Aboriginal people) and has high biodiversity values. Hope Vale Aboriginal Community is situated about 45 km north of Cooktown on the east coast and 26 km by road from the coast adjacent to the GBR World Heritage Area (Fig. 9 1).

Hope Vale is the hometown of the Guugu Yimithirr people. The community has a population of about 1200 people. Their average income is low by Australian standards: 63% of Aboriginal people on Cape York have annual incomes of less than US$8500. Altman (2004) points out that this value ignores the contribution of the customary sector of the economy, which is not monetized and therefore not quantified or recognized in mainstream terms. Most people are dependent on social welfare payments from the Australian government that are largely delivered by the Commonwealth Government's Community Development Employment Program (CDEP), which is administered by the Hope Vale Aboriginal Council as part of its local government responsibilities (Altman and Gray 2000). Scheme participants generally work part-time for wages on community-orientated jobs (Arthur 1999). Several part-time community rangers are employed under this program in an attempt to undertake the day-to-day management of the community's land and sea country.

CDEP participants combine this 'working for the dole', with other activities including customary hunting and fishing as they see fit. Dugong and turtle hunting are very important customary practices. Most dugong hunting and butchering occurs north of Lookout Point (Fig. 9.1). Turtle hunting is more widespread, mostly occurring on inshore and offshore reefs. Turtles are usually brought back alive and butchered near beach camps on the coast near the community.

The management context

Prior to 1870, the Guugu Yimithirr people maintained one language but traditionally occupied land that was divided into about 37 separate family groups, within 15 clan estates, where each was protected by a strict set of laws governing belief and land title. Today the clans and family groups are still responsible for their traditional land and sea countries and thus are the potential partners in co-management with the managing agencies.

The Commonwealth (national) and Queensland (provincial) governments share statutory responsibility for this region. GBRMPA determines the policies that govern the operation of the GBR Marine Park, which ends at the low water mark. The extensive intertidal region is under the jurisdiction of the Queensland government and is in the GBR Coast Marine Park. The land is under state (local government) jurisdiction. The complexity of these jurisdictional arrangements have made it difficult to develop a satisfactory legal framework to manage indigenous hunting, especially as the legal situation surrounding the rights of indigenous Australians has been (and still is) extremely complex, controversial, and volatile.

In 1992, the Australian High Court made the landmark decision to recognize the potential existence of native title (Mabo v The State of Queensland No 2 (1992) 173 CLR 1). This decision provided a new framework for the recognition of indigenous rights to land, sea and wildlife, and gave the indigenous peoples some bargaining power over the management of their traditional lands (Davies et al. 1999). The High Court of Australia confirmed the potential for native title to exist over customary sea country in October 2001 (Commonwealth of Australia v Yarmirr (2001) HCA56 (11 October 2001). The High Court determined that native title over the sea, in the case on which their judgment was based, was not exclusive, and that interests of importance specifically include traditional hunting of dugongs and turtles. This precedent has been confirmed in a subsequent sea-country determination. Although the existence of native title in the sea country of the Guugu Yimithirr people has not been confirmed at the time of writing this review, the potential for such recognition has complicated negotiations about the management of dugong and turtle hunting since the early 1990s.

Towards cooperative management

GBRMPA initiatives: 1983–1996

As part of the process of drafting plans to zone or rezone the GBR Marine Park, the GBRMPA is required to formally consult with stakeholders. In their submission, regarding the initial zoning of the section of the park spanning the sea country of the Guugu Yimithirr people, Marsh and Heinsohn (1982) expressed concern over the possible overexploitation of dugongs in the Starcke River region (Fig. 9.1). As a result, GBRMPA staff met with members of the Hope Vale community on several occasions to explain the zoning plan with its requirements for permits for traditional hunting (Smith and Marsh 1990). There were problems with these negotiations resulting from: (1) the inherent problems of public meetings in Aboriginal communities; (2) a lack of understanding of the community dynamics by agency staff; and (3) the hunters' reluctance to voice their concerns about the marine park. As a result, most of the negotiations were conducted between GBRMPA staff and members of the Hope Vale Community Council, not all of whom were conversant with or representative of the hunters' perspective (Smith and Marsh 1990).

GBRMPA first implemented a dugong permit system in 1983 (Smith and Marsh 1990). On the basis of scientific advice, 20 individual permits were issued prior to a four week open hunting season in January 1984. The permit conditions were: (1) one dugong per hunter, or a quota of 20 for the community for the season; (2) no female dugongs with attendant calves to be taken; (3) no firearms to be used; (4) catch data sheets to be completed and returned; (5) the permits to be available for inspection within the park; and (6) the permits to be valid only north of the Endeavour River (Fig. 9.1). The permits were allocated on a first come, first served basis. No permits were issued for turtles; their harvest continued to be unregulated. GBRMPA advised the community of possible penalties for infringement of permit conditions under the relevant legislation.

The introduction of the dugong hunting permit system at Hope Vale by this managing agency resulted in widespread apprehension, confusion, and misconception in the community regarding the existence, function, and regulation of the GBR Marine Park (Smith and Marsh 1990). The Hope Vale people felt victimized as at that time, GBR Marine Park regulations on dugong hunting were applied to them but not to members of the other east coast Aboriginal communities. In addition, at that time, there was limited control over other anthropogenic effects on dugongs such as commercial gill netting and shark meshing for bather protection (Marsh

2000; Marsh et al. 2005). Community members considered the permits to be a serious infringement of their traditional hunting rights. There was dissatisfaction with the number of dugongs allowed per permit and the permit allocation arrangements. Some non-hunters received permits while known hunters missed out, exacerbating pre-existing tensions within the community. There was also general dissatisfaction with the manner in which some management officers interacted with community members. Thus, the introduction of a dugong permit system at Hope Vale resulted in the community being hostile towards the management agencies (Marsh et al. 1984), a bad start from which it has been difficult to recover.

Although minor alterations were made in the method of distributing the permits for the 1985 and 1986 hunting seasons, the general discontent remained. These management developments raised the awareness of Hope Vale residents to the government's concern for the management and conservation of dugongs. They also resulted in a disproportionate amount of attention being focused on dugong hunting, so that the quota became a target (Smith and Marsh 1990).

After an extensive period of research in the community, Smith recommended a hierarchical list of management options (Smith 1987; Smith and Marsh 1990). In increasing severity, they were: (1) community dugong hunting permits; (2) declaring current dugong hunting areas as official hunting areas; (3) closed seasons; and (4) quotas. This broad management system was designed to allow each indigenous community hunting in the GBR Marine Park to be covered by the same scheme, but had the flexibility to cater for both community circumstances and specific circumstances. It also allowed for different options as circumstances changed.

The Marine Park Authority adopted these recommendations but in practice applied them only to the Hope Vale community. Between 1986 and 1991, GBRMPA issued permits for dugong hunting to the Hope Vale Aboriginal Council for the whole community. The area used by the community for hunting was informally designated as the hunting area. This designation served two functions: (1) recognition of Aboriginal hunting rights for the area; and (2) a disincentive for the community to expand hunting into other areas (should the means become available) until the status of the dugong population was confirmed. The day-to-day management agency was responsible for maintaining catch records for the community. Provision was made for the collection of dugong skulls, or at least the tusks, and any available capture information. The first permit for turtle hunting was issued in December 1986 and four community permits had been issued to the community council by 1996, each for approximately one year and for an unspecified number of turtles.

No permits for dugong hunting were issued between February 1991 and November 1997 when the practice of issuing permits was resumed at the request of the Hope Vale Community. This request was prompted by an incident in late 1996 when a community resident allegedly killed up to 11 dugongs in one hunting session. Reports of this incident caused concern in the community (and amongst the management agencies). The incident was investigated but there was insufficient evidence to mount a prosecution (Smith 1998).

Cooperative initiatives: 1997–1999

The community approached GBRMPA after this alleged incident, specifically requesting assistance to develop a plan to enable the community to manage their turtle and dugong hunting. While this incident was a catalyst, the Guugu Yimithirr people, and especially the elders, had become increasingly aware over several years of the need to manage hunting. Their concern was prompted by both their perception of reduced numbers in local populations of both turtles and dugongs, and developments elsewhere in the GBR region. Scientific evidence that dugong numbers had declined along the urban coast of the GBR (Marsh et al. 1996) led to a ban on the issue of permits for traditional hunting south of Cooktown (Fig. 9.1) by the Great Barrier Reef Ministerial Council in 1997 and the declaration of a series of dugong protection areas (Marsh 2000). As one elder from Cape York put it (Williams 1996): 'You know my boy them dugong are in trouble down south, it can happen up here if I don't start looking after them now'.

The Hope Vale Natural and Cultural Resource Management Office, through its coordinator, became a driving force behind the development of a hunting management plan. The community invited the managing agencies for the GBR region (the GBRMPA and the Queensland Parks and Wildlife Service) to become involved. This stimulated in-kind support, including the implementation of the dugong information and education strategy in indigenous communities (Hunter and Williams 1997), training for rangers in permit allocation (Hunter 1998), and a contractual agreement to fund a consultant to collate all previous work undertaken into a draft plan for community feedback.

As explained by Nursey-Bray (2000), the consultation process for the plan involved key hunters from the relevant clan estates, key traditional owner representative organizations, and members of the local Turtle and Dugong Hunting Management Council, which was established during the planning process to:

- Ensure that sustainable hunting practices are maintained and implemented
- Ensure that traditional knowledge about turtle and dugong hunting and species are maintained and handed down to the youth of the community
- Ensure that the Turtle and Dugong Hunting Management Plan is implemented and practised (HVAC 1999)

The local native title representative body, the Cape York Land Council, was integrally involved, allowing one of its key staff members to work on the plan intermittently for almost a year (Nursey-Bray 2000). The plan was reviewed by the council's lawyers and was read and commented on by the council's senior staff, including traditional owners of country in the region that had sufficient qualifications and expertise to comment on native title considerations. The Cape York Land Council also funded the production and printing of the plan. During the planning process, a workshop was held to explain the draft plan to key non-indigenous stakeholders, including representatives of government agencies, NGOs and researchers.

Box 9.1 Principles underlying the Hope Vale Turtle and Dugong Hunting Management Plan

(1) Recognition of traditional owner and historical owner custodianship of marine estates.
(2) Recognition of sea rights.
(3) Recognition and commitment to community responsibility for sea management.
(4) Recognition of cultural importance of indigenous hunting.
(5) That any curtailment of hunting will not be accepted if it diminishes native title in any way.
(6) Recognition of the knowledge of the elders past and present about both turtle and dugong and traditional hunting.
(7) Recognition of traditional knowledge and its incorporation and acknowledgement where possible into the plan.
(8) That the plan reflect and be part of the tradition of transmission of hunting knowledge over time.
(9) That hunting traditions are maintained.
(10) That this plan is part of a wider planning process both indigenous and non-indigenous. This includes processes of sub-regionalization, land and sea management, plan development and legislative and other policy directions.
(11) Whilst this community will accept responsibility for managing its impact upon these species, it expects other sectors, which impact on the turtle and dugong to do likewise.

Staff members of the managing agencies were closely involved in the process (Smith 1998; Nursey-Bray 2000).

A Guugu Yimithirr Bama Wii – Girrbithi and Ngawiya – A Turtle and Dugong Hunting Management Plan (HVAC 1999) was formally endorsed by the signatures of 38 community members including: the chair of the key local corporations and the local council; 19 elders; a representative of the local Elders Justice Network; the chair and several members of the local Turtle and Dugong Hunting Management Council; the project and plan convenor; and the community rangers.

The plan's underlying principles indicated that it was an important step forward by the community to secure outside acceptance of its ability to enter into the marine management arena and to demand to be respected and involved on its own terms (Box 9.1).

These aspirations assert the indigenous rights of use that provided the momentum for the plan. They are also the embodiment of the assumption of the Guugu Yimithirr of their right and ability to manage their sea country and its wildlife resources effectively (Nursey-Bray 2003). The vision of the Hope Vale plan is: 'To develop and implement controlled and sustainable hunting practices that will minimize the impact on, and contribute to the protection and survival of dugong (Girrbithi) and turtle (Ngawiya) species for the enjoyment and use of future generations' (HIVAC 1999).

This vision is accompanied by a supporting statement that the plan would in no way compromise native title. The plan clearly states (HVAC 1999) that it is to be considered a living document that will: 'Remain flexible and open to amendment at all times, subject to community, review, evaluation and appropriate consultations in order to anticipate and incorporate changes in native title issues, legislation and community needs as appropriate and desired by the Hope Vale community'.

The plan has five objectives (Box 9.2). Together with the vision, they articulate the community's aspirations. The plan opens with a dedication to the elders and subsequently a number of statements from traditionally-affiliated and historically-associated individuals from across the community and clans that express support for the document. These statements are a strong affirmation that the community drove the planning process (Nursey-Bray 2003). They also express the key priorities that informed the planning at all times. These have included the wish for community by-laws, the need to maintain traditional and historical associations with culture and identity, the important role of elders, the importance of sharing responsibility, and the need to work together and maintain community ownership of the plan.

The establishment of the Turtle and Dugong Hunting Management Committee was a very important step in plan development and was an acknowledgment of the need for the indigenous community to be involved in management. The committee's

Box 9.2 Objectives of the Hope Vale Turtle and Dugong Hunting Management Plan (HVAC 1999)

(1) To develop a controlled indigenous hunting regime for Dugong (Girrbithi) and Turtle (Ngawiya) through careful planning, monitoring, and management.
(2) To protect dugong and turtle habitat by managing the activities carried out on the land and sea by both Traditional Owners and visitors according to the desires of the Traditional Owners.
(3) To maintain the activity, knowledge, and skill of traditional hunting for turtle and dugong ensuring that this important cultural activity is continued through future generations.
(4) To assist the community to develop and/or reinstate customary laws to manage traditional hunting in conjunction with State and Commonwealth legislation.
(5) To revitalize respect for the law and sea management aspirations of individual clan groups and identify ways in which these groups can work together to ensure the survival and prosperity of dugong and turtles.

membership aimed to include all representative agencies and bodies at Hope Vale, to be across clan groups, and to reflect the whole of community aspirations (male and female) about traditional hunting. This group decided to take responsibility for the plan and its implementation and worked together to develop the vision statement and definition of roles (Nursey-Bray 2000).

The community decided to use the same management tools that had characterized the earlier initiatives by the managing agencies including: a hunting region that was the same as that defined under the community permits issued by GBRMPA; a defined hunting season over the summer holidays (15 December to 15 February); community hunting permits as issued by GBRMPA for 20 dugongs and 25 turtles (similar numbers to those authorized by the original GBRMPA community permits that reflected contemporary scientific advice). The plan also reflected important new developments that asserted community responsibility and control including:

- A body representative of hunting interests in the community, the Turtle and Dugong Hunting Management Council that includes representatives of traditional owners, hunters, community rangers, and the Elders Justice Network to determine who receives hunting authorities under the community permits, who should get punished for breaches, and the resultant penalties.
- A community license framework with hunting authorities (to be provided by GBRMPA) to be distributed through the Natural and Cultural Resource Management Office by the rangers and coordinator.
- Penalties for infringement including a mix of Aboriginal justice and prosecution by the managing agencies (for third or serious offences).
- Hunting authorities limited to one per immediate family with disputes over allocation referred to mediation and resolution by the Turtle and Dugong Hunting Management Council.
- A ban on non-indigenous people participating in hunting expeditions.
- Provision of special hunts for funerals but
- A ban on hunts for weddings and birthdays.
- Provision of a special hunt each year to provide meat for the old people in the community.
- Butchering rules for turtles that address animal welfare concerns.
- Bans on taking turtle eggs, and removing a turtle shell from the butchering site.
- A recommendation that two rangers be employed each year to monitor the hunt and enforce the plan.
- A requirement for hunters to check in with rangers on return from a hunt to ensure that catch information is recorded.
- A requirement that the plan apply to all people living within the Hope Vale local government region as well as to relevant groups living outside this jurisdiction.

The plan did not designate specific responsibilities and timelines for its implementation. It did, however, make a series of 17 recommendations that were designed to facilitate the implementation process.

The Hope Vale primary school was involved throughout the project. All artwork included in the final document was done by the children as part of an art competition featuring turtles and dugongs. This competition was run in conjunction with an education program about the values of turtles and dugongs. The final launch of the plan took place at the school assembly, and concluded with the children singing a song about turtle and dugong hunting protection and management, and a community barbeque. This whole process: (1) promoted an awareness amongst the children (the future hunters) about the plan and the importance of hunting management; (2) created a sense of ownership by the children through the inclusion of their artwork in the final document; and (3) involved the adults as parents, facilitating community pride in the project (Nursey-Bray 2000). This pride was strengthened six months later when the community's plan was awarded the Australian Prime Minister's Community Environment Award 2000.

The post-planning period: 1999–2001

After the plan's launch, Hope Vale community faced many challenges including the death of key elders and individuals, who had championed the plan in the community, personnel changes in the coordinators and rangers employed in their Natural and Cultural Resource Management Office, a complete turnover of staff at the Indigenous Policy Liaison Unit at GBRMPA, who had championed the plan in that agency, and a lack of resources for implementation. Despite being a member of the government that had awarded the Australian Prime Minister's Community Environment Award 2000 to the plan, the then Commonwealth Minister for the Environment effectively prevented GBRMPA from issuing community hunting permits as required by the plan, allegedly because he feared backlash from conservation groups. Changes to the case law regarding native title also altered the community's political landscape and the legal position of traditional marine hunting. In addition, new national environmental legislation (Commonwealth Environmental Protection and Biodiversity Conservation Act 1999) strengthened and clarified GBRMPA's position with respect to the protection of the GBR region's World Heritage values, including dugongs and turtles.

As a result of these developments, the Hope Vale Turtle and Dugong Hunting Management Plan was not implemented effectively, despite the enthusiasm evident at its launch in late 1999, and the subsequent accolade from the Australian Prime Minister. Several meetings involving both the community and staff of the managing agencies were convened in 2000 and 2001 to address the issue of implementation. None of the 13 relevant recommendations were fully addressed.

Subsequent developments

Although western scientists have been concerned about the sustainability of the indigenous harvests of the stocks of dugongs and turtles in the Cape York region and adjacent Torres Strait for many years, the scientific evidence of overharvest has strengthened recently (Limpus 1999; Chaloupka 2002, 2003; Marsh 2003; Heinsohn et al. 2004; Marsh et al. 2004, 2005). Despite numerous attempts to present the scientific information in an accessible format, many indigenous people still do not accept this evidence. In addition, a survey of indigenous fishing in northern Australia (Henry and Lyle 2003) produced controversial estimates of the magnitude of the take of dugongs and turtles. These estimates raised the public profile of the issue and polarized the positions of some conservationists and indigenous groups.

In 2003, the Australian Minister of Environment instructed that the matter of this overharvest be addressed with high priority. The Marine and Coastal Committee of the Natural Resource Management Ministerial Council established a cross-jurisdictional task force on dugong and marine turtle populations and asked that it develop a national approach to assist in ensuring that the indigenous harvests of dugongs and marine turtles is sustainable and legal. The task force included representatives from Commonwealth, State and Territory government departments and agencies responsible for environment, natural resource management, and indigenous issues. The Ministerial Council subsequently endorsed the National Partnership Approach, which was released in November 2005. The approach contains five goals supported by a number of objectives and also outlines possible activities that could be used to implement it (http://www.deh.gov.au/coasts/species/turtles/national-approach.html). These goals are:

(1) Improve the information base available to indigenous communities for managing the sustainable harvest of turtles and dugongs.
(2) Respect for indigenous and non-indigenous knowledge and management.
(3) Improve education and awareness.
(4) Identify the economic, social, and cultural factors that may contribute to unsustainable harvest levels and identify and implement measures to address them.
(5) Protect sea country resources.

Both Commonwealth and State management agency staff are currently consulting with traditional owners about the management of the dugongs and green turtles in relation to all anthropogenic effects, not just indigenous hunting. In addition, considerable government funding from the National Heritage Trust has recently been earmarked for community-based management and catch monitoring initiatives for dugongs and green turtles across some of the remote regions of northern Australia, potentially assisting with the problem of resources that was one of the impediments to the implementation of the Hope Vale plan and of most co-management initiatives with indigenous peoples in Australia (Davis et al. 1999).

Within the GBR region, the relevant management agencies are attempting to address all known human effects on sea turtles and dugongs (Marsh et al. 2005). The 2003 GBRMP Zoning Plan (GBRMPA 2004), which is a statutory document, specifically makes provision for Traditional Use of Marine Resource Agreements (TUMRAs) (Havemann et al. 2005). Guidelines for establishing TUMRAs are being established in consultation with indigenous representative bodies. TUMRAs are envisaged as the primary mechanism for reconciling traditional use of the marine environment including hunting with the conservation of biodiversity by encouraging indigenous peoples to exercise their stewardship role in conservation in a culturally appropriate and scientifically robust manner. TUMRAs will also be consistent with the Commonwealth Native Title Act 1993. Each TUMRA will describe how an individual indigenous group would like to manage the traditional use of marine resources in its sea country. Once such an agreement is finalized, parties to a TUMRA can apply to GBRMPA to have it accredited. As part of the TUMRA process, communities such as Hope Vale will be encouraged to develop management arrangements for dugong and turtle hunting. The Hope Vale Turtle and Dugong Management Plan will presumably be reviewed and revised as necessary as part of the process of accrediting the two TUMRAs foreshadowed for the sea country of the Guugu Yimithirr people.

Discussion and conclusions

Over the last 20 years, members of Hope Vale community and the agencies with statutory responsibility for that community's sea country have come a long way both together and separately in their attempts to address the difficult issue of sustainable marine hunting of dugongs and green turtles (Marsh et al. 1984; Baldwin, 1985, 1988; Smith and Marsh 1990; Williams 1996; Benzaken et al 1997; Hunter and Williams 1997; Hunter 1998; Nursey-Bray 2000, 2003; Marsh 2003). Considerable cash and in-kind investments by government agencies (particularly GBRMPA), the community and other sources (such as the Pew Charitable Trust Foundation and James Cook University) have underwritten this progress. However, most of this support has been spent in negotiations with the community and in scientific research rather than in helping the community with on-ground management initiatives.

The process of producing A Guugu Yimithirr Bama Wii – Girrbithi and Ngawiya – the Hope Vale Turtle and Dugong Management Plan, was a significant achievement by Hope Vale community. Hope Vale was the first major dugong and turtle hunting community in northern Australia to have produced a comprehensive plan for managing marine hunting. Some other remote indigenous communities are now viewing this plan as a prototype for their own initiatives.

The present national approach to the problem, the development of TUMRAs in the GBR region and the re-zoning of the GBRMP (Marsh et al. 2005) potentially overcame several of the major concerns of the Hope Vale community: (1) hunting regulations were applied to them but not to members of other indigenous communities; (2) their plan was not recognized in law; (3) resources to implement their plan were extremely limited and short-term; (4) there was a perceived emphasis by the managing agencies on indigenous hunting rather than all anthropogenic effects; (5) there was a focus on indigenous hunting rather than more generally on the management of sea country.

Many lessons have been learned as a result of this complex journey towards the cooperative management of indigenous marine hunting by Hope Vale Aboriginal community and the policy-makers and day-to-day managers who are responsible for the management of the adjacent Great Barrier Reef World Heritage Area. These lessons need to be distilled to inform other attempts at managing marine hunting and fishing, especially of threatened species of cultural significance. My list of such lessons follows:

- Management of the use of marine resources of indigenous peoples should address the dual imperatives of social justice and biodiversity conservation. In a developed country such as Australia, the major concern is long-term employment and capacity building rather than food security.
- Management of the sea country of indigenous peoples should not concentrate on hunting or fishing in isolation. Indigenous peoples wish to be involved in all aspects of the management of their country in a manner that they consider culturally appropriate and that uses their traditional knowledge.
- It is important for managing agencies to spend time in culturally appropriate education and extension programs before management initiatives are negotiated. Communities must understand a problem if they are to respond effectively.
- When dealing with a matter as symbolic of cultural identity as indigenous hunting of megafauna, it is important to develop a management framework at a regional scale. If individual communities are treated as a prototype they will resent an initiative that does not apply to other communities. Communication between communities, even isolated communities in remote areas with a low socio-economic base, is now fast and efficient. Another imperative for implementing management at a regional scale is that both dugongs and sea turtles undertake extensive movements (Marsh et al. 2004).
- Community-based initiatives must have the potential for formal legal recognition. Communities are unlikely to be able to enforce community-based management initiatives today without such support.
- Indigenous communities are understandably wary of initiatives that they consider threatening to their aspirations. In Australia, this includes native title aspirations. Thus management initiatives should acknowledge these concerns.
- Management initiatives should be culturally appropriate and developed by the community and the managing agencies together. Successful initiatives are an iterative mixture of both top-down and bottom-up strategies. Top-down initiatives from managing agencies are important for statutory support, capacity building and resources. Bottom-up initiatives from the community are an essential prerequisite for community ownership and control. Features of effective indigenous community wildlife management in Australia have been identified by Davies and colleagues (1999).

- Management arrangements need to be flexible and adaptive, allowing the management partnerships to grow iteratively in a distinct cycle, which treats management intervention as a hypothesis for testing and evaluation. This approach allows all parties to test their assumptions systematically and to measure how well the arrangements have worked and improve them if necessary. Robinson and colleagues (in press) outline a suggested process. Thus successful co-management initiatives are likely to take many years to develop.
- The problem and the proposed solution have to be owned by many people in both the community and the managing agencies, rather than a few champions. An initiative must have the capacity to survive changes in personnel.
- Negotiations about management of sea country must be conducted in a manner that allows all parties to participate on an equal footing, and to respect each other's aspirations. Jones (2004) provides practical advice on how to set an effective negotiation table in such circumstances.
- Planning management initiatives is much easier than implementation, which requires resources and community capacity. The resources required for implementation should ideally be available before planning is completed to capitalize on the momentum generated during the planning phase and as an incentive during the negotiation process.
- Although, as explained above, some funds have been provided by the Australian Government to support community-based management of dugongs and sea turtles in northern Australia, this funding is project-based and short-term. A secure long-term funding base is required. Altman (2003) notes that: 'supporting indigenous economic futures on country has the potential to generate benefits not just for Aboriginal people but also for a variety of national stakeholders'. The development of local indigenous natural resource management of land and sea country in the remote regions of northern Australia is a potentially cost-effective option (Smith 2004). Such an initiative has the potential to provide employment opportunities in communities that are at present largely dependent on social security to manage remote areas that are currently managed largely by benign neglect. Using locals should be much more cost-effective than meeting the logistical challenge and expense of relocating management agency staff from other areas, either permanently or on an expeditionary basis.
- Australian indigenous communities do not trust western research or researchers. Nonetheless, because research is conducted over many years, it provides a valuable opportunity to build trust by following agreed protocols and should be considered as part of the community education and extension process. Meaningful indigenous participation even extending to 'co-managed' research at all levels of the research process from framing and prioritizing the questions, designing the research protocol, collecting and analyzing the data, and disseminating the results helps to build trust in the outcomes of the research process, especially if the results can be set in the context of relevant traditional knowledge (Innes and Ross 2001).

References

Anonymous (1992) *Agenda 21, Rio Declaration on Environment and Development*. http://www.un.org/documents/ga/conf151/aconf15126-1annex1.htm

Altman, J.C. (2003) *Peoples on Country, Healthy Landscapes and Sustainable Indigenous Economic Futures: the Arnhem Land Case*. Paper presented at the International Association for Landscape Ecology World Conference 2003.

Altman, J.C. (2004) Development options on Aboriginal land: sustainable indigenous hybrid economies in the twenty-first century. In: G. Ward, S. Davies and G. Gray (Eds) *The Power of Knowledge, the Resonance of Tradition*. Aboriginal Studies Press, Canberra.

Altman, J.C. and Gray, M.C. (2000) *The Effects of the CDEP Scheme on the Economic Status of Indigenous Australians: Some Analyses Using the 1996 Census*. Centre for Aboriginal Economic Policy Research, Australian National University, Discussion Paper 195.

Arthur, W.S. (1999) *Careers, Aspirations and the Meaning of Work in Remote Australia: Torres Strait*. Centre for Aboriginal and Economic Policy Research, Australian National University, Discussion Paper 190.

Baldwin, C. (1985) *Management of Dugong: an Endangered Marine Species of Traditional Significance. The Need for User Involvement in an Integrated Program of Research, Management and Education*. Great Barrier Reef Marine Park Authority, Townsville, Technical Report 1.

Baldwin, C. (1988) Management of dugong: an endangered marine food species of traditional significance: the need for user involvement in an integrated program of research, management and education. In: F. Gray and L. Zann (Eds) *Workshop on Traditional Knowledge of the Marine Environment in Northern Australia*. Great Barrier Reef Marine Park Authority, Townsville, Workshop Series 8.

Benzaken, D., Smith, G. and Williams, R. (1997) A long way together; the recognition of Indigenous interests in the management of the Great Barrier Reef World Heritage Area. In: *State of the Great Barrier Reef World Heritage Area Workshop*. Great Barrier Reef Marine Park Authority, Townsville.

Bomford, M. and Caughley J. (Eds) (1996) *Sustainable Use of Wildlife by Aboriginal Peoples and Torres Strait Islanders*. Australian Government Publishing Service, Canberra.

Chaloupka, M. (2002) Stochastic simulation modelling of southern Great Barrier Reef green turtle population dynamics. *Ecological Modelling*, **148**: 79–209.

Chaloupka, M. (2003) *Phase 2: Development of a Population Model for the Southern Great Barrier Reef Green Turtle Stock*. Great Barrier Reef Marine Park Authority, Research Publication 81.

Davies, J., Higginbottom, K., Noack, D., Ross, H. and Young, E. (1999) *Sustaining Eden: Indigenous Community Based Wildlife Management in Australia*. International Institute for Environment and Development, London.

Freeman, M.M.R., Bogoslovskaya, L., Caulfield, R.A., Egede, I., Krupnik, I.I., Stevenson, M.G. (1998) *Inuit, Whaling and Sustainability*. Altamira Press, Lanhan MD.

GBRMPA. (1981) *Nomination of the Great Barrier Reef by the Commonwealth of Australia for Inclusion on the World Heritage List*. Great Barrier Reef Marine Park Authority, Townsville.

GBRMPA (2004) *Great Barrier Reef Marine Park Zoning Plan 2003*. Great Barrier Reef Marine Park Authority, Townsville.

Havemann, P., Thieret, D., Marsh, H. and Jones, C. (2005) Decolonising conservation? Traditional use of marine resource agreements and dugong hunting in the Great Barrier Reef World Heritage Area. *Environmental and Planning Law*, **22**: 258–80.

Heinsohn, R., Lacy, R.C., Lindenmayer, D.B., Marsh, H., Kwan, D. and Lawler, I. (2004) Unsustainable harvest of dugongs in Torres Strait and Cape York (Australia) waters: two case studies using population viability analysis. *Animal Conservation*, **7**: 417–25.

Henry, G.W. and Lyle, J.M. (Eds) (2003) *The National Recreational and Indigenous Fishing Survey, July 2003*. Australian Government Department of Agriculture, Fisheries and Forestry, Canberra. http://www.affa.gov.au/corporate_docs/publications/pdf/fisheries/final_recsurvey_report.pdf

Hilton-Taylor, C. (2000) *The 2000 Red List of Threatened Species*. IUCN, Gland, Switzerland.

Hunter, B. (1998) *Community Based Management Initiatives for Marine Turtle and Dugong*. Report to Great Barrier Reef Marine Park Authority, Townsville.

Hunter, B. and Williams, R. (1997) *Sustainable Hunting: In Search of a Solution*. Sea Turtle Management and Conservation in Northern Australia Workshop.

HVAC (1999) *A Guugu Yimithirr Bama Wii – Girrbithi and Ngawiya – A Turtle and Dugong Hunting Management Plan*. Hope Vale Aboriginal Community, Hope Vale, Australia.

Innes, J., and Ross, H. (2001) *Co-managed Research as a Strategy for Informing the Development of Indigenous and Government Management Partnerships over the Great Barrier Reef*. International Association for the Study of Common Property Inaugural Pacific Regional Meeting, Brisbane, QLD. Available at Digital Library of the Commons, http://dlc.dlib.indiana.edu/documents/dir0/00/00/07/03/index.html

Johannes, R.E. and MacFarlane, J.W. (1991) *Traditional Fishing in the Torres Strait Islands*. CSIRO Division of Fisheries, Hobart.

Jones, C. (2004) Is there a future for coexistence? A question of strategy. *Indigenous Law Bulletin*, **6**: 15–18.

Limpus, C. (1999) *Green Turtle*, Chelonia mydas *in Queensland: Population Status and Sustainable Harvest, September 1999 Review*. Report to Queensland Parks and Wildlife Service, Brisbane.

Marsh, H. (2000) Evaluating management initiatives aimed at reducing the mortality of dugongs in gill and mesh nets in the Great Barrier Reef World Heritage Area. *Marine Mammal Science*, **16**: 684–94.

Marsh, H. (2003) *Progress Towards the Cooperative Management of Marine Hunting by a Remote Indigenous Community and Government Agencies in the Great Barrier Reef Region*. GBRMPA/JCU Final Report, Townsville.

Marsh, H., and Heinsohn, G.E. (1982) *Zoning the Cairns Section of the Great Barrier Reef Marine Park for the Conservation and Management of Dugongs* (Dugong dugon). Report to Great Barrier Reef Marine Park Authority, Townsville.

Marsh, H., Smith, A. and Kelly, G. (1984) *Dugong Hunting by Members of the Hopevale Aboriginal Community and the Initial Reaction to the GBRMP and the Dugong Permit System*. Report to the Great Barrier Reef Marine Park Authority, Townsville.

Marsh, H., Corkeron, P., Lawler, I.R., Lanyon, J.M. and Preen, A.R. (1996) *The Status of the Dugong in the Southern Great Barrier Reef Marine Park*. Great Barrier Reef Marine Park Authority, Townsville, Research Publication 41.

Marsh, H., Lawler, I., Kwan, D., Delean, S., Pollock, K. and Alldredge, M. (2004) Aerial surveys and the potential biological removal technique indicate that the Torres Strait dugong fishery is unsustainable. *Animal Conservation*, **7**: 435–43.

Marsh, H., De'ath, G., Gribble, N. and Lane, B. (2005) Historical marine population estimates: triggers or targets for conservation? The dugong case study. *Ecological Applications*, **15**: 481–92.

Nursey-Bray, M. (2000) *A Guugu Yimmithirr Bama Wii Ngawiya and Girrbithi, The Turtle and Dugong Hunting Management Plan – Hopevale, a Background Report*. Consultancy report to the GBRMPA and Hope Vale Aboriginal Council.

Nursey-Bray, M. (2003) Fishing for answers: the incorporation of indigenous knowledge in Australia: developing cross cultural literacy. In: N. Haggan, C. Brignall and L. Wood (Eds) *Putting Fishers Knowledge to Work, A Fisheries Research Centre Report*, Vol. 11. VUBC, Vancouver, Canada.

Ponte, F., Marsh, H. and Jackson, R. (1994) Indigenous hunting rights – ecological sustainability and the reconciliation process in Queensland. *Search*, **25**: 258–61.

Robinson, C.J., Ross, H. and Hockings, M. (in press) *Development of Co-operative Management Arrangements in the Great Barrier Reef; an Adaptive Management Approach*. CRC Reef Research Centre, Technical Report 55.

Smith, A. (1987) An Ethnobiological Study of the Usage of Marine Resources by Two Aboriginal Communities on the East Coast of Cape York Peninsula, Australia. PhD thesis, James Cook University, Townsville.

Smith, A.J. and Marsh, H. (1990) Management of traditional hunting of dugongs (*Dugong dugon* (Muller, 1776)) in the Northern Great Barrier Reef, Australia. *Environmental Management*, **14**: 47–55.

Smith, B.R. (2004) *Some Natural Resource Management Issues for Indigenous People in Northern Australia*. Paper presented at the Arafura Timor Research Facility Forum: Research in the Arafura Timor Seas, Australian National University, Canberra. http://www.anu.edu.au/nie/events/2004_atrf/paper_smith.html

Smith, G. (1998) *The Hope Vale Hunting Management Plan, Guugu Yimmithirr Bama Wii, Community Strategy for the Management of Traditional Hunting*. Report to Great Barrier Reef Marine Park Authority, Townsville.

Stokes, T. and Dobbs, K. (2002) *Fauna and Flora of the Great Barrier Reef World Heritage Area*. Report to Great Barrier Reef Marine Park Authority, Townsville.

Williams, R. (1996) Who's listening and who's learning? Aboriginal and Torres Strait Islander knowledge of turtle and dugong in the Great Barrier Reef Marine Park region In: *Ecopolitics IX, 1–3 September 1995, Northern Territory University, Darwin*.

Part III

Meso-scale Fisheries

Chapter 10

Partnerships Towards Sustainable Use of Australia's Northern Prawn Fishery

Catherine M. Dichmont, Neil R. Loneragan, David T. Brewer, and Ian R. Poiner

Introduction

The Northern Prawn Fishery (NPF) in Australia has a 6000 km coastline and a total area of some 250 000 km^2 (Fig. 10.1). At present, the fishery catches nine species of prawns (shrimp). The original vessels in the fishery were mostly small wooden otter trawlers, with a modal length of ~16 m that towed four nets and stored the catch in brine tanks (Fig. 10.2). In contrast, the vessels in use today are efficient steel trawlers of a modal length of about 22 m, towing two large nets (Fig. 10.2). They are equipped with color sounders, satellite-based tracking systems, and large freezers that reduce their down time by offloading their product to barges. Spotter planes are used to find concentrations of banana prawns. A considerable effort to improve the quality of prawns has led to the introduction of snap freezers as well as grading and packing at sea. The fishery now lands between 6000 and 8000 metric tons (MT) of prawns annually. The fleet was reduced from more than 280 vessels in the 1980s to the present fleet of 85 vessels after many management changes (Fig. 10.3). From 1992/93 to 2001/02, the real revenue for operators in the NPF fluctuated between US$90 and US$143 million, with an average of US$113 million (Galeano et al. 2004), making the NPF one of Australia's top five most valuable fisheries. (Figures based on the 2005 exchange rate from AU$, in which the money was earned, to US$.)

Exploratory fishers found banana prawns in the Gulf of Carpentaria in the 1950s, but at densities too low to justify a fishery (Pownall 1994). The Gulf Prawn Survey of 1963–1965 created history as being the first properly-conducted scientific investigation in Australia to lead directly to the development of a major fishery. The survey was a joint operation of Commonwealth and state government departments, a private fishing company (Craig Mostyn and Company) and the

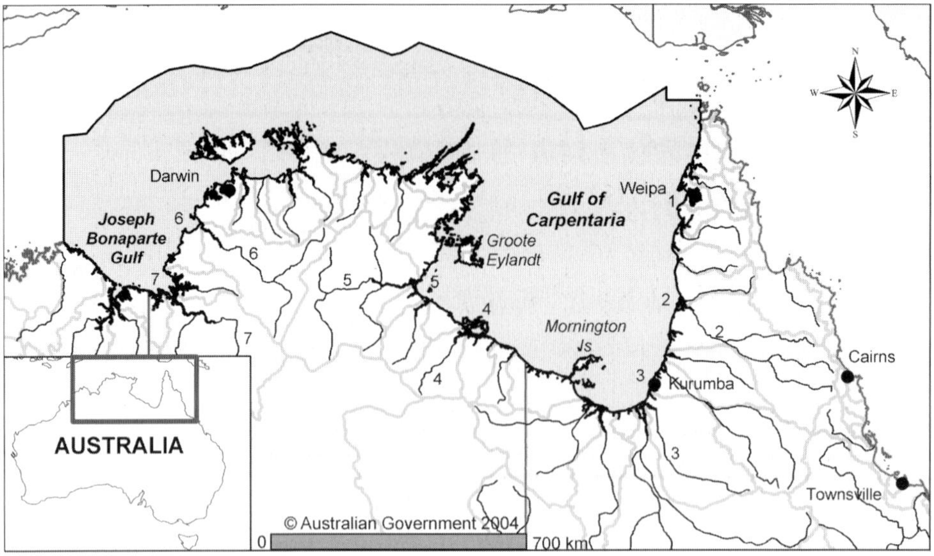

Figure 10.1 Map showing the extent of the Northern Prawn Fishery. Insert shows its position in the northern tropics of Australia (Australian Fisheries Management Authority). Also shown are catchments and some of the main river systems adjacent to the Northern Prawn Fishery shaded in grey. 1 = Embley River, 2 = Mitchell River, 3 = Norman River, 4 = McArthur River, 5 = Roper River, 6 = Daly River, 7 = Victoria River.

Commonwealth Scientific and Research Organization (CSIRO) Division of Fisheries and Oceanography.

When commercial prawn fishing began in the Gulf of Carpentaria in the early 1960s, it initially targeted only the banana prawn (*Penaeus merguiensis*) – a species that forms dense aggregations that stir the sediment, making visible mud 'boils'. Catches of *P. merguiensis* peaked at more than 12 000 MT in 1974 (Fig. 10.3) and have remained high, although they fluctuate from year to year. These large catches, together with open access to the fishery that attracted large foreign vessels in the 1960s and 1970s and government boat-building subsidies, resulted in the fishing fleet growing to about 280 vessels in the early 1980s. Furthermore, because *P. merguiensis* is short-lived and aggregates, the fishery made substantial investments in processing and targeting equipment. The increased efficiencies led to the allowable number of fishing days being reduced from year-round in the 1960s, to a few months in the 1970s, and to just over a month from the 1990s (Somers and Wang 1997).

The area fished in the NPF increased greatly during the late 1970s and early 1980s when a different suite of prawn species was captured: two valuable tiger prawn species (*Penaeus esculentus* and *P. semisulcatus*); the less valuable endeavour prawns (*Metapenaeus endeavouri* and *M. ensis*); and a minor fishery for blue-legged king prawns (*P. latisulcatus*) (see Baldwin et al. 1998 and Lavery et al. 2004 for a discussion on nomenclature). This tiger prawn fishery expanded

Figure 10.2 Typical vessels built in the 1970s (top) and in the 1980s or 1990s (bottom). Both vessels are still active in the fishery today, although the top vessel is uncommon. (Source: Bob Pendrey, CSIRO (top) and David Vance, CSIRO (bottom)).

rapidly until it, too, began to suffer from excessive capacity and declining catches in the mid-1980s. A further fishery that targets a different banana prawn species (*P. indicus*) started in the Joseph Bonaparte Gulf, about 1900 km west of the main fishing ground in the Gulf of Carpentaria (Fig. 10.1), during the early 1980s. As a result, from the late 1980s, the Northern Prawn Fishery covered the full range of commercial prawn species available in the northern region.

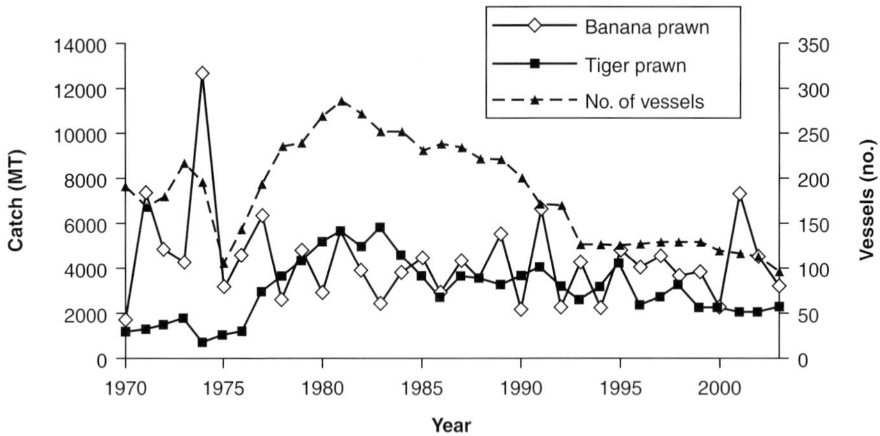

Figure 10.3 Annual banana and tiger prawn landings (MT) and the number of vessels fishing in the Northern Prawn Fishery.

Management

Input controls such as size of gear and number of fishing days, rather than output controls such as catch quotas, are the basis for management of the NPF. The two states and the territory whose waters include the NPF, initially managed the fishery. Given the different priorities of each state or the territory, there was little possibility of a unified management response for resources that occurred within more than one jurisdiction.

An Offshore Constitutional Settlement (OCS) between the governments of the States of Queensland and Western Australia, and the Northern Territory, and the Australian Federal (or 'Commonwealth') Government was signed for the NPF in 1988. The Australian Federal Government became responsible for many of the resources in the NPF that straddle state or international boundaries. The Australian Fisheries Service, which was restructured in 1992 to form the present-day Australian Fisheries Management Authority (AFMA), became responsible for the management of the NPF.

AFMA is responsible for the day-to-day management of fisheries that fall under the jurisdiction of the Australian Federal Government. The Fisheries Management Act 1991 includes five specific legislative objectives:

(1) Implementing efficient and cost-effective management on behalf of the Commonwealth (Australian Federal Government).
(2) Ensuring that the exploitation of fisheries resources and the carrying on of any related activities are conducted in a manner consistent with the principles of ecologically sustainable development, in particular, regarding the effects of fishing activities on non-target species and the marine environment.
(3) Maximizing economic efficiency in the exploitation of fisheries resources.

(4) Ensuring accountability to the fishing industry and to the Australian community in the Authority's management of fisheries resources.
(5) Achieving government financial targets in relation to the recovery of the costs of the Authority.

AFMA is a statutory authority with a board consisting of a chairperson, government director, managing director, and five nominated directors. No more than two directors can be currently engaged in fishing or fish processing (Smith et al. 1999).

The partnership approach

The Fisheries Management Act 1991 emphasizes a 'partnership approach' among fishery managers, scientists, industrial sectors such as processors and owners, conservation groups, and other stakeholders. This partnership involves close consultation, but also direct input into, and responsibility for, providing advice to the AFMA board. Two of the committees established to facilitate this process are fundamental to the successful management of the NPF: the Northern Prawn Management Advisory Committee (NORMAC), and the Northern Prawn Resource Assessment Group (NPRAG). Although these committees interact and provide advice to each other, they report directly and independently to the board of AFMA, thereby maintaining the independence of scientific and management advice. NORMAC has an independent chairperson and eight members, of which five are from industry, one from a conservation group, a scientist, and an AFMA representative. There are also two permanent observers: one representing the Northern Territory and the States of Queensland and Western Australia and, recently, one from the Commonwealth Department of Environment and Heritage; the Department responsible for independently auditing Australia's export fisheries. The NPRAG has about ten members; mostly stock-assessment, biological, and economic scientists; fisheries managers; and industry members. It is the technical committee that provides scientific advice on the status of stocks, by-catch, and the ecosystem. Enforcement is the responsibility of AFMA and is based on risk–benefit analyses that rely, amongst other methods, on satellite tracking of all vessels in the NPF, and some at-sea observer and enforcement coverage.

The second major piece of legislation affecting the operation of fishers in the NPF is the Environment Protection and Biodiversity Conservation (EPBC) Act 1999. Under this act, every export fishery in Australia must be certified every five years as being ecologically sustainable to obtain a license to export product from Australia. In its first attempt, the NPF was successfully certified under the EPBC Act in 2004. The two main principles against which the sustainability of a fishery is assessed are:

(1) A fishery must be conducted in a manner that does not lead to overfishing, or for those stocks that are overfished, the fishery must be conducted such that there is a high degree of probability the stocks will recover. In the case of the

NPF, overfishing is defined as a stock that is below the spawning stock size that corresponds to maximum sustainable yield (S_{MSY}) and is therefore treated as a limit reference point below which the resource should not fall.

(2) Fishing operations should be managed to minimize their effect on the structure, productivity, function, and biological diversity of the ecosystem.

A complex ecosystem

Prawns grow rapidly and most reach commercial size and reproductive maturity at the age of six months, although they take 9 to 12 months to reach the larger, more valuable, sizes. In the life cycle of many of the commercial penaeids in northern Australia, the subadults and adults live in offshore waters, while the postlarvae and juveniles are found in the intertidal and shallow subtidal waters of estuaries and coastal waters (Fig. 10.4; Dall et al. 1990). The combination of tidal currents and the behavior of the larvae and postlarvae provides the mechanism for the larvae to move to the inshore and estuarine nursery habitats (Rothlisberg et al. 1996; Condie et al. 1999), while the juvenile emigration from these habitats is size-related in some species, but in banana prawns is stimulated by the amount of rainfall and reduction of salinity in estuaries (Staples and Vance 1986).

Rainfall in northern Australia is dominated by the summer monsoons, with about 90% of rainfall occurring between December and April and effectively no rainfall

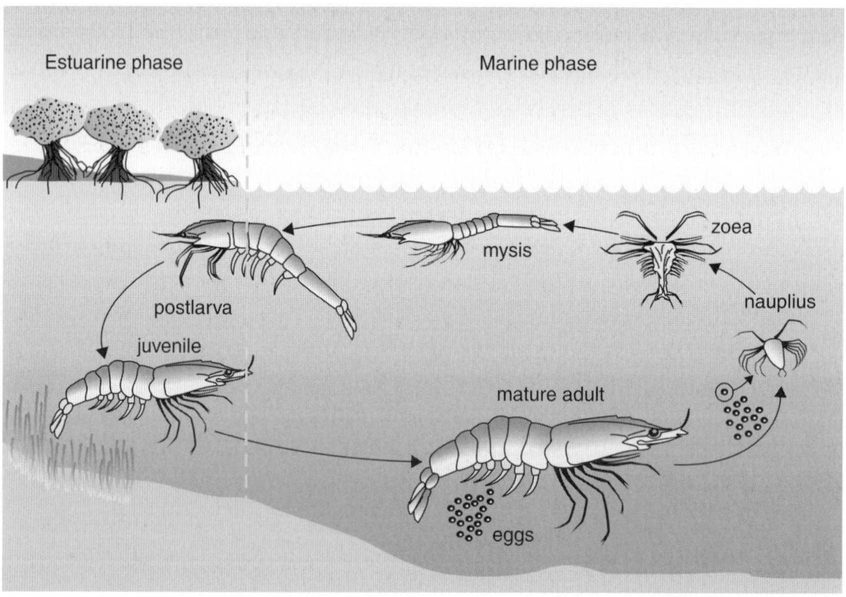

Figure 10.4 The life cycle of the prawns caught commercially in the Northern Prawn Fishery showing inshore nursery habitats and offshore spawning grounds and larval habitat. (Drawing by Louise Bell, CSIRO Marine Research.)

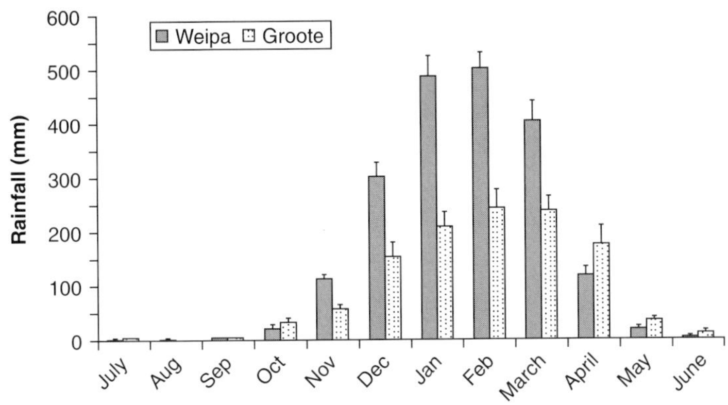

Figure 10.5 Mean annual rainfall (± 1 SE) for Weipa in the northeastern Gulf of Carpentaria and for Groote Eylandt in the western Gulf of Carpentaria for the period between 1970 and 2000. (Data provided by the Bureau of Meteorology.)

between June and September (Fig. 10.5; Vance et al. 1985; Gillanders and Kingsford 2002). The influence of rainfall on the estuaries and coastal waters is moderated by the size of the catchments, and can vary greatly around the Northern Prawn Fishery (Fig. 10.1). For example, rivers in the northeastern Gulf of Carpentaria are relatively short, with small catchments. The size of both the river systems and catchments increases towards the southeast. Consequently, in high rainfall years, the effects of rainfall on salinity in the estuaries and coastal waters of the Gulf of Carpentaria are greater and last longer in the southeast (Vance et al. 1998). The distribution of seagrass, the critical settlement and juvenile habitat for tiger prawns, may also be affected by the interaction of catchment size and rainfall. It has been shown in studies elsewhere that seagrass abundance has decreased following increased freshwater streamflow, possibly mediated by increased turbidity and therefore decreased light penetration (Gillanders and Kingsford 2002). Thus seagrasses have a very limited distribution along the eastern Gulf of Carpentaria and increase in coverage and diversity in the western Gulf, where rainfall is low and catchments are smaller (Fig. 10.6) (Poiner et al. 1987, 1989; Poiner and Peterken 1995). There is a link between high rainfall and banana prawn catch, although this connection is moderated by the interaction between rainfall and catchment size (Vance et al. 1998).

Coastal habitats and prawn production

In the NPF region, seagrass and algae provide the critical nursery habitat for tiger prawns *Penaeus esculentus* and *P. semisulcatus* (Loneragan et al. 1998), whereas juvenile banana prawns, *P. merguiensis* and *P. indicus*, are found along mangrove-lined mudbanks (Vance et al. 1996; Kenyon et al. 2004a). Western king prawns *P. latisulcatus* are associated with sandy substrates (Potter et al. 1991). In general,

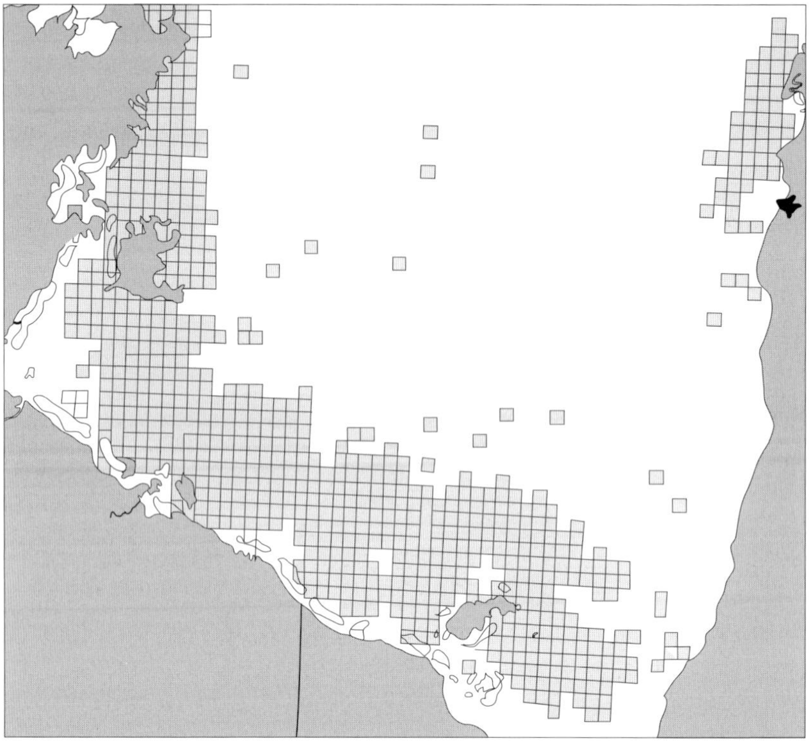

Figure 10.6 Distribution of seagrass in the Gulf of Carpentaria, shown by lines close to the coast, (Poiner et al. 1987) and the grids where the average annual catch of tiger prawns exceeds 150 kg.

most postlarvae and juveniles recruit to inshore waters between October and December, and February and April. The juveniles recruit into the fishery fairly continuously over many months except during the winter months from June to September (Dall et al. 1990; Vance et al. 1996, 1998).

The significance of coastal habitats to the productivity of the NPF was recognized very early in the fishery by researchers, managers, and fishers. As a result, trawling has been banned in the rivers of Queensland since the 1960s and over seagrass beds in the inshore waters since the early 1970s (Taylor 1994). Research on understanding the relationship between coastal habitats and prawn populations has been supported and encouraged by industry and managers since the 1970s and is continuing. The managers of the fishery have valuable information on the distribution of juvenile prawns in relation to different coastal habitats (Loneragan et al. 1998; Vance et al. 2002), the effect of the loss of coastal habitats on juvenile prawn populations and the fishery (Rothlisberg et al.1988) and the processes that affect juvenile prawns in their coastal nursery habitats (Vance et al. 2002). The partnership approach and the appreciation of the significance of coastal habitats to fishery production led to closure of seagrass beds and inshore waters to trawling from the early 1970s

(Kenyon et al. 2004b). Closures are enforced using a satellite tracking system known as a vessel monitoring system.

The environment of the fishery

Depth and sediment type influence the distribution and abundance of commercial-sized penaeids (Dall et al. 1990). In general, most catches of *P. merguiensis, P. esculentus*, and *M. endeavouri* are taken in depths from 20–40 m, those of *P. semisulcatus* and *M. ensis* in 30–50 m, while *P. indicus* is fished at 50–80 m in the Joseph Bonaparte Gulf (Somers 1994a, b). The proportion of mud in the sediments influences distribution: *Penaeus indicus, P. merguiensis*, and *P. semisulcatus* are found in areas with a higher proportion of mud (65–100%) than *P. esculentus* and *M. endeavouri* (20–80% mud); *M. ensis* spans a range of sediment types (30–95% mud); and *P. latisulcatus* is found on sandier substrates (<50% mud) (Somers 1994a, b).

The inshore waters of the Gulf of Carpentaria are mainly turbid and muddy. The offshore waters are generally clear over soft sediments, where patches of species-rich hard-bottom communities dominated by sponges and corals are found. The hard substratum prevents these patches in the NPF from being trawled (Hill et al. 2002). A survey of the Gulf of Carpentaria in May 2004 located several of these patches in the southeastern Gulf and found extensive epibenthic communities of sponges, corals, and other species living at depths of about 20 m. The plotters that fishers use to record information on trawls show that the distribution of these untrawlable grounds varies greatly throughout the Gulf, ranging from about 2% of the total area in the Weipa region to about 40% in the north of Mornington and Vanderlin Islands in the southern Gulf (Die et al. 2001). The geological history of these hard-bottom areas is unknown. Apart from these untrawlable grounds, trawlers target areas of high prawn abundance. As a result, only about 25% of the NPF is trawled and a much smaller area is trawled heavily (>100 days of trawling each year) (Die et al. 2001).

Demonstrating the sustainability of the fishery requires knowledge of the effects of trawling on the prawns, the environment, and other species caught in trawl nets. The NPF covers about 6000 km of coastline and 15 marine bioregions identified in the Interim Marine Comprehensive Representative Area process (IMCRA 1998). Little is known of the benthic flora and fauna, particularly outside the Gulf of Carpentaria. Of 16 research surveys carried out in the NPF (Hill et al. 2002), all except two were targeted at prawns and not the rest of the marine community. NORMAC has encouraged research in 15 of the bioregions and to assess the areas potentially designated as marine protected areas, particularly their effect on the fishery. This support resulted in a major survey in March 2005 to collect simultaneous data on the benthic habitats, associated fauna, and on the hydrographic environment in the Gulf of Carpentaria.

Sustainability of target species

Assessment advice

Simulation modeling techniques have been used from the 1970s to identify the opening dates of the banana prawn fishing season such that the value per recruit is maximized (Lucas et al. 1979; Somers 1985, 1990; Somers and Wang 1997). Much of the field research to support this model was based on extensive interannual pre-season sampling in several stock areas of the NPF. This work would not have been possible without cooperation of industry, which provided vessels at no charge (Somers and Taylor 1990). Stock assessments of the tiger prawns showed that it was overexploited from the 1990s (Table 10.1; Wang and Die 1996; Dichmont et al. 2003a). Subsequent assessments of the tiger prawn resource changed from being simplistic to being very complex and computer intensive. The early model used for management was an equilibrium surplus production model (Somers 1990), which was replaced by a fully age-based (Wang and Die 1996) or biomass dynamic model (Haddon 2001), and this ultimately led to a model that keeps track of spawning stock and recruitment size (without estimating the full age structure) through a delay-difference model (Dichmont et al. 2003a). Even though the models have different assumptions and levels of complexity, they consistently found that the tiger prawn resource was overexploited.

Perhaps the most influential scientific discovery for management was the finding that there is a strong stock–recruitment relationship for tiger prawns. This means that the expected level of recruitment declines with declining spawning-stock size (Wang and Die 1996). This was unexpected, for the prevailing view was that prawn fishing, no matter how intense, would not affect recruitment and the future productivity of the resource. This finding in the NPF combined with the collapse of the tiger prawn resource in Exmouth Gulf in 1982–1983 (Penn and Caputi 1986), increased the acceptance of the stock–recruitment relationship, particularly by the scientific community. A recent meta-analysis of published data on tiger prawn populations throughout the world found that prawn recruitment was in fact related to the abundance of spawners, and prawn fisheries should, therefore, be managed to maintain sufficient mature adults to yield high recruitment (Ye 2000).

History of effort reductions

Management of the NPF since the 1980s has been characterized by attempts to restructure the fishery and reduce effort. This was driven by the substantial increases in effort when new technologies were adopted as well as by declining catches. Several approaches to reducing the size of the fleet (Table 10.1) led to a marked drop in the number of vessels in the fishery (Fig. 10.3). This did not, however, equate to a reduction in fishing mortality (Bishop et al. 2000; Cartwright 2005), because of the inequity in vessel efficiency and the fleet's increased ability to locate and catch prawns in a day, through for example, increases in on-board technology

Table 10.1 Scientific advice and management development (with intended target reduction if stated) in the Northern Prawn Fishery, 1980–2004.

Year	Scientific advice and management development	Target reduction (if any)
1980	Introduction of limited entry.	
1985	CSIRO presents data showing a decline of the brown tiger prawn stocks and recommends a 25% reduction in effort. Statutory fishery rights are granted in the form of A-units (a combination of hull dimensions and engine horse power). Voluntary buy-back scheme introduced (tends to reduce only latent effort).	
1987	Reduction from quad to twin gear, mid-year closure, ban on daylight trawling during the tiger prawn season.	30% reduction of effort
1988	Restriction on head-rope length of nets.	
1990	Voluntary industry-funded buy-back scheme with loans from the government (failed to reduce the fleet to the target; actual reduction by target date was 72 000 units (a reduction of 172 trawlers).	50 000 A-units (i.e. fewer than 130 vessels) by 1 April 1993
1993	Compulsory, industry-funded, buy-back scheme (reduces the fleet to 137 (128 active) vessels). Removal of net-size restrictions (but use of double gear only remains).	Across the board reduction of 30% of the remaining A-units
1995	Start of annual assessments with an age-based model developed by CSIRO for consideration by the NPRAGs. NPRAG declares effective effort to be too high.	
1996	Fishing power estimated to be increasing at 2–5% per year (the stock assessment used 5% p.a. as agreed by NORMAC). A Gear Units Workshop (attended by industry, scientists and managers) considers the idea of gear units as a management tool. First attempt by NPRAG to introduce biological reference points; deferred by the NORMAC 39 meeting to 1997 for further discussion.	
1997	NPRAG advises that both tiger prawn stocks are overexploited and recommends immediate reduction of effort by at least 10%. Independent tiger prawn stock assessment confirms NPRAG advice.	Season change intended to decrease effort by 10% (i.e. a net of 5% after the 5% increase in fishing power) and similarly for 1998
	NORMAC recommends that the fishery be closed 3 weeks earlier at the end of the year and during the mid-season closure for 1998; also recommends issuing 15% fewer gear units when they are introduced in 1999. AFMA board accepts advice (although gear units are introduced in 2000).	

Table 10.1 *Continued.*

Year	Scientific advice and management development	Target reduction (if any)
1998	Mandatory introduction of satellite-based vessel monitoring system across the fleet. NPRAG advises that spawning stocks are well below target levels and that rebuilding of stocks 'requires significant and urgent efforts'. Some sectors of industry dispute the advice. The end of season closure reverts to the end of November and a large area closure is implemented on 1 November. Effort at maximum sustainable yield (EM_{SY}) becomes the target reference point when gear units are introduced.	
1999	NPRAG advises that effort in 1998 was 35% greater than EM_{SY}, that both tiger prawn species remain overexploited, and that effort needs to reduced by 35%. NORMAC recommends replacing the spatial closure implemented in 1998 by larger mid-year and end of year closures. Allocation Advisory Panel investigates the translation formula used in the gear unit system. Bycatch Action Plan is released.	15% reduction
2000	Australian Senate Inquiry endorses gear-based management. Gear-based management starts in July and gear is reduced by 15%. NPRAG reports that seasonal closures have reduced effort, but that stocks are still overexploited and declining. AFMA Board and Minister write to NORMAC insisting on reductions in effort, TEDs become compulsory in the tiger prawn fishery.	10% reduction in effort
2001	CSIRO develops a new model for tiger prawn stocks. Technical review of stock assessment models. NPRAG considers tiger prawn stocks are overexploited (brown tiger prawns are below 50% of their target level). Minister calls meeting of fishers and local politicians to demand action. NORMAC sets a new target to rebuild tiger prawn stocks by 2006. NORMAC agrees to reduce gear units by 25% and institute seasonal changes in July 2002. TEDs made compulsory in banana prawn fishery.	40% effort reduction on brown tiger prawns and 25% on grooved tiger prawns
2003	Stocks of both grooved and brown tiger prawn show signs of recovery. Effort-neutral trade-offs between gear, season and spatial closures are investigated but do not result in management changes.	
2004	NPRAG declares that both grooved and brown tiger prawn stocks are recovered, although with caution, as survey results do not clearly support this.	
2004/05	NORMAC changes the target to maximum economic yield; a lower effort level than the recovery target set in 2001 and further reduces gear units by 25%.	

and vessel size (Dichmont et al. 2003b). As early as 1987, Buckworth (1987) quantified substantial increases in vessel fishing power, which continued through the 1990s, particularly with the advent of color sounders and global positioning systems (GPS) (Robins et al. 1998; Bishop et al. 2000; Dichmont et al. 2003b). In the early years, total effort did not decline when vessels left the fishery as those remaining simply increased the number of hours they fished. Later, the introduction of new technologies increased fishing effort. Certain sections of the industry were reluctant to accept these findings. Considerable research was done to quantify the extent of the change.

In 1985, each vessel was assigned a number of 'A-units', which is a transferable statutory fishing right (a form of individual property right) based on vessel volume and engine horse power, and effort reductions were based on these units. This system was inflexible, however, because even small A-unit reductions could result in boats being unable to fish and actually impeded restructuring the fleet. Eventually, both industry and management agreed that the A-unit management system had to be changed. Consequently, NORMAC formed a working group in 1992 to investigate alternative management systems, and AFMA and NORMAC embarked on extensive consultations in 1993 to determine the new form of management.

Scientists expressed concern about the status of the tiger prawn stocks in the NPF throughout the late 1990s, declaring that the tiger prawn resources were biologically overexploited every year from 1997 (Die et al. 1997) and noting that fishing power was increasing rapidly (Dichmont et al. 2003b). For example, the use of GPS and plotters to target prawns was shown to increase fishing power by 12% over three years (Robins et al. 1998). Scientists expressed their concerns at NORMAC meetings from 1995 (Table 10.1). The reductions in fishing effort implemented in the 1990s were, however, less than those advised by the NPRAG and were often partly offset by increases in vessel fishing power. Advice, based on stock assessments, included effort reductions required to reach NORMAC's 1990s maximum sustainable yield (MSY) target. Decisions on major effort reductions were delayed pending the change from one management system to another. For example, the NPRAG advised that effort on tiger prawns should be reduced immediately by 35% in 1998, but the suggested three week closure reduced fishing effort by only an estimated 15%.

There was general agreement in the industry – and unanimous agreement at NORMAC in November 1996 – that management should change to a system based on tradable gear rather than A-units, where the number of gear units a vessel has defines the amount of head-rope and therefore roughly the width of the trawl net it can use. The industry's consensus disappeared, however, because of opposition by operators who considered that they would be disadvantaged by the new formula (Cartwright 2005). These operators, who owned smaller vessels or smaller fleets, also disputed the results of the stock assessment; especially that fishing power was increasing. Instead, they argued that changes in the catch rates were largely attributable to environmental factors, predation, or the reduced fleet being too small to search all the prawn grounds effectively (Cartwright 2005). The dispute within the industry led to a senate inquiry that ultimately supported the shift to gear units and the simultaneous reduction in effort. An international review of the stock assessment methods supported the conclusion based on the stock-assessment advice that the resource was overexploited (Deriso 2001).

In the early 2000s, three major changes were made to the management of the fishery:

(1) The gear-unit system, based on head-rope length, was implemented and immediately reduced effort by 15%.
(2) Scientific advice on the need for a further substantial reduction in fishing mortality was accepted – the effort on *P. esculentus* was cut by 40% and on *P. semisulcatus* by about 25%.
(3) An annual independent monitoring program that was financed mainly by the industry aimed at creating a recruitment and spawning index of abundance was implemented (Dichmont et al. 2004).

These reductions were implemented in 2001 by shortening the total head-rope length by 25%. In addition, changes were made in the length of fishing seasons to reduce fishing effort on tiger prawns. Today, there are indications of a recovery of the tiger prawn stock. Management also changed its objectives from setting the level of fishing effort to achieve the MSY, to managing so that there is at least a 70% probability that the size of the spawning stock is above that required to achieve the MSY. Furthermore, the fishery invested heavily in a large-scale annual independent monitoring program (Dichmont et al. 2004). In 2005, management again changed its objectives to achieving maximum economic yield, which has resulted in a further gear reduction of 25%.

Prawns are not the only catch

The effect of fisheries on their incidental and unwanted catch, or by-catch, has become a growing and major issue in the last few decades (Andrew and Pepperell 1992; Alverson 1994; Kennelly 1995). All trawl nets used to catch prawns have small mesh and are towed along a biologically-diverse seabed. Consequently, this unselective gear often catches a large and diverse suite of non-target species. Alverson (1994) estimated that around one-third of the world's discards come from prawn trawl fisheries. The by-catch consists mostly of small fish with no value to an industrial fishery. The resulting large quantities of discards and the effect of trawls on endangered or vulnerable and often charismatic species, including turtles, sharks, rays, sawfish and seahorses, have given prawn trawl fisheries a poor reputation.

The NPF has a variety of by-catch species, usually making up to 90% of the catch (Brewer et al. 1998; Stobutzki et al. 2001a). These species include:

- Large quantities of small teleost fish that are between 87.5% and 95.2% of the catch by weight and include >300 species (Pender et al. 1992; Blaber et al. 1997)
- 6 species of marine turtles

- >50 species of elasmobranchs (Stobutzki et al. 2001a)
- 13 species of sea snakes (Milton 2001)
- 230 benthic invertebrate taxa (Stobutzki et al. 2000)

Many of these species are dead when discarded, or have a poor chance of survival (Hill and Wassenberg 1990, 2000). Some are endangered, including five of the six species of marine turtles caught, or are protected by Australian law, such as sea snakes, sea horses, and pipefishes. Others, mostly elasmobranchs, have life histories that make them highly vulnerable to population declines with any increase in mortality rates (Last and Stevens 1994).

The high volume and diversity of the by-catch of this fishery continue to present fishery operators, managers, and researchers with significant challenges. The Australian Fisheries Management Act 1991 and the Environmental Protection and Biodiversity Conservation Act 1999 require that negative effects on endangered species are prevented, catches of non-target species are reduced, and the long-term sustainability of by-catch and by-product populations is demonstrated. The scale and complexity of these challenges has required a forward-thinking, innovative and coordinated approach by researchers, managers, and fishers.

By-catch research and management

The partnership approach to decision-making in this fishery has meant that managers have actively sought knowledge and implemented by-catch avoidance initiatives. The balance of views from the different NPF committees has provided a well-considered, logical, and forward-thinking approach to by-catch and other ecological issues. These initiatives include establishing a broadly-representative NORMAC subcommittee to advise specifically on by-catch issues, currently known as the Fishing Impacts and Solutions Working Group. This group's support has resulted in the successful funding of a range of programs to improve knowledge of by-catch and related issues (Table 10.2), and the instigation of a new blueprint – the NPF's by-catch action plans – to guide the direction of by-catch issues into the future. The use of by-catch action plans is now widespread in Australian fisheries, with another nine Commonwealth-managed fisheries adopting similar plans.

Some initiatives aimed at managing prawn stocks have indirectly, but significantly, reduced the effects on by-catch. For example, effort reductions since the mid-1980s have reduced the annual number of boat days from 34 000 to the current level of 12 500. There has been a concurrent reduction in area fished from 1240 six-minute grids in 1984 to 804 in 2000. Although not quantified, this greatly reduced effort is likely to have improved the sustainability of by-catch populations in the Gulf of Carpentaria and other regions in the NPF. Perhaps the most significant management initiative relating directly to by-catch was the mandatory introduction of turtle excluder devices (TEDs) and by-catch reduction devices (BRDs) in 2000 (Fig. 10.7). TEDs are, in their most common form, inclined hard grids (Fig. 10.8) that guide any large object or organism out of the trawl before it reaches the net's

Table 10.2 Recently completed or current projects focused on by-catch and other non-target species that have been supported by the NPF's Management Advisory Committee, and the involvement of the NPF industry.

Project title	Industry involvement
Monitoring the catch of turtles in the Northern Prawn Fishery, 1998–2002.	Data collected by voluntary industry observers
Effects of trawl design on by-catch and benthos in prawn and finfish fisheries; 1993–96.	
An assessment of the TED performance in the NPF banana prawn fishery – final report.	Data collected by scientific observers and industry crews on industry vessels
Quantifying the effects of trawling on the seabed biota of the NPF; 2002–2005.	
Surrogates I – Predictors, impacts, management and conservation of the benthic biodiversity of the Northern Prawn Fishery; 2000–2002.	GPS data from industry volunteers to define untrawlable ground
Commercialization of by-catch reduction strategies and devices within northern Australian prawn trawl fisheries; 1996–1999.	Gear technologist providing assistance on TED and BRD operation on NPF vessels
Ecological sustainability of by-catch and biodiversity in prawn trawl fisheries; 1996–1999.	Data collected by scientific observers on industry vessels
Assessment and improvement of BRDs and TEDs in the NPF; a cooperative approach by fishers, scientists, fisheries technologists, economists and conservationists; 2000–2002.	Data collected by scientific observers on industry vessels
Design, trial and implementation of an integrated, long-term by-catch-monitoring program, road tested in the NPF. 2002–2005	Data collected by voluntary industry observers

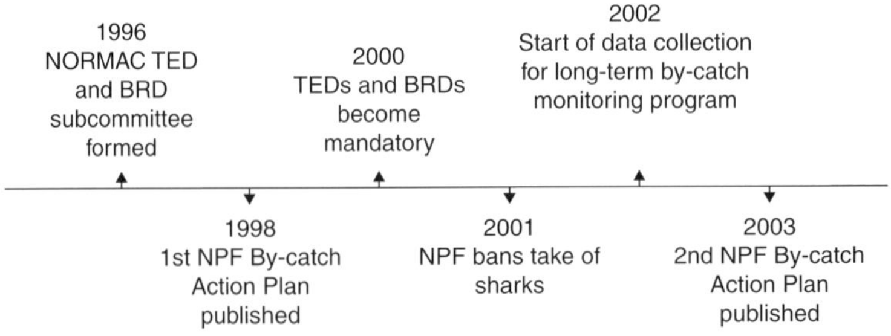

Figure 10.7 Time series of management actions to reduce the NPF's effect on by-catch species.

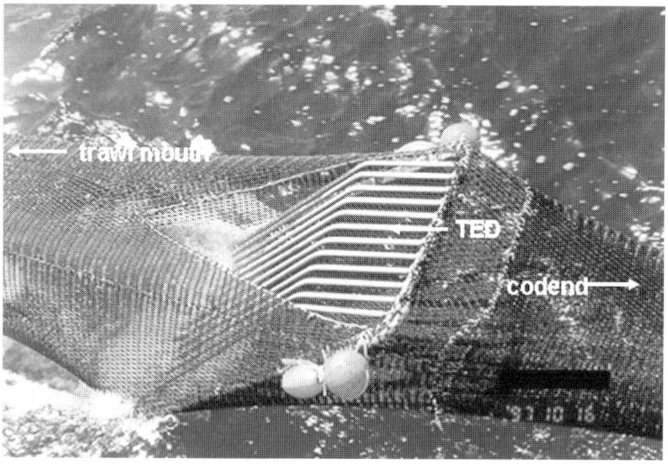

Figure 10.8 Turtle excluder devices (TEDs) have been mandatory in trawls in Australia's Northern Prawn Fishery since 2000. They have dramatically reduced catches of endangered marine turtles and vulnerable species of sharks, rays, and sawfish.

end. Their introduction has resulted in a 99% reduction in catches of marine turtles from 5730 ± 1907 per year from 1979–1988, to fewer than 100 in 2001, and of turtle drownings from 344 ± 125 to fewer than 10 (Poiner et al. 1990; Brewer et al. 2004). The catch of large elasmobranchs (sharks >1 m long or rays >1 m wide) has also been reduced by 91% since the introduction of TEDs (Brewer et al. 2004).

The implementation of this gear change followed two large NORMAC research projects that investigated the performance of different TEDs and BRDs to determine the most effective devices (Brewer et al. 1998) and, in particular, to help with the installation and effective use of TEDs (Robins et al. 2000) (Table 10.2). This research ensured that the prawn catch was reduced by only 3–6% when TEDs and BRDs became mandatory in 2000, which would have been the major contributing factor to the high compliance in the installation and use of TEDs by NPF fishers. The United States of America has placed import embargos on countries not using TEDs and the early success of NPF in 2001 has ensured that prawn exports have not been embargoed.

Other projects supported by NORMAC have increased the understanding of the interaction between the fishery and incidentally-affected populations. For example, studies have included: (1) the effects of trawling on the sessile seabed biota undertaken between 2002 and 2005; (2) a study of the sustainability of the by-catch and biodiversity undertaken between 1996 and 1999; and (3) baseline descriptions comparing by-catch composition of the tiger and banana prawns (Table 10.2) (Stobutzki et al. 2001a). Research such as this has enabled the managers and the industry to make informed environmental decisions. The research has also led to the development of new methods to assess the risk to the sustainability of by-catch species populations from trawling (Stobutzki et al. 2001b, 2002), which have been adopted by other fisheries (Hobday et al. 2004).

Some of the key decisions made by the NPF have been based on information from other fisheries and other sources of scientific data. For example, in 2001 NORMAC banned the retention of any elasmobranchs, and as with other management changes in the NPF, the compliance by fishers to this ban has been very high. This decision was based mainly on the growing body of evidence that these species are highly vulnerable to many forms of fishing pressure (Last and Stevens 1994) and that the thriving trade in their fins was seriously depleting shark populations (NOAA 2002). The NPF's decisions and actions to reduce by-catch have allayed the criticism of the industry, paved the way for positive assessments of the NPF and led to a product export approval by the Australian government's primary environmental auditing agency in 2004.

Cooperation, participation, and compliance

The partnership approach to management that includes fishing company managers and fishers on committees has promoted an active role for industry in research and monitoring programs. Participation in programs with an ecological focus has helped NPF fishers to understand modern fisheries management and lowered barriers to their participation in sampling and monitoring. As an example, many of the research projects listed in Table 10.2 have depended on the investment and voluntary participation of fishers, which has often included attending training programs to learn how to identify by-catch species and collect scientific data.

The value of industry participation in management is also demonstrated in the fishery's high rate of compliance with management changes to their fishing operation. For example, initially, the introduction of TEDs in 2000 was not widely welcomed by NPF fishers. Nonetheless, despite relatively little enforcement, the industry has almost unanimously and effectively implemented these devices (Robins et al. 2003).

Future challenges

The NPF's management processes and responsible attitude towards ecological issues has helped to develop a proactive and ecologically responsible culture amongst its fishers. Perhaps the most impressive feature of this progress has been the willingness to look for solutions and to fund programs that are ground-breaking. Current initiatives aim to establish new, more accurate, risk assessments for by-catch species and a long-term program that will monitor the status of by-catch populations. Fishers are currently working in a program to continually improve BRDs to reduce the quantity of incidental small fish and invertebrates. There remain challenges: to establish marine protected areas; to gain a better understanding of the unseen effects of trawling on sessile benthic communities; and the management of species at significant risk, notably sawfish and sea snakes. The NPF has a good record in dealing with difficult issues and the management processes it has developed should ensure future success.

Summary

Management performance

The partnership approach of AFMA has been a great success, but it has been hard and challenging work. It has helped produce science that is relevant to industry and management and is used internationally. It has responded effectively to environmental issues, such as protecting major nursery areas, protecting recruits and spawning prawns, and also reducing the by-catch of many vulnerable species. In addition, the partnership approach has produced clearly defined objectives for the fishery (NPF Strategic Plan and the By-catch Action Plan) and strategic directions for research (Five-year Strategic Research Plan) that has resulted in a good balance between tactical and strategic research and management.

There can be a downside to the partnership approach. When management decisions affect the economic viability and an individual fisher's tenure in the industry, debate can become heated and decisions can be delayed or stalled altogether. For example, there was about a seven year lag between the times when concerns about tiger prawn stocks were first raised to the time when decisive management action was taken. Even so, while there was still no consensus on the status of the stocks, the decision to reduce effort was implemented. The delay in the implementation of effort reductions had the unfortunate consequence of large effort reductions that were disruptive to the industry. Should similar cases occur, AFMA should not expect consensus, but should either give NORMAC a clear ultimatum or take decisive action itself.

Vision for the future

The future will be strongly linked to the partnership approach. Long-term prawn and by-catch monitoring programs have been established and our understanding of ecosystem function is steadily improving. Ecosystem-based fisheries management is difficult for scientists, managers, and industry to implement, but is more likely to be implemented successfully within the partnership approach. In reality, the ultimate goals of fishers, managers, and industry are not always very different. For example, achieving independent, international certification for ecological sustainability requires that management be ecosystem-based, backed by science, and proactive decision-making is made. Management by a partnership approach of all stakeholders is most likely to achieve a viable industry with a strong future.

References

Alverson, D.L., Freeberg, M.H., Murawski, S.A. and Pope, J.G. (1994) *A Global Assessment of Fisheries Bycatch and Discards.* FAO Fisheries Technical Paper, 339.

Andrew, N.L. and Pepperell, J.G. (1992) The by-catch of shrimp trawl fisheries. *Oceanography and Marine Biology: Annual Review*, **30**: 527–65.

Baldwin, J.D., Bass, A.L., Bowen, B.W. and Clark, W.H. (1998) Molecular phylogeny and biogeography of the marine shrimp *Penaeus. Molecular Phylogenetics and Evolution*, **10**: 399–407.

Bishop, J., Die, D. and Wang, Y-G. (2000) A generalized estimating equations approach for analysis of the impact of new technology on a trawl fishery. *Australian and New Zealand Journal of Statistics*, **42**: 159–77.

Blaber, S., Brewer, D., Burridge, C., Farmer, M., Milton, D., Salini, J., Wang, Y-G., Wassenberg, T., Buxton, C., Cartwright, I., Eayrs, D., Rawlinson, N., Buckworth, R., Gill, N., MacCartie, J., Mounsey, R. and Ramm, D. (1997) *Effects of Trawl Design on Bycatch and Benthos in Prawn and Finfish Fisheries.* FRDC Project 93/179, Final Report.

Brewer, D.T., Rawlinson, N., Eayrs, S. and Burridge, C.Y. (1998) An assessment of Bycatch Reduction Devices in a tropical Australian prawn trawl fishery. *Fisheries Research*, **36**: 195–215.

Brewer, D.T., Heales, D.S., Eayrs, S.J., Taylor, B.R., Day, G., Sen, S., Wakeford, J., Milton, D.A., Stobutzki, I. C., Fry, G.C., van der Velde, T.D., Jones, P.N., Wang, Y-G., Dell, Q., Austin, M., Carter, D., Nelson, C., Nichols, J. and Gofton, T. (2004) *Assessment and Improvement of TEDs and BRDs in the NPF: a Co-operative Approach by Fishers, Scientists, Fisheries Technologists, Economists and Conservationists.* FRDC Project 2000/173, Canberra.

Buckworth, R.C. (1987) Changes in fishing effort and catching power in the DMZ tiger prawn fishery. *Northern Prawn Information Notes, February 1987*, CSIRO Division of Fisheries, Cleveland, Australia.

Cartwright, I. (2005) The Australian Northern Prawn Fishery. In: S. Cunningham and T. Bostock (Eds) *Successful Fisheries Management: Issues, Case Studies and Perspectives*. Eburon Academic Publishers, Netherlands.

Condie, S.A., Loneragan, N.R., Die, D.J. (1999) Modeling the recruitment of tiger prawns (*Penaeus esculentus* and *P. esculentus*) to nursery grounds in the Gulf of Carpentaria, northern Australia: implications for assessing stock-recruitment relationships. *Marine Ecology Progress Series*, **178**: 55–68.

Dall, W., Hill, B.J., Rothlisberg, P.C. and Staples, D.J. (1990) The biology of the Penaeidae. *Advances in Marine Biology*, **27**: 1–489.

Deriso, R. (2001) *A Review of the 2001 Assessment of Tiger Prawns in the Northern Prawn Fishery*. Report to AFMA.

Dichmont, C.M., Punt, A.E., Deng, A., Dell, Q. and Venables, W. (2003a) Application of a weekly delay-difference model to commercial catch and effort data for tiger prawns in Australia's Northern Prawn Fishery. *Fisheries Research*, **65**: 335–50.

Dichmont, C.M., Bishop, J., Venables, W.N., Sterling, D., Rawlinson, N. and Eayrs, S. (2003b) *A New Approach to Fishing Power Analyses and its Application in the Northern Prawn Fishery*. AFMA Research Fund R99/1494.

Dichmont, C.M., Vance, D., Burridge, C., Pendrey, B., Deng, A., Ye, Y. and Loneragan, N. (2004) *Designing, Implementing and Assessing an Integrated Monitoring Program for the NPF.* FRDC Project Report 2002/101.

Die, D., Wang, Y-G. and Bishop, J. (1997) *1996 Tiger Prawn Stock Assessment*. NPFAG Working Paper, 97/2.

Die D., Loneragan, N., Haywood, M., Vance, D., Manson, F., Taylor, B. and Bishop, J. (2001) *Indices for Recruitment and Effective Spawning for Tiger Prawns Stocks in the Northern Prawn Fishery*. FRDC Project 1995/014.

Galeano, D., Langenkamp, D., Shafron, W. and Levantis, C. (2004) *Australian Fisheries Surveys Report 2003*. ABARE, Canberra.

Gillanders, B.M. and Kingsford, M.J. (2002) Impact of changes in flow of freshwater on estuarine and open coastal habitats and the associated organisms. *Oceanography and Marine Biology: an Annual Review*, **40**: 233–309.

Haddon, M. (2001) *Modeling and Quantitative Methods in Fisheries*. Chapman and Hall/CRC Press, Boca Raton.

Hill, B.J. and Wassenberg, T.J. (1990) Fate of discards from prawn trawlers in Torres Strait. *Australian Journal of Marine and Freshwater Research*, **41**: 53–64.

Hill, B.J. and Wassenberg, T.J. (2000) The probable fate of discards from prawn trawlers fishing near coral reefs. A study in the northern Great Barrier Reef, Australia. *Fisheries Research*, **48**: 277–86.

Hill, B.J., Haywood, M.D.E., Gordon, S.R., Condie, S.A., Ellis, N., Tyre, A.J., Vance, D.J., Dunn, J.R., Mansbridge, J., Moeseneder, C., Bustamante, R., Pantus, F. and Venables, W. (2002) *Surrogates 1 – Predictors, Impacts, Management and Conservation of the Benthic Biodiversity of the Northern Prawn Fishery*. FRDC Project 2000/160, CSIRO Division of Marine Research, Cleveland, Queensland.

Hobday, A., Smith, A., Stobutzki, I., Bruce, B., Bustamante, R., Butler, A., Chesson, J., Deng, R., Dennis, D., Garvey, J., Hartmann, K., Knuckey, I., Ling, S., Manson, F., Milton, D., Pitcher, R., Sainsbury, K., Stevens, J., Walker, T., Wayte, S., Webb, H., Wise, B. and Young, J. (2004) *Ecological Risk Assessment for Australian Commonwealth Fisheries: Final Report – Stage 1. Hazard Identification and Preliminary Risk Assessment.* AFMA Report R01/0934.

IMCRA (1998) Interim Marine and Coastal Regionalisation for Australia: an ecosystem-based classification for marine and coastal environments. In: R. Thackway and I.D. Cresswell (Eds) *Environment Australia*. Department of Environment, June 1998.

Kennelly, S.J. (1995) The issue of bycatch in Australia's demersal trawl fisheries. *Reviews in Fish Biology and Fisheries*, **5**: 213–34.

Kenyon, R.A., Loneragan, N.E., Manson, F.J., Vance, D.J. and Venables, W.N. (2004a) Distribution of juvenile red-legged banana prawns (*Penaeus indicus*) and juvenile white banana prawns (*Penaeus merguiensis*) among nursery habitats, and proximity of the offshore fishery, in the Joseph Bonaparte Gulf, north-west Australia. *Journal of Experimental Marine Biology and Ecology*, **309**: 79–108.

Kenyon, R.A., Jarrett, A., Bishop, J., Taranto, T., Dichmont, C. and Zhou, S. (2004b) *Documenting the History of and Providing Protocols and Criteria for Changing Existing or Establishing New Closures in the NPF*. Report to the Australian Fisheries Management Agency, AFMA Report R02/0881.

Last, P.R. and Stevens, J.D. (1994) *Sharks and Rays of Australia*. CSIRO, Australia.

Lavery, S., Chan, T.Y., Tam, Y.K. and Chu, K.H. (2004) Phylogenetic relationships and evolutionary history of the shrimp genus *Penaeus s.l.* derived from mitochondrial DNA. *Molecular Phylogenetics and Evolution*, **31**: 39–49.

Loneragan, N.R., Kenyon, R.A., Staples, D.J., Poiner, I.R. and Conacher, C.A. (1998) The influence of seagrass type on the distribution and abundance of postlarval and juvenile tiger prawns in the western Gulf of Carpentaria, Australia. *Journal of Experimental Marine Biology and Ecology*, **228**, 175–96.

Lucas, C., Kirkwood, G. and Somers, I. (1979) An assessment of the stocks of the banana prawn *Penaeus merguiensis* in the Gulf of Carpentaria. *Australian Journal of Marine and Freshwater Research*, **30**: 639–52.

Milton, D. M. (2001) Assessing the susceptibility to fishing of populations of rare trawl bycatch: sea snakes caught by Australia's Northern Prawn Fishery. *Biological Conservation*, **101**: 281–90.

NOAA (2002) United States extends ban on shark finning to Pacific Ocean. http://www.publicaffairs.noaa.gov/releases2002/feb02/noaa02018.html

Pender, P.J., Willing, R.S. and Cann, B. (1992) NPF bycatch a valuable resource? *Australian Fisheries*, **51**: 30–1.

Penn, J.W. and Caputi, N. (1986) Spawning stock recruitment relationships and environmental influences on the tiger prawn (*Penaeus esculentus*) fishery in Exmouth Gulf, western Australia. *Australian Journal of Marine and Freshwater Research*, **37**: 491–505.

Poiner, I.R. and Peterken, C. (1995) Seagrasses. In: L.P. Zann and P. Kailola (Eds) *The State of the Marine Environment Report for Australia, Technical Annex 1. The Marine Environment*. Department of Environment, Sport and Territories, Australia.

Poiner, I.R., Staples, D.J. and Kenyon, R. (1987) The seagrass communities of the Gulf of Carpentaria, Australia. *Australian Journal of Marine and Freshwater Research*, **38**: 121–32.

Poiner, I.R., Walker, D.I. and Coles, R.G. (1989) Regional studies – seagrasses of tropical Australia. In: A.W.D. Larkum, A.J. McComb and S.A. Shepherd (Eds) *The Biology of Seagrasses – An Australian Perspective*. Elsevier, New York.

Poiner, I.R., Buckworth, R.C., Harris, A.N.M. (1990) Incidental capture, and mortality of sea turtles in Australia's Northern Prawn Fishery. *Australian Journal of Marine and Freshwater Research*, **41**: 97–110.

Potter, I.C., Manning, R.J.G. and Loneragan, N.R. (1991) Size, movements, distribution and gonadal stage of the western king prawn (*Penaeus latisulcatus*) in a temperate estuary and local marine waters. *Journal of Zoology, London*, **223**: 419–45.

Pownall, P. (Ed) (1994) *Australia's Northern Prawn Fishery: the First 25 Years*. NPF25, Cleveland, Australia.

Robins, C.M., Wang, Y-G. and Die, D. (1998) The impact of global positioning system and plotters on fishing power in the northern prawn fishery, Australia. *Canadian Journal of Fisheries and Aquatic Sciences*, **55**: 1645–51.

Robins, J., Eayrs, S., Campbell, M., Day, G. and McGilvray, J. (2000) *Commercialisation of Bycatch Reduction Strategies and Devices within Northern Australian Prawn Trawl Fisheries*. Department of Primary Industries Report QO00006, Queensland.

Robins, C.M., Goodspeed, A.M., Poiner, I. and Harch, B.D. (2003) *Monitoring the Catch of Turtles in the Northern Prawn Fishery*. FRDC Report, Canberra.

Rothlisberg, P., Staples, D., Poiner, I. and Wolanski, E. (1988) Possible impact of the greenhouse effect on commercial prawn populations in the Gulf of Carpentaria. In: G.I. Pearman (Ed.) *Greenhouse: Planning for Climate Change*. CSIRO Publications, East Melbourne.

Rothlisberg, P.C., Craig, P.D. and Andrewartha, J.R. (1996) Modelling penaeid prawn larval advection in Albatross Bay, Australia: defining the effective spawning population. *Marine Freshwater Research*, **47**: 157–68.

Smith, A.D.M., Sainsbury, K.J. and Stevens, R.A. (1999) Implementing effective fisheries-management systems – management strategy evaluation and the Australian partnership approach. *ICES Journal of Marine Science*, **56**: 967–79.

Somers, I.F. (1985) Maximising value per recruit in the fishery for banana prawns, *Penaeus merguiensis*, in the Gulf of Carpentaria. In: P.C. Rothlisberg, B.J. Hill and D.J. Staples (Eds) *Second Australian National Prawn Seminar 85*. NPS2, Cleveland, Queensland.

Somers, I.F. (1990) Manipulation of fishing effort in Australia's penaeid prawn fisheries. *Australian Journal of Marine and Freshwater Research*, **41**: 1–12.

Somers, I.F. (1994a) No resource, no fishery: prawns and their habitat. In: P.C. Pownall (Ed.) *Australia's Northern Prawn Fishery: the First 25 Years*. NPF25, Cleveland, Australia.

Somers, I.F. (1994b) Species composition and distribution of commercial penaeid prawn catches in the Gulf of Carpentaria, Australia, in relation to depth and sediment type. *Australian Journal of Marine and Freshwater Research*, **45**: 317–35.

Somers, I. and Taylor, B. (1990) Industry co-operation saves on NPF research bill. *Australian Fisheries*, **179**: 46–7.

Somers, I. and Wang, Y. (1997) A simulation model for evaluating seasonal closures in Australia's multispecies Northern Prawn Fishery. *North American Journal of Fisheries Management*, **17**: 114–30.

Staples, D.J. and Vance, D.J. (1986) Emigration of juvenile banana prawns *Penaeus merguiensis* from a mangrove estuary and recruitment to offshore areas in the wet–dry tropics of the Gulf of Carpentaria, Australia. *Marine Ecology Progress Series*, **27**: 239–52.

Stobutzki, I.C., Blaber, S.J.M., Brewer, D.T., Fry, G.C., Heales, D.H., Miller, M.J., Milton, D.A., Salini, J.P., van der Velde, T.D., Wassenberg, T.J., Jones, P.N., Wang, Y-G., Dredge, M., Courtney, A.J., Chilcott, K. and Eayrs, S.J. (2000) *Ecological Sustainability of By-catch and Biodiversity in Prawn Trawl Fisheries*. FRDC Project 96/257.

Stobutzki, I.C., Miller, M.J., Jones, P.N. and Salini, J.P. (2001a) Bycatch diversity and variation in a tropical Australian penaeid fishery; the implications for monitoring. *Fisheries Research*, **53**: 283–301.

Stobutzki, I., Farmer, M.J. and Brewer, D.T. (2001b) Sustainability of fisheries bycatch: a process for assessing highly diverse and numerous bycatch. *Environmental Conservation*, **28**: 167–81.

Stobutzki, I.C., Miller, M.J., Heales, D.H. and Brewer, D.T. (2002) Sustainability of elasmobranchs caught as bycatch in a tropical prawn (shrimp) trawl fishery. *Fishery Bulletin*, **100**: 800–21.

Taylor, B. (1994) A delicate balancing act: management of the fishery. In: P.C. Pownall (Ed.) *Australia's Northern Prawn Fishery: the First 25 Years*. NPF25, Cleveland, Australia.

Vance, D.J., Staples, D.J. and Kerr, J.D. (1985) Factors affecting year-to-year variation in the catch of banana prawns (*Penaeus merguiensis*) in the Gulf of Carpentaria, Australia. *Journal Conseil International pour l'Exploration de la Mer*, **42**: 83–97.

Vance, D.J., Haywood, M.D.E., Heales, D.S. and Staples, D.J. (1996) Seasonal and annual variation in abundance of postlarval and juvenile grooved tiger prawns *Penaeus semisulcatus* and environmental variation in the Embley River, Australia: a six-year study. *Marine Ecology Progress Series*, **133**: 43–55.

Vance, D.J., Haywood, M.D.E., Heales, D.S., Kenyon, R.A. and Loneragan, N.R. (1998) Seasonal and annual variation in abundance of postlarval and juvenile banana prawns, *Penaeus merguiensis*, and environmental variation in two estuaries in tropical northeastern Australia: a six-year study. *Marine Ecology Progress Series*, **163**: 21–36.

Vance, D.J., Haywood, M.D.E., Heales, D.S., Kenyon, R.A., Loneragan, N.R. and Pendrey, R.C. (2002) Distribution of juvenile penaeid prawns in mangrove forests in a tropical Australian estuary, with particular reference to *Penaeus merguiensis*. *Marine Ecology Progress Series*, **228**: 165–77.

Wang, Y-G. and Die, D. (1996) Stock-recruitment relationships of the tiger prawns (*Penaeus esculentus* and *Penaeus semisulcatus*) in the Australian Northern Prawn Fishery. *Marine and Freshwater Research*, **47**: 87–95.

Ye, Y. (2000) Is recruitment related to spawning stock in penaeid shrimp fisheries? *ICES Journal of Marine Science*, **57**: 1103–9.

Chapter 11
The Western Rock Lobster Fishery in Western Australia

Bruce F. Phillips, Roy Melville-Smith, and Nick Caputi

Introduction

The western rock lobster, *Panulirus cygnus*, occurs on the western seaboard of Australia and is found in commercial quantities from Augusta in the south to Shark Bay in the north (Fig. 11.1), an area along approximately 1000 km of coast. The fishery is Australia's most valuable single-species fishery; landings are currently valued at more than US$185–250 million per year, with a flow-on value of approximately US$500 million per year (Department of Fisheries, Western Australia 2005). The commercial fishery dates back to the 1890s, when rock lobsters were caught for the local market. Before World War II, comparatively small quantities were caught and canned for export. However, the fishery expanded greatly after the war both in terms of effort and fishing area, leading to the catches climbing from <500 metric tons (MT) to over 8000 MT by the end of the 1950s (Gray 1999). Annual commercial catches since the mid-1980s have fluctuated between about 8000 and 14 400 MT, but on average have yielded around 11 000 MT. Currently, 540 boats in the commercial fishery operate some 57 000 pots (traps) over the season (mid-November to the end of June) totaling an estimated 10.8 million pot lifts each season.

There is also a large recreational catch component to this fishery. Licensed recreational fishers are permitted to catch lobsters by diving, or by using up to two pots per person per day over the same season as commercial fishers and can keep up to eight legal-sized lobsters per person per day. Since the late 1990s, recreational fishers have accounted for between 250 and 500 MT each season, which is 2–4% of the total western rock lobster catch. This catch has been growing in recent years as increasing numbers of recreational fishers take up the unrestricted access to the recreational fishery (Melville-Smith et al. 2005).

Because of the long history and importance of the fishery, numerous reviews have examined aspects of the biology of the species and its management (Hancock 1981; Phillips 1981; Phillips and Brown 1989; Brown 1991; Gray 1992, 1999; Caputi et al. 2000, 2003; Hall and Brown 2000; Melville-Smith and Anderton 2000; Phillips et al. 2000; Phillips and Melville-Smith 2005). The object of this

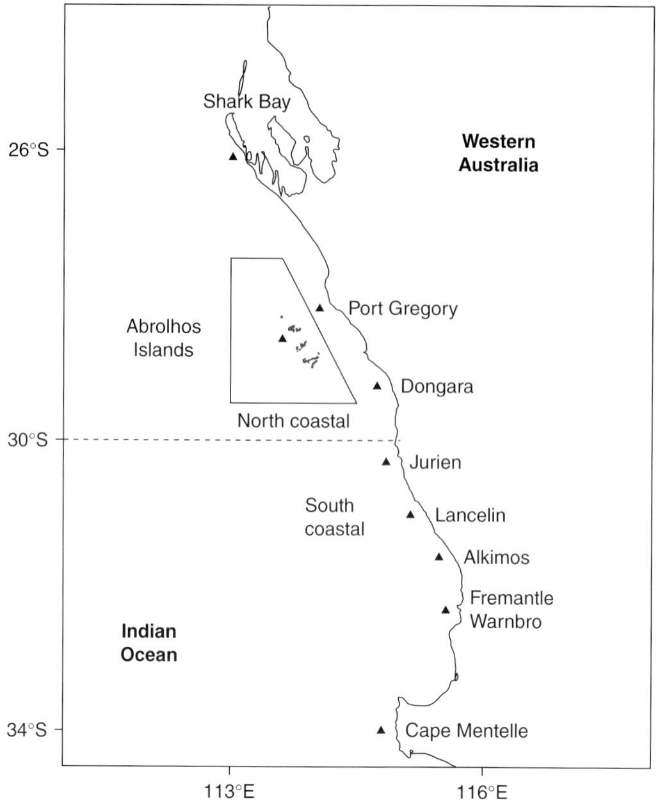

Figure 11.1 The western rock lobster (*Panulirus cygnus*) commercial and recreational fishery off Western Australia, showing the north and south coastal and the Abrolhos Islands areas. The sites where the puerulus collectors used for catch prediction are located are marked by a ▲.

chapter is to build on these earlier reviews by documenting the progress that has been made in maintaining the sustainability of the Western Rock Lobster Fishery, and to identify future challenges. Contemporary Western Rock Lobster Fishery management concentrates on controlling the exploitation rate through fishing effort and maintaining a specified breeding-stock size. Environmentally sustainable development (ESD) indicators are a recent initiative, and while some of these more general sustainability issues will be discussed, this review focuses on issues specific to management of the resource rather than the effects of its fishing on the environment and socio-economic elements.

Threats to sustainability

Since at least the 1970s, management of the Western Rock Lobster Fishery has been guided by five principles formulated into actions that can be summarized as:

(1) Acquiring reliable catch, effort, and length-frequency data to monitor the effects of fishing pressure on the exploited population

(2) Setting a legal minimum size, protection of females, and gear restrictions with adequate inspection and legislative backing
(3) Controlling fishing effort to protect the breeding stock and maximize the benefit from the resource
(4) Understanding the stock, recruitment, and environment interrelationships; and
(5) Communicating effectively and regularly with both commercial and recreational fishers to ensure the objectives and methods of management are well understood

The management arrangements have been developed with these principles as a guide. Together with a strongly directed and comprehensive research program, compared to other lobster fisheries in Australia and around the world, this approach has resulted in the recognition that the Western Rock Lobster Fishery is one of the best-managed fisheries in the world.

Research database

Reliable catch, effort, and size-frequency data

The Western Rock Lobster Fishery has a database extending more than 40 years that includes commercial catch returns, voluntary logbooks, processor returns; abundance estimates of pre-recruits, breeding stock, and size-frequency data from on-board monitoring of the commercial catch and recreational catch surveys. Extensive databases such as these were vital components to Walters and colleagues' (1993) spatio-temporal model and in more recent times, Hall's (2000) and Hall and colleagues' (2000) size-structured model. The models have been used at different times to assess the status of the stock and to forecast the effects of proposed management strategies on the breeding stock.

Technology change

An ongoing core program is determining the rate of increase of fishing power over the last 20 years, associated with changes in boats, gear, and fish-finding technology. The introduction of global positioning systems and other technologies, including changes in vessel size and power and components of gear power and efficiency are measured and monitored. Estimates of fishing efficiency are used to adjust fishing effort data (pot lifts) to remove the increased efficiency bias from industry-based catch data, and this provides the basis for population abundance estimates and stock assessment.

Size and gear restrictions

Legal size

The legal minimum size for the fishery was set in 1900 at 76 mm carapace length (CL). It has essentially remained unchanged since that time, although since the

1993/94 season it was increased by 1 mm to 77 mm for the first 2.5 months of the 7.5 month fishing season. This was one of a number of management measures introduced in the early 1990s aimed at reducing fishing mortality and improving the brood stock (Hall and Chubb 2001).

Compliance

An obvious threat to the sustainability of any commercial fishery, namely compliance of regulations, was addressed early on in the development of the fishery. In the 1960s, effort limitations were introduced and it was probably not a coincidence that illegal fishing activity became widespread. Subsequently, undersized animals and egg-bearing lobsters stripped of their brood were being processed and this black market spread from the locality to markets in eastern Australia (Gray 1999; Caputi et al. 2000).

Management introduced measures to counteract the problem, specifically legislation to restrict processing to licensed establishments, more inspectors, increased fines for offences, convictions were recorded against the vessels rather than their skippers, and licenses could be cancelled after the third offence. These measures helped to make this a highly compliant industry. Even today, much effort goes into ensuring compliance and research into improving the efficiency of compliance (McKinlay and Millington 1999; McKinlay 2002). Currently, for every 1000 lobsters that are checked by enforcement officers, as few as one to two individuals are illegal (McKinlay 2002). An attempt has been made to adjust the less reliable historic catch data for illegal activities and under-reporting of catch (Caputi et al. 2000).

Mortality and damage to the discarded catch

During the 1980s, it was recognized that large numbers of undersized lobsters were being killed unnecessarily, exposed to predation, or having their growth affected by poor handling practices on fishing vessels. These practices involved needless and lengthy exposure of undersized animals to the air before they were sorted and released. Additionally, mortality was high due to excessive handling and associated limb loss and the release of undersized animals in unsuitable habitats away from their home reefs (Brown and Caputi 1983, 1984, 1986; Brown and Dibden 1987). In response, the number of escape gaps in pots was increased from one to three in 1986. Furthermore, codes of fishing practice have been introduced into the industry and are continually being updated and communicated verbally on coastal tours of the fishery, and printed material sent to stakeholders (Stevens et al. 1995).

Despite the progress, there is more room for improvement in reducing the loss of appendages by poor or prolonged handling. Researchers are exploring the possibility of stunning the catch with cold water before it is sorted, in the expectation that this will minimize appendage loss (Davidson and Hosking 2001).

Controlling fishing effort

Effort in the fishery has been controlled since 1963, when the maximum total number of pots was fixed but transferable (Phillips et al. 2000). In more recent years, when other rock lobster fisheries in Australasia turned to catch quotas, the Western Rock Lobster Fishery management considered adopting the catch quota approach (Bowen 1994). Nevertheless, both fishers and managers have strongly supported retention of input control in the form of total allowable effort (TAE) and transferable effort units in the past; a second review of input and output controls is currently being undertaken.

Life-history relationships

Life-history relationships of *Panulirus cygnus* have been studied extensively (Caputi et al. 2003) and are complex (Fig. 11.2 and 11.3). Of special interest is that at about three to four years after settlement, premature animals undergo a synchronized molt in November and migrate offshore. The majority of these pale-shelled animals, known locally as 'whites', migrate into deeper water (>40 m). By their next molt they have assumed their normal red coloration and are once again known as 'reds'.

Stock and recruitment relationships

Although there are very significant peaks and troughs in the annual catch, they are not due to fishing pressure. Rather, they are mainly due to environmental

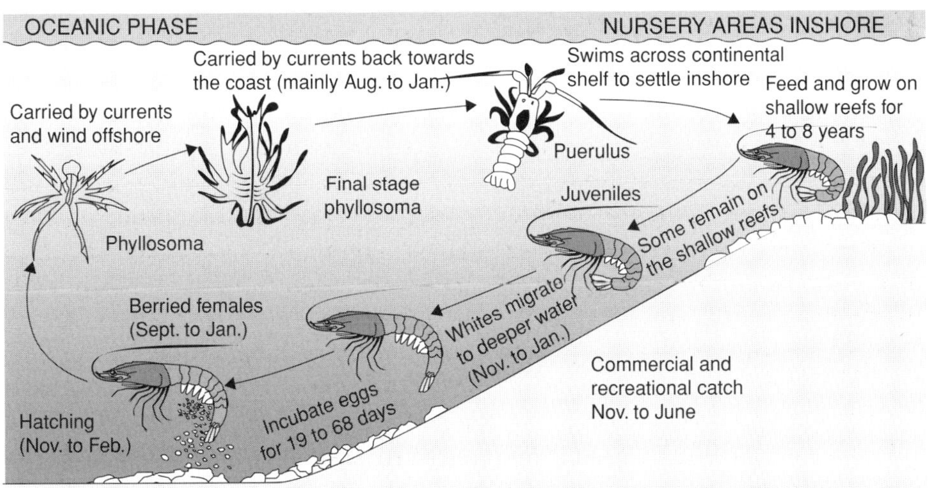

Figure 11.2 Schematic life history of the western rock lobster, *Panulirus cygnus* (from Kailola et al. 1993).

Figure 11.3 Life-history stages of the western rock lobster, *Panulirus cygnus*: (a) berried female; (b) late-stage phyllosoma larvae; (c) puerulus; (d) red and white juveniles.

perturbations that drive the success of the postlarval settlement stage (puerulus). This, in turn, affects recruitment into the capture size of the fishery (Pearce and Phillips 1988; Caputi et al. 1995a).

Environment and catch prediction

The El Niño Southern Oscillation (ENSO) influences the strength of the Leeuwin Current and the frequency and strength of westerly winds, which is responsible for determining the level of puerulus settlement (Pearce and Phillips 1988; Caputi et al. 1995a). Studies of puerulus settlement on collectors (Fig. 11.4) along the coast of Western Australia during the 1980s showed a good correlation between the numbers settling and the catches in the fishery three to four years later (Phillips 1986; Caputi et al. 1995b, c). These findings have been developed into a predictive system that is recognized by industry, government, and fishers (Sharp 2000). This is separate from the population models that have been developed for the fishery, but not used to predict future catches (Walters et al. 1993; Hall et al. 2000).

Figure 11.4 A Phillips collector used to collect the puerulus stage of the western rock lobster, *Panulirus cygnus* (Phillips 1972).

One of the major benefits of this ability to predict catches up to four years in advance is that management can be adapted to a future specific year class before fishing takes place. By contrast, other fisheries without this predictive ability can only react to the effects of a poor year class after it has been fished. For example, measures that were introduced in the 1993/94 season were specifically introduced then because of the prediction that that season was one of good recruitment, whereas the following seasons were predicted to have poor recruitment.

Protecting egg production

One of the most serious threats to the fishery in the past – and one that still requires vigilant attention – is the potential overexploitation of the brood stock. Western rock lobsters live along many hundreds of kilometers of coastline (Fig. 11.1) and, although a single stock, they do have regional differences in their biological characteristics. For example, females and probably males mature at a much smaller size in the offshore Abrolhos Islands than near the coast, and even on the coast (Grey 1979) there are substantial differences in size at maturity. Fishing a population below or around the size at maturity during the rapid somatic growing phase generally optimizes the catch. Unless properly managed this is, however, a risky management strategy because it could overexploit the parent stock, which in turn could lead to overfishing of recruits.

The size of the western rock lobster brood stock was estimated to have shrunk to low levels in the early 1990s (Walters et al. 1993). The postulated reasons were that more intense and efficient fishing caused higher exploitation rates. Fishers could target isolated deepwater reefs in the offshore breeding grounds more efficiently when electronic fishing aids, notably global positioning systems (GPS), were introduced (Brown et al. 1995; Caputi et al. 2000). Results of the model developed by Walters and colleagues (1993) suggested that egg production had been reduced to 15–20% of their pristine levels (Anon. 1993). Furthermore, a decline in Abrolhos Island puerulus settlement from 1984 was possibly due to the reduction in spawning stock, as environmental effects could not easily explain the decline. Consequently, the first signs of recruitment overfishing appeared to be emerging (Caputi et al. 1995a).

Scientists, industry, and managers were unified in their concern about the state of the brood stock and the signs that recruitment was being affected. This led management to introduce a suite of measures aimed at rebuilding the brood stock to the levels of about 1980. These management changes, which were supported by industry, were introduced in 1993. The changes included an 18% reduction in the number of pots allowed by the commercial fishery. Nominal fishing effort is measured as trap lifts unadjusted for increases in fishing efficiency and latent effort are pot allocations that are not fully utilized by, for example, not being fished for the full fishing season (Caputi et al. 2000). Nominal fishing effort had increased through the 1980s with uptake of latent effort in the fishery, combined with greater efficiency using electronic and other fishing aids. Consequently, fishing mortality had increased substantially. A 10% pot reduction was imposed through the mid- to late 1980s at a rate of 2% per season, but this stabilized rather than decreased the effort. Pot reduction was aimed at reducing the fishing effort and the pressure on the breeding animals.

Second, what has become known as the 'setose' rule was introduced, which required fishers to return female lobster bearing ovigerous setae or a tar spot, which is an external spermatophore. This was in addition to existing legislation that required fishers to return egg-bearing animals back to the sea. The commercial and recreational fishing seasons start in mid-November. Mature females, while not necessarily carrying eggs, generally have ovigerous setae during the spawning season (September–December) and until February or March, when many molt into a non-setose phase. They remain non-setose until May, when they molt back into the setose state in preparation for the mating that starts in June and the breeding season, which starts around September. Therefore, while the setose rule has not excluded all mature females from capture, it has largely limited their exploitation to the proportion that molt to the non-setose stage in the period between February and May.

The third management change was the introduction of a legal maximum size for female lobsters of 115 mm carapace length (CL) in the southern zone (south of 30° S, Fig. 11.1) and 105 mm CL in the northern zone. This rule was aimed at protecting the large females in the population. Although these animals make up only a small part of the population, they are thought to make a large contribution to egg production. Chubb (1991) for example, showed that large female western rock lobsters are capable of producing multiple broods each season and can carry hundreds of thousands more eggs than small females.

The fourth change was an increase in the legal minimum size by 1 mm CL from 76 to 77 mm during 15 November to 31 January. Most western rock lobster around the legal minimum size migrate offshore around the start of the commercial fishing season in December and January and are targeted by the commercial fishery, which takes large numbers of them. Once in the deeper offshore waters, fewer are caught, and it was therefore expected that increasing the minimum size during the migrating phase would enable more animals to survive to become part of the breeding population.

The obvious question that follows is to ask how successful the management changes have been in rebuilding the brood stock. The state of the brood stock is currently measured by two different methods. The first, described by Chubb and colleagues (1995), uses commercial monitoring data from the northern and southern coastal management zones. The weakness of this index is that it does not automatically account for increases in fishing efficiency, which are inherent in the use of commercial catch-and-effort data. Although these data can be adjusted to make up this deficiency (Brown et al. 1995), there is inevitable uncertainty about the modified data. In the case of the Western Rock Lobster Fishery the monitoring data are adjusted by about 1–2% per year to account for increased efficiency.

The fishery-independent method of establishing an egg-production index was established for the Western Rock Lobster Fishery in the early 1990s (Melville-Smith et al. 1998). In brief, a survey is conducted each year at the same GPS positions on a number of sites covering the full extent of the commercial fishery. Standardized fishing gear, bait, and lunar phase are used to limit the variables that could influence catch rates. Regional egg production is calculated from the number and size of breeding females caught in the pots, taking into account size–fecundity relationships. The weakness of this approach is that uncontrolled environmental factors such as temperature and swell at the time of sampling can affect catch rates. The overall effect of the above management on the state of the brood stock by both fishery-dependent and fishery-independent methods is a substantial increase in egg production within the last decade (Fig. 11.5).

When the management changes were introduced, they were expected to increase egg production in the fishery and assessment by an age-structured model has shown their effectiveness (Hall and Chubb 2001). The model suggests that by 1999/2000, egg production in the fishery was 134% above the 1980/81, or target level. It also estimates that the catch in 1999/2000 was 91% of what might have been taken had the above management measures not been in place. Furthermore, the model

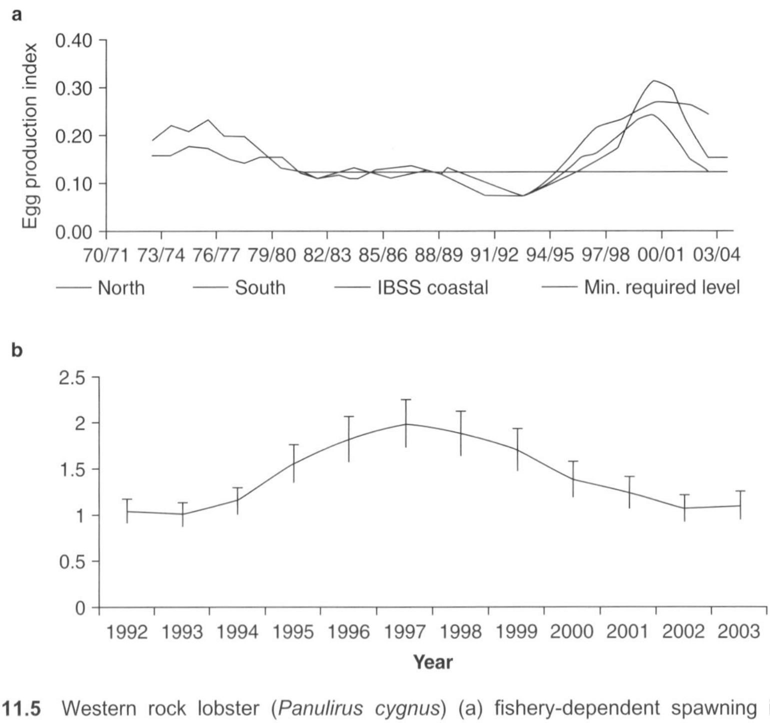

Figure 11.5 Western rock lobster (*Panulirus cygnus*) (a) fishery-dependent spawning index for the north and south coastal regions and a combined coastal fishery-independent index (IBSS) and (b) fishery-independent spawning-stock indices for the Abrolhos Islands. A three year moving average has been used to smooth all indices.

calculates that if each of the four management measures had been applied separately and singularly, egg production in the 1999/2000 season would have been 76, 84, 94, and 102% of the target level for the minimum size, maximum size, pot reduction, and setose regulations, respectively (Hall and Chubb 2001).

While management measures such as those introduced in the 1993/94 season are in place to protect the brood stock, there seems to be little future likelihood of declining recruitment due to overfishing. Currently, managers have no wish to increase the size of the brood stock beyond the target level at the beginning of the 1980s (Fig. 11.5a). When landings in the 2001/02 fishing season were predicted to be below average, the maximum size rule for females was temporarily rescinded to bolster catches (Chubb 2002). The effect of this management change was considered an acceptable risk to the fishery and was instituted to maintain a constant supply to the markets in Asia (Department of Fisheries, Western Australia 2002a). With the breeding stock at levels considered adequate for the fishery, it is likely that environmental effects will be the major factor driving future population and catch variation.

Communication with fishers and stakeholders

The principal method of consultation in the fishery is through the Rock Lobster Industry Advisory Committee (RLIAC), a statutory ministerial advisory committee. The committee includes fishers, processors, recreational fishing representatives, and managers. Other representatives including scientists, conservationists, and specialist representatives from industry may attend, reliant on permission of the chair of the committee. The functions of the RLIAC, as set out in the Western Australian Fish Resources Management Act, 1994, are: (1) to identify issues that affect the rock lobster fishery; (2) to advise the minister on matters relating to the management, protection, and development of the rock lobster fisheries; and (3) to advise the minister on matters relating to rock lobster fisheries where the advice of the advisory committee is sought by the minister.

Since its establishment in 1965, RLIAC has made communication and consultation with industry its first priority. RLIAC has been the forum where issues, particularly those related to sustainability, have been debated and recommendations made to the minister. The committee takes problems that affect the fishery and the industry generally and through discussion, management papers, and meetings with industry, initiates ways to resolve them.

The fishery management group has a very successful process of communicating with stakeholders. This includes an annual coastal tour that includes a series of public meetings for fishing communities and the general public and publication of the western rock lobster Commercial Fisheries Production Bulletin, which is produced two or three times a year. Commercial fishers' involvement in management was, until recently, by way of RLIAC and the West Australian Fishing Industry Council. This involvement was enhanced in 2002 by the creation of an industry-representative organization: the Western Rock Lobster Council.

Threats to sustainability in the future

With egg production in the fishery being monitored, and with the previously-described ways of adjusting exploitation rates to maintain breeding stock levels, the main biological threat to sustainability of the resource is under control. However, there are always other issues arising that if left unattended, could cause future problems.

Regional contributions to egg production

One of the risks associated with the management of egg production is the disproportionately large egg production by lobsters in the Abrolhos Islands. For example, Chubb (1991) calculated that these islands contributed between 45 and 60% of the stock's egg production between 1984 and 1989 – a direct result of their high densities of lobsters and their small size at maturity. The large quantities of eggs produced in this relatively small and isolated region, and therefore the presumed importance of that brood stock to the fishery, does leave the fishery vulnerable in the unlikely event of a serious disaster, such as a major oil spill or localized disease outbreak.

The effect of increased fishing efficiency

The effect of effort creep – increase in the catching power of a rock lobster pot due to improvement in fishing technology – will need constant vigilance in this input-regulated fishery. The use of nominal fishing effort to plot changes in catch per unit effort over time (Caputi et al. 2000), indicates that this index has been stable since the 1970s. When catch rates are adjusted for increases in fishing efficiency due to gear technology improvements, such as color echo sounders, GPS, and better designed and faster boats, the catch rates with standardized effort have, however, gradually declined.

An assessment of the increase in the efficiency has also been undertaken using a depletion analysis during the non-migrating part of the fishery (March to June) (Wright et al. 2006). This technique provides estimates of the residual legal biomass at the end of the fishing season and an estimate of the harvest rate. In recent years, it has been estimated that efficiency increases, especially by fishers in the northern zones of this fishery, may be leading to a decline in egg production in this region. In response to this, the minister has approved a 15% effort reduction in the northern zone and a smaller 5% reduction in the south. These effort reductions are not straightforward pot reductions, but are instead a combination of relatively small (10%) pot reductions over certain periods of the season; fishing closures for extended periods of up to three weeks, as well as particular days, including Christmases and New Years, Sundays, and for three days over the full moon in certain months; and even extending legal minimum size changes from 76 mm to

77 mm CL for an additional two weeks. The changes are variable depending on fishing region (northern or southern fishing zones) and are outlined more fully elsewhere (Anon. 2005). To assist in the decision-making process, the economic effect on fishers of the different management measures was evaluated to assess the short-term (1–2 years) effect on the net revenue (Thomson and Caputi 2005).

Growth of recreational fishing

One of the issues to be addressed is the expansion of the recreational fishery. Since 1986, the recreational license sales have increased by ~150% (Fig. 11.6) and the increase in the percentage of the total landings has been even greater (Fig. 11.7). Over 90% of the recreational rock lobster catch is made in the shallow inshore grounds in depths of <20 m (Melville-Smith and Anderton 2000) – well outside of the depths where the brood stock of the coastal regions of the fishery are found. Therefore, while recreational fishers are competing with commercial fishers for a greater share of the inshore catch, particularly around large metropolitan areas, they are not directly affecting the brood stock.

Strategies and recommendations for developing an integrated approach for managing the recreational, aboriginal, ecotourism, and commercial rock lobster sectors were developed and released by a minister-appointed Integrated Fisheries

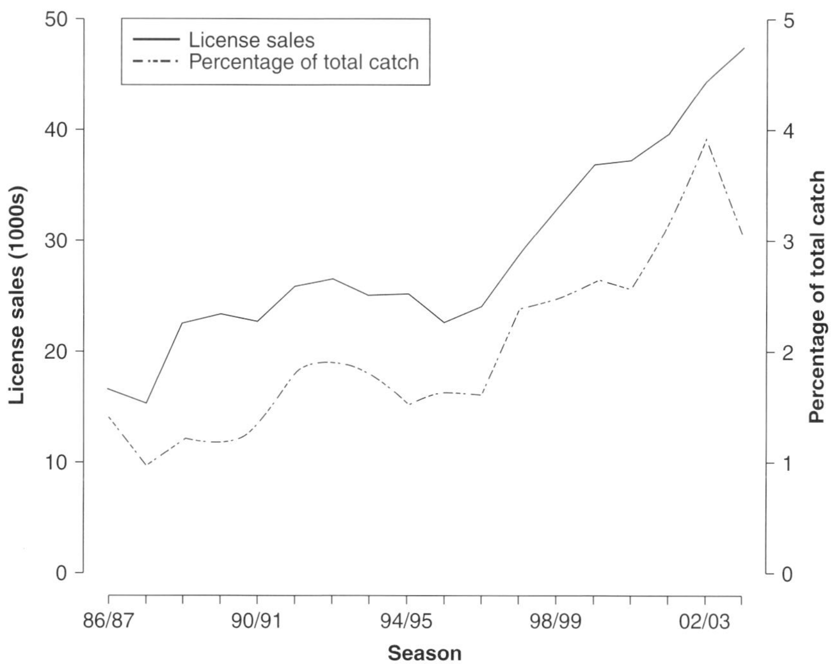

Figure 11.6 Recreational western rock lobster (*Panulirus cygnus*) license sales and catch, as a percentage of the total landings made by the commercial and recreational sector between the 1986/87 and 2003/04 fishing seasons (adapted from Melville-Smith et al. 2004).

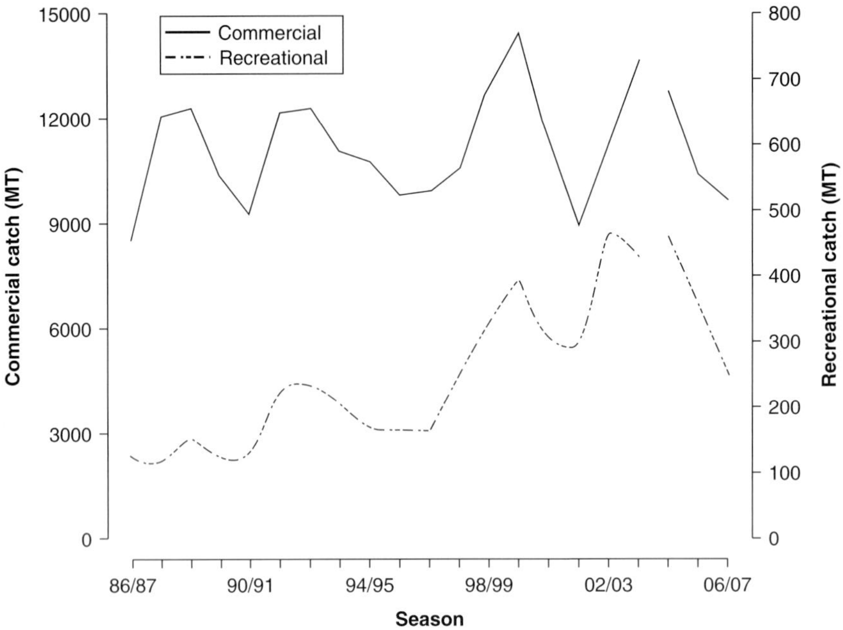

Figure 11.7 Time series of recreational and commercial western rock lobster (*Panulirus cygnus*) landings and predicted landings. Recreational catches are estimates from mail surveys that began in the 1986/87 season. Commercial catches are from fishers' monthly returns.

Management Committee in mid-2003 (Department of Fisheries, Western Australia 2002b). These recommendations included the establishment of an Integrated Fisheries Allocation Advisory Committee (IFAAC) that will advise on the allocation issues between groups, such as the recreational and commercial sectors of the Western Rock Lobster Fishery (Department of Fisheries, Western Australia, 2002b; IFAAC 2005).

A report has been issued by the Department of Fisheries, Western Australia, where the executive director has approved sustainability of the western rock lobster fishery and outline harvest levels upon which the IFAAC will be able to base their recommended allocations among the various user groups (Department of Fisheries, Western Australia, 2005). The Minister of Fisheries, on consideration of the IFAAC's advice, will determine the allocations to each sector.

Harvesting pueruli for aquaculture

Currently, the propagation of *P. cygnus* lobsters from egg to marketable-sized animals seems unlikely given their long and complicated larval life. Harvesting of pueruli and growing them out is, however, a relatively simple operation that has been successful in a number of countries, most notably Vietnam (Tuan et al. 2000). The possibility that western rock lobster pueruli will be harvested from the wild fishery for commercial sale, for grow-out at sea or onshore is, if not adequately controlled, seen as a potential threat to both the sustainability and economics of the fishery. A model developed to determine the effect of removing different numbers of pueruli on the subsequent catch in the wild fishery found that effects are likely to be slight unless many millions of pueruli are removed (Phillips et al. 2003a). Potential losses to the wild fishery could be countered by reductions in effort that takes into account the number of pueruli harvested and their settlement strength. Methods for capturing the pueruli for aquaculture using artificial seaweed collectors have been developed (Phillips et al. 2001).

Research towards the development of rock lobster aquaculture in Australasia is receiving substantial funding through the Fisheries Research and Development Corporation, an Australian federal funding body. Licenses to capture and grow-out pueruli of rock lobsters have been issued in New Zealand and Tasmania and it is only a question of time before western rock lobster is farmed in Australia.

Enhancement of stocks

Plans for translocation of juveniles, reseeding areas with aquacultured pueruli and postpueruli, or increasing survival of naturally-settling pueruli or juveniles by introducing artificial structures, have all been proposed at different times. A research study assessing the possibilities for enhancing the natural settlement of the western rock lobster is currently being conducted.

Reproductive biology issues

The western rock lobster catch has become male-dominated since females have received greater protection with the maximum size and setose rules. There has, however, been no evidence of sperm limitation (MacDiarmid and Butler 1999) or of other reproductive biology issues related to strongly biased sex ratios. Research to monitor these issues is undertaken as part of the fishery-independent breeding stock survey and the at-sea commercial monitoring program (Chubb 1991; Chubb et al. 1995).

Economics and social interactions

Winning the 1999 Premier's Award 'For Excellence in Public Sector Management' in the category of 'Economic Development' for the Western Rock Lobster Fishery increased the recognition of the work of the Western Australian Fisheries Department (Western Australian Fisheries Department 2000). The award recognized the contribution made by the western rock lobster fishery in economic and social benefits to Western Australia. These include: contributing an estimated 28% of the 1997/98 total export value of Australian fisheries products; 79% of the State's seafood exports creating the equivalent 2000 full-time jobs; and contributing to the support of 6000 other jobs in processing, boat building, and service industries.

The Western Australian Government recently made a commitment to review management arrangements for the Western Rock Lobster Managed Fishery. This review will be completed in 2007 and will include an investigation into alternative management strategies that are expected to achieve an optimal mix of economic and social benefits and ecological sustainability. Under this review, the current management arrangements are to be viewed as the benchmark against alternative management. In particular, the review is to identify management approaches that provide the best long-term socio-economic return to the State from the ecologically sustainable use of the western rock lobster resource.

As a first step the review will assess three management scenarios, namely: (1) retention of existing management; (2) individually transferable effort; and (3) a total allowable catch system. The second option will examine limitations on the number of pots fished, but allow flexibility for individual operators to choose when they fish and how much gear they use at different times of the year. In other words, a management system with a quota on effort, but the ability to set the total allowable effort based on the choices of the fisher. The third option will examine a system whereby units have a value in total allowable kilograms of lobster, but biological controls remain to ensure the breeding stock is not at risk (Thomson and Caputi 2005).

Increasing awareness

The Western Rock Lobster Fishery was the first fishery in the world to be awarded certification as a well-managed fishery by the Marine Stewardship Council (MSC), an international organization set up to assess and certify sustainable fisheries. Accreditation in March 2000 followed an extensive review conducted by an independent panel that examined the operation of the fishery, taking into account its sustainability and effect on the marine environment (Phillips et al. 2003b). The MSC assessment is based on the criteria outlined in the document entitled 'Principles and Criteria for Sustainable Fishing'. For the purposes of MSC certification, a sustainable fishery is defined as one that is conducted in such a way that it:

(1) Can be continued indefinitely at a reasonable level
(2) Maintains and seeks to maximize ecological health and abundance
(3) Maintains the diversity, structure, and function of the ecosystem on which it depends, as well as the quality of its habitat, minimizing the adverse effects that it causes
(4) Is managed and operated in a responsible manner, in conformity with local, national, and international laws and regulations
(5) Maintains present and future economic and social options and benefits; and
(6) Is conducted in a socially and economically fair and responsible manner

The certification is a positive indication of the quality of the management of this fishery. In 2004, the process of a second MSC certification of the fishery commenced as the first certification expired in August 2005. An extension into 2006 has been given while the review is being conducted.

A new development in Australia is a federal government initiative concerned with sustainability of fish stocks. The Department of the Environment and Heritage, which is the Australian federal department responsible for environmental issues, has determined that from December 2003, only marine products from fisheries that are environmentally sustainable can be exported. The assessment of such fisheries is a similar process to the MSC assessment (Fletcher 2001) and the Western Rock Lobster Fishery has received export certification.

There are recommendations to establish marine national parks in several regions of Western Australia that will influence western rock lobster fishing operations (Report of the Marine Parks and Reserves Selection Working Group 1994). Some recommendations of this working group have been promulgated, including the creation of a marine reserve at Jurien Bay, and other planned marine national parks or refuge areas.

Additionally, a code of practice has been adopted. This includes boat and gear safety in the operation of vessels, gear, bait, and waste disposal, which has enhanced the status of the industry. The procedures for handling of lobster ensure that they are delivered to the factories alive and missing as few limbs as possible, and this ensures that they are in the best condition for live export.

Conclusions

The attention of managers of the Western Rock Lobster Fishery is shifting away from the fishery's biological sustainability, which was believed to be under threat in the late 1980s, toward increasing the value and reducing the costs. A number of strategies are being considered; for example, mechanisms to reduce within and between-season catch variability caused by different seasonal catch rates and inter-annual recruitment pulses. An example of this was the removal of the maximum size prior to the 2001 season. In this case, it was predicted to have catches well below the average, and as the breeding stock was well above the target levels, the maximum size was removed for one year to increase the catch for that year. The current shift in focus towards maximizing economic returns has generated new controversies, notably whether the industry's management committees should be advising government on short-term market benefits or concentrating on stock sustainability issues alone (Donohue 1998a, b).

The Standing Committee on Ecologically Sustainable Development (ESD) of the Legislative Council of the Western Australian Parliament recently undertook an examination of the management and sustainability of the Western Rock Lobster Fishery. The committee concluded that this fishery 'is a major success story in terms of sustainability of its yield and the infrastructure managing the fishery is operating within ESD principles' (and the only criticism was that 'to date, it has focused primarily on sustainable yields rather than on the ecosystems supporting the western rock lobster resource' (Sharp 2000)). The gap is currently being addressed as a part of the ongoing MSC accreditation obligations. The high catch of recent years resulting from favorable environmental conditions in 1999–2000 have begun to have a significant effect on marketing and wholesale prices, suggesting that the fishery is becoming an economic victim of its biological success (Morgan 2001).

Contemporary global issues such as 11 September in the USA, a slow economic recovery in major far eastern markets, and most recently the SARS virus, raised economic concerns and stimulated attempts to develop new markets (Department of Fisheries 2001). More recently (2005) economic concerns have risen about the effect of increases in fuel costs on the fishery.

The Western Rock Lobster Fishery is well positioned through the efforts of many years of collaborative management, supported by well-directed research and monitoring programs, such that it will be able to enjoy future prosperity. While some future questions in the Western Rock Lobster Fishery have been identified, it is inevitable that other unforeseen issues will arise, as fisheries sustainability is a dynamic process. The challenge for fisheries researchers, managers, and industry, is to have the foresight to identify these potential problems and to have the means and ability to address them.

Summary

The Western Rock Lobster Fishery has 540 boats operating about 57 000 pots. The average annual catch of 11 000 MT is valued in excess of US$300 million. In addition to the commercial catch, recreational fishers take about 400 MT a year. Sustainability in this fishery is maintained by analysis of a comprehensive fisheries database, some of which dates back to the 1960s, and includes factors such as catch, effort, length-frequencies, fishery-dependent and fishery-independent breeding-stock surveys, puerulus settlement monitoring, and recreational catch monitoring. This data combined with an extensive set of management controls that include limited fishing season and legal minimum and maximum sizes, and an effective compliance program have resulted in the above described success. Effort in the fishery is controlled by input restrictions on the number of pots allowed and number of days fishing, which are implemented after considerable consultation with industry. The principal method of ensuring the sustainability of the fishery is by monitoring the size of the breeding stock, using data from both a commercial at-sea monitoring program and an annual fishery-independent breeding-stock survey. When the breeding stock fell to low levels in the early 1990s, management initiatives succeeded in returning it to what are considered to be safe levels. Catches are currently high, but fishers have acquired sufficient scientific knowledge to understand that catches will fluctuate for environmental reasons and to take this into account in their fishing operations. Environmental effects have been shown to drive the level of settlement in a particular season. These settlement levels are in turn highly correlated with catches three to four years later, and this information provides predictions of future catches and improves management of the fishery. The fishery was awarded Marine Stewardship Council certification in March 2000, the first in the world to receive this imprimatur.

References

Anonymous (1993) *Chairman's Report to the Minister of Fisheries on Management Recommendations for 1993/94 and 1994/95 Western Rock Lobster Seasons*. Western Australian Department of Fisheries Management Paper 55.

Anonymous (2005) New package to protect lobsters. *ProWest*, September/October.

Bowen, B.K. (1994) *Long Term Management Strategies for the Western Rock Lobster Fishery: Evaluation of Management Options*. Western Australian Department Fisheries Management Paper 67.

Brown, R.S. (1991) A decade (1980–1990) of research and management of the western rock lobster (*Panulirus cygnus*) fishery of Western Australia. *Revista Investigaciones Marinas*, **12**: 204–22.

Brown, R.S. and Caputi, N. (1983) Factors affecting the recapture of undersize western rock lobster *Panulirus cygnus* George returned by fishermen to the sea. *Fisheries Research*, **2**: 103–28.

Brown, R.S. and Caputi, N. (1984) Factors affecting the growth of undersize western rock lobster, *Panulirus cygnus* George, returned by fishermen to the sea. *Fisheries Bulletin*, **83**: 567–74.

Brown, R.S. and Caputi, N. (1986) Conservation of recruitment of the western rock lobster (*Panulirus cygnus*) by improving survival and growth of undersize rock lobsters captured and returned by fishermen to the sea. *Canadian Journal of Fisheries and Aquatic Sciences*, **43**: 2236–42.

Brown, R.S. and Dibden, C.J. (1987) *A Re-examination of the Methods Used for Handling Undersize Western Rock Lobsters*. Western Australian Department Fisheries Report 78.

Brown, R.S., Dibden, C.J. and Barker, E.H. (1995) A preliminary assessment of the effect of increases in fishing power on stock assessment and fishing effort of the western rock lobster (*Panulirus cygnus* George 1962) fishery in Western Australia. *Crustaceana*, **68**: 227–37.

Caputi, N., Chubb, C.F. and Brown, R.S. (1995a) Relationships between spawning stock, environment, recruitment and fishing effort for the western rock lobster, *Panulirus cygnus*, fishery in Western Australia. *Crustaceana*, **68**: 213–26.

Caputi, N., Brown, R.S. and Chubb, C.F. (1995b) Regional prediction of the western rock lobster (*Panulirus cygnus*) catch in Western Australia. *Crustaceana*, **68**: 246–56.

Caputi, N., Brown, R.S. and Phillips, B.F. (1995c) Prediction of catches of the western rock lobster (*Panulirus cygnus*) based on indices of puerulus and juvenile abundance. *ICES Marine Science Symposium*, **199**: 287–93.

Caputi, N., Chubb, C.F., Hall, N.G. and Brown, R.S. (2000) Measurement of catch and fishing effort in the western rock lobster fishery. In: B.F. Phillips and J. Kittaka (Eds) *Spiny Lobsters: Fisheries and Culture*. Blackwell Science Ltd., Oxford.

Caputi, N., Chubb, C.F., Melville-Smith, R., Pearce, A. and Griffin, D. (2003) Relationships between life history stages of the western rock lobster, *Panulirus cygnus* fishery in Western Australia. *Fisheries Research*, **65**: 47–61.

Chubb, C.F. (1991) Measurement of spawning stock levels for the western rock lobster, *Panulirus cygnus*. *Revista Investigaciones Marinas*, **12**: 223–33.

Chubb, C.F. (2002) West coast rock lobster managed fishery status report. In: J.W. Penn (Ed.) *State of the Fisheries Report 2001/2002*. Western Australian Department Fisheries, Perth.

Chubb, C.F., Caputi, N. and Brown, R.S. (1995) Stock and recruitment indices for the western rock lobster (*Panulirus cygnus*) fishery – biological and other considerations. In: A.J. Courtney and M.G. Cosgrove (Eds) *Proceedings of the Workshop on Spawning Stock-Recruitment Relationships (SRRs) in Australian Crustacean Fisheries*. Department Primary Industries, Brisbane.

Davidson, G. and Hosking, W. (2001) Nice legs, shame about the waste! Ways of controlling handling – induced appendage loss. In: R. van Barneveld (Ed.) *Developments in Rock Lobster Enhancement and Aquaculture III*. Wellington, New Zealand.

Department of Fisheries, Western Australia (2001) *Western Rock Lobster Fishery 2000/01 Season*. Commercial Fisheries Production Bulletin 24.

Department of Fisheries, Western Australia (2002a) West coast rock lobster managed fishery management summary. In: J.W. Penn (Ed.) *State of the Fisheries Report 2001/2002*. Western Australian Department Fisheries, Perth.

Department of Fisheries, Western Australia (2002b) *Report to the Minister for Agriculture, Forestry and Fisheries by the Integrated Fisheries Management Review Committee*. Western Australia Department Fisheries Management Paper 165.

Department of Fisheries, Western Australia (2005) *Integrated Fisheries Management Report, Western Rock Lobster Resource*. Western Australia Department Fisheries Management Paper 192.

Donohue, K. (1998a) *Western Rock Lobster Management – Options and Issues*. Western Australia Department Fisheries Management Paper 101.

Donohue, K. (1998b) *Future Directions for the Rock Lobster Industry Advisory Committee and the Western Rock Lobster Managed Fishery*. Western Australia Department Fisheries Management Paper 123.

Fletcher, W.J. (2001) *Policy for the Implementation of Ecologically Sustainable Development for Fisheries and Aquaculture within Western Australia*. Western Australia Department Fisheries Management Paper 157.

Gray, H. (1992) *The Western Rock Lobster* Panulirus cygnus. *Book 1: A Natural History*. Westralian Books, Geraldton, Australia.

Gray, H. (1999) *The Western Rock Lobster* Panulirus cygnus. *Book 2: A History of the Fishery*. Westralian Books, Geraldton, Australia.

Grey, K.A. (1979) Estimates of the size of first maturity of the western rock lobster, *Panulirus cygnus*, using secondary sexual characteristics. *Australian Journal of Marine and Freshwater Research*, **30**: 785–91.

Hall, N.G. (2000) Modelling for Fishery Management, Utilizing Data for Selected Species in Western Australia. PhD thesis, Murdoch University, Murdoch, Perth, Australia.

Hall, N.G., and Brown, R.S. (2000) Modelling for management: the western rock lobster fishery. In: B.F. Phillips and J. Kittaka (Eds) *Spiny Lobsters: Fisheries and Culture*. Blackwell Science Ltd., Oxford.

Hall, N.G. and Chubb, C. (2001) The status of the western rock lobster, *Panulirus cygnus*, fishery and the effectiveness of management controls in increasing the egg production of the stock. *Australian Journal of Marine and Freshwater Research*, **52**: 1657–67.

Hall, N.G., Liangyue, C., Chubb, C., Caputi, N., Cheng, H., Melville-Smith, R. and Shanks, S. (2000) *Modelling to Explore Management Strategies to Optimize the Value of the Rock Lobster Fishery of Western Australia*. Final Report, Fisheries Research Development Corporation Project 97/104, Canberra.

Hancock, D.A. (1981) Research for management of the rock lobster fishery of Western Australia. *Proceedings Gulf and Caribbean Fisheries Institute*, **33**: 207–29.

IFAAC (Integrated Fisheries Allocation Advisory Committee) (2005) *Draft Allocation Report*. Government of Western Australia, Perth.

Kailola, P.J., Williams, M.J., Stewart, P.C., Reichelt, E., McNee, A. and Grieve, C. (1993) *Australian Fisheries Resources*. Bureau of Resource Sciences, Canberra.

MacDiarmid, A.B. and Butler, M.J., IV (1999) Sperm economy and limitation in spiny lobsters. *Behavior Ecology and Sociobiology*, **46**: 14–24.

Marine Parks and Reserves Selection Working Group (1994) *A Representative Marine Reserve System for Western Australia*. Western Australian Conservation and Land Management, Como.

McKinlay, J.P. (2002) *Optimising the Efficiency and Effectiveness of Enforcement to Achieve Compliance in the Western Rock Lobster Fishery*. Fisheries Research and Development Corporation Report 1998/156.

McKinlay, J.P. and Millington, P.J. (1999) Fisher obligations in co-managed fisheries: the case for enforcement. In: R. Shotton (Ed.) *Use of Property Rights in Fisheries Management. Proceedings of FishRights 99 Conference, Fremantle, Western Australia.* FAO Fisheries Technical Paper No. 404/2.

Melville-Smith, R. and Anderton, S.M. (2000) *Western Rock Lobster Mail Surveys of Licenced Recreational Fishers 1986/87 to 1998/99*. Western Australian Department of Fisheries Research Report 122.

Melville-Smith, R., Chubb, C.F., Caputi, N., Cheng, Y.W., Christianopoulos, D. and Rossbach, M. (1998) *Fishery Independent Survey of the Breeding Stock and Migration of the Western Rock Lobster* (Panulirus cygnus). Final Report, Fisheries Research and Development Corporation Project 96/108. Canberra.

Melville-Smith, R., Thomson, A.W. and Caputi, N. (2004) Improved forecasts of recreational western rock lobster (*Panulirus cygnus*) catches in Western Australia, by predicting licence usage. *Fisheries Research*, **68**: 203–8.

Melville-Smith, R., Thomson, A.W. and Readhead, P. (2005) Licenced recreational rock lobster fishery status report. In: J.W. Penn, W.J. Fletcher and F. Head (Eds) *State of the Fisheries Report 2004/05*. Department of Fisheries, Western Australia.

Morgan, G.R. (2001) Initial allocation of harvesting rights in the rock lobster fishery of Western Australia. In: R. Shotton (Ed.) *Case Studies on the Allocation of Transferable Quota Rights in Fisheries*. FAO Fisheries Techical Paper 411.

Pearce, A.F. and Phillips, B.F. (1988) ENSO events, the Leeuwin Current, and larval recruitment of the western rock lobster. *Journal du Conseil International pour l'Exploration de la Mer*, **45**: 13–21.

Phillips, B.F. (1972) A semi-quantitative collector of the puerulus larvae of the western rock lobster *Panulirus longipes cygnus* George (Decopoda, Palinuridae). *Crustaceana*, **22**: 147–54.

Phillips, B.F. (1981) The circulation of the southeastern Indian Ocean and the planktonic life of the western rock lobster. *Oceanography Marine Biology Annual Review*, **19**: 11–39.

Phillips, B.F. (1986) Prediction of commercial catches of western rock lobster *Panulirus cygnus* George. *Canadian Journal of Fisheries and Aquatic Sciences*, **43**: 2126–30.

Phillips, B.F. and Brown, R.S. (1989) The West Australian rock lobster fishery: research for management. In: J.F. Caddy (Ed.) *Marine Invertebrate Fisheries: their Assessment and Management*. Wiley, New York.

Phillips, B.F. and Melville-Smith R. (2005) Sustainability of the Western Rock Lobster Fishery: a review of past progress and future. *Bulletin of Marine Science*, **76**: 485–500.

Phillips, B.F., Chubb, C.F. and Melville-Smith, R. (2000) The status of Australia's rock lobster fisheries. In: B.F. Phillips and J. Kittaka (Eds) *Spiny Lobsters: Fisheries and Culture*. Blackwell Science Ltd., Oxford.

Phillips, B.F., Melville-Smith, R., Cheng, Y.W. and Rossbach, M. (2001) Testing collector designs for commercial harvesting of western rock lobster (*Panulirus cygnus*) puerulus. *Marine Freshwater Research*, **52**: 1465–73.

Phillips, B.F., Melville-Smith, R. and Cheng, Y.W. (2003a) Measuring the effects of pueruli removals and habitat improvement in assessing sustainability of spiny lobster populations. *Fisheries Research*, **65**: 89–101.

Phillips, B.F., Ward, T. and Chaffee, C. (2003b) *Eco-labelling in Fisheries: What is it all About?* Blackwell Science Ltd, Oxford.

Sharp, C. (2000) *Report of the Standing Committee on Ecologically Sustainable Development in Relation to the Management and Sustainability of the Western Rock Lobster*. Western Australia Legislative Council Report 6.

Stevens, R., Turner, S. and Whisson, G. (1995) *Fifteen Minutes: Code of Practice for Handling Live Rock Lobster*. Western Australian Fishing Industry Council, Perth, Australia.

Thomson, N. and Caputi, N. (2005) An economic analysis of management options in the Western Rock Lobster Fishery of Western Australia. *Proceedings of the 3rd Conference of the North American Association of Fisheries Economists, University of British Columbia, Vancouver, British Columbia, Canada, 2005.*

Tuan, L.A., Nho, N.T. and Hambrey, J. (2000) Status of cage mariculture in Vietnam. In: I.C. Liao and C.K. Lin (Eds) *Cage Aquaculture in Asia: Proceedings 1st International Symposium of Cage Aquaculture in Asia*. Asian Fisheries Society, Manila, and World Aquaculture Society – Southeast Asian Chapter. Bangkok, Thailand.

Walters, C.J., Hall, N., Brown, R. and Chubb, C. (1993) Spatial model for the population dynamics and exploitation of the Western Australian rock lobster, *Panulirus cygnus*. *Canadian Journal of Fisheries and Aquatic Sciences*, **50**: 1650–62.

Western Australian Fisheries Department (2000) *Annual Report Highlights 1999–2000*. http://www.fish.wa.gov.au/annualreport/ar2000/ar2000high.pdf

Wright, I, Caputi, N. and Penn, J.W. (2006) Depletion-based population estimates for western rock lobster (*Panulirus cygnus*) fishery in Western Australia. *New Zealand Journal of Marine and Freshwater Research*, **40**: 107–22.

Chapter 12
Successes and Challenges in the Hoki Fishery of New Zealand

Mary Livingston and Kevin Sullivan

Introduction

Hoki (*Macruronus novaezelandiae*) (Fig. 12.1) was one of 29 commercial fish species first incorporated into New Zealand's quota management system (QMS) using individual transferable quotas (ITQs) when the QMS was first introduced in 1986 (Annala et al. 1991). With annual total allowable catch (TAC) levels averaging 200 000–250 000 metric tons (MT) until 2003, hoki has been New Zealand's largest and most successful fishery, and remains a major export earner for the fishing industry (http://www.seafood.co.nz). Much of the product is currently exported to European and North American markets as fillets, but some is also sold in New Zealand.

Under New Zealand legislation, the TAC is based on the maximum sustainable yield (MSY) of the fishery, qualified by economic, environmental, and social factors. A formal stock assessment process to estimate MSY in terms of current annual yield (CAY) is followed annually, as defined in the 'Guide to Biological Reference Points for the 2005–07 Yield Fishery Assessment Meetings' (Ministry of Fisheries 2006). The stock assessment includes an estimate of the yield, risk to the stock, and sources of possible errors and uncertainties. The assessment also determines if the stock is currently above, below, or at a level that can produce MSY.

When the QMS was introduced, hoki was thought to consist of a single stock (Patchell 1982), so annual catches were allocated to a single management area, known as HOK1[1]. The stock assessment for hoki has become increasingly complex over time, as more has been discovered about the life history and biology of the fish, including the discovery of a second major spawning ground in 1987 (Livingston 1990). Fish on the two spawning grounds were found to have different growth rates and different morphologies (Horn and Sullivan 1996; Livingston and Schofield

[1] 'HOK' is the shorthand code used for this species, and '1' defines the part of the 200 mile exclusive economic zone (EEZ) where the quota is allocated, which in the case of hoki, is virtually the entire EEZ.

Figure 12.1 Hoki spawning at 400 m depth in Cook Strait Canyon, New Zealand (courtesy of National Institute of Water and Atmospheric Research – NIWA).

1996) and, on this basis, hoki have been assessed as two stocks since 1990, western and eastern, even though the single management area HOK1 has been retained (Annala et al. 1990).

The introduction of the QMS system and a formal stock assessment process led to a major improvement in fisheries management in general and, for hoki, enabled research and monitoring requirements for stock assessment to be clearly identified and implemented. Different methodologies were used to track stock abundance, and state of the art assessment models were developed and modified to utilize as many data sources as possible.

The hoki fishery developed after the boom–bust of the deepwater orange roughy (*Hoplostethus atlanticus*) fishery, and was the first large-scale fishery to become largely operated by New Zealand owned vessels. Until 1987, foreign vessels, operating under joint-venture arrangements with New Zealand companies, took most of the hoki catch. With the introduction of the QMS and a stock assessment indicating an annual yield of 250 000 MT or more in 1986 (Patchell and Coombs 1986), the New Zealand fishing industry expanded and invested in the purchase of vessels to capitalize on their own fish resources. They were, however, cautious and anxious not to exploit the stock beyond sustainable levels, as there was considerable uncertainty about the stock size (Yates and Russell 1987). During the early years of the development of the fishery, TAC levels were easily adjusted more or less by consensus between parliamentary ministers, the Ministry of Fisheries and stakeholders using information from the stock assessment.

Annual catch levels fluctuated between 200 000 and 250 000 MT from 1986 to 2002, and the fishery was regarded as an example of successful and sustainable management strategies within New Zealand's QMS system. In 2001, the Hoki Management Company successfully achieved Marine Stewardship Council (MSC) certification. The MSC is a non-profit organization originally developed by the WWF and Unilever in 1997 to certify seafood products obtained from sustainably-managed fisheries in an environmentally acceptable manner (http://www.msc.org;

Chapter 11). A comprehensive, lengthy, and rigorous evaluation process of the fishery was followed by independent experts contracted by MSC, and certification of the hoki fishery was considered a major success since it was the first large white-fish fishery to achieve certification in the world.

MSC certification raised the profile of environmental issues associated with trawling in New Zealand, and the risk assessment followed by the MSC process identified seabird and marine mammal capture, followed by the effects of trawling on bottom habitats, as being the highest areas of environmental risk posed by the fishery (http://www.msc.org). Following a number of corrective action requirements, some of the more pressing environmental issues were addressed by the industry, as outlined in the website reports. As in other parts of the world, environmental concerns about fishing have been increasingly raised in New Zealand, and the hoki fishery has been taking much of the lead on this, in part because the fishery is large, and in part because of the requirements of the MSC process. The certification period runs for five years, and the hoki fishery is currently (2006) undergoing re-evaluation for recertification.

There are however other challenges facing the hoki fishery. Stock abundance and catch rates have dropped significantly in the past four years, and the assessment in 2005 indicated that the western stock was at or possibly below a sustainable level (Francis 2005). Major cuts to the TAC were implemented in 2004 and 2005, bringing the TAC down from its peak of 250 000 MT to 100 000 MT. This has caused widespread disruption and financial hardship to the industry, with many vessels being sold and fishermen losing jobs. It is clear that poor recruitment of young hoki for a number of years led to the current situation, and management responses to cut quota levels have ensured that exploitation levels have been adjusted appropriately. Nevertheless, some might argue that the cuts did not happen rapidly enough, and the cause of the poor recruitment is not well understood.

In this chapter, we give a broad overview of the biology of hoki, the history of the fishery and we describe how the fishery is monitored. We also discuss some of the confusion that arose from conflicting abundance indices and lessons learnt from a scientific viewpoint in interpreting stock status. The challenge now is to ensure that stocks rebuild to a satisfactory level, and that fishery managers remain sensitive to the uncertainties associated with stock assessment, including fish population and environmental processes, while maintaining a viable business where escalating fuel costs and unfavorable currency exchange rates pose profit limits.

Biology

Life history and productivity

The New Zealand hoki is a deepwater merluccid hake common to New Zealand and South Australian waters and is closely related to the South American species *Macruronus magellanicus*. Other names include whiptail, blue hake, and blue

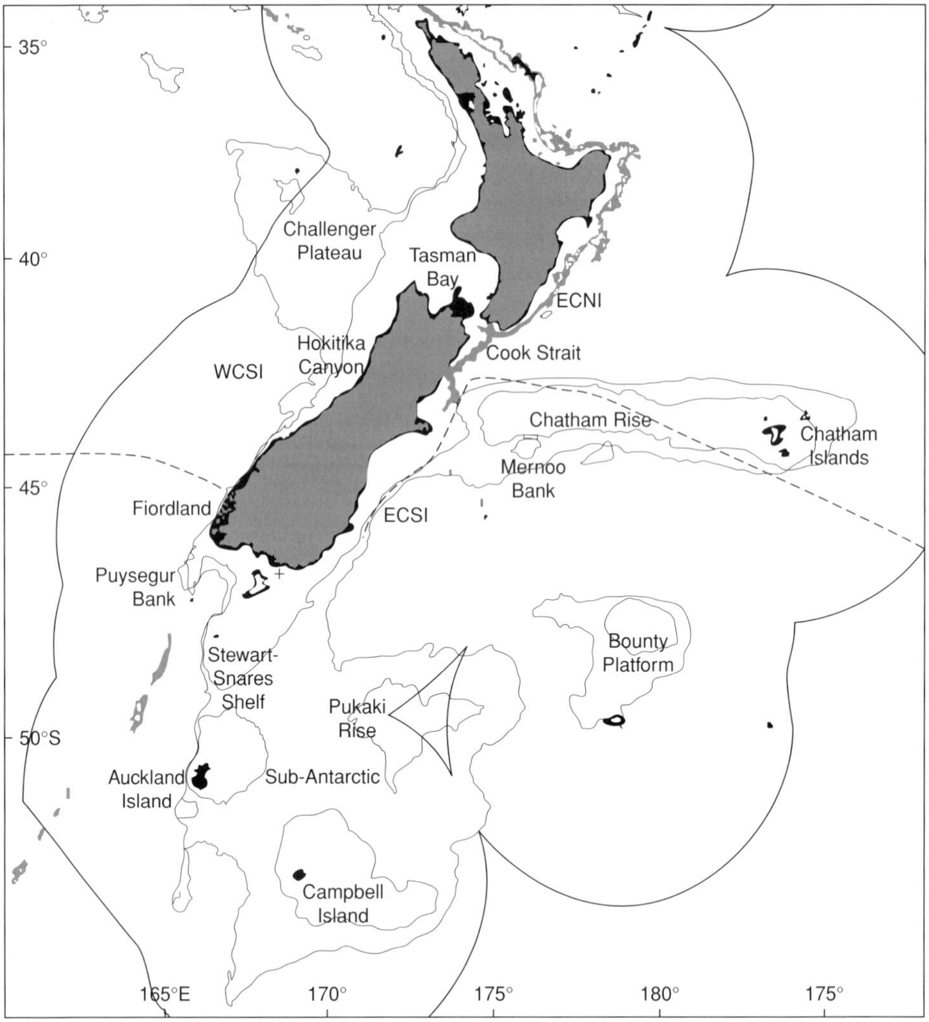

Figure 12.2 Map of New Zealand showing locations mentioned in text. ECNI = east coast of North Island; ECSI = east coast of South Island; WCSI = west coast of South Island. Solid line is the position of the Exclusive Economic Zone boundary. Bathymetry lines are 500 m and 1000 m depths. Dashed line is the approximate position of the Subtropical Front with Subtropical Water to the north and Sub-Antarctic Water to the south.

grenadier. The name 'hoki' is a Maori word meaning 'to return' but it does not appear to bear any relation to the naming of the Hokitika Canyon, where hoki happen to spawn off the west coast of the South Island (Fig.12.2).

Hoki generally dominate bottom-dwelling fish communities in depths of 200–800 m (Francis et al. 2002), and are particularly abundant in cool waters associated with the Subtropical Front and Sub-Antarctic Water south of the front (Fig. 12.2). These waters bathe two bathymetric features, the Chatham Rise and the Campbell Plateau, that lie to the east and south of the South Island respectively (Fig. 12.2) and hoki are widely dispersed across both features. Immature hoki are more usually found around the coast of the South Island as well as in shallow waters across the Chatham Rise (Bull and Livingston 2000). Mature size hoki greater than 80 cm total length are generally found in commercial quantities throughout the year in waters deeper than 400 m, either on the Chatham Rise and Campbell Plateau, or off the west coast of the South Island or in Cook Strait during the spawning season. Large hoki are sometimes caught in mid-water feeding aggregations over deep seamounts east of the Chatham Rise, but do not spawn there (Livingston et al. 2004).

Spawning generally occurs in association with submarine canyon features found around the South Island, particularly Hokitika Canyon off the west coast of the South Island and in Cook Strait (Fig. 12.2). The fish undergo long migrations during May and June from their feeding grounds on the Chatham Rise and Campbell Plateau to their spawning grounds where they form dense spawning schools from July to September. Spawning therefore takes place during the austral winter, and like northern hemisphere temperate spawners, occurs in time for the hatched larvae and juvenile fish to capitalize on the higher productivity and food availability that typically occurs in spring.

Hoki have moderately high fecundity with an average size spawning female of 90 cm total length (TL) releasing a total of about 1 million eggs over a series of spawning events within a season (Langley 1993; Schofield and Livingston 1998). Not all mature-size hoki spawn every year. Winter surveys of both the Chatham Rise and sub-Antarctic have found significant numbers of large hoki with no gonad development at times when spawning is occurring in other areas (Livingston et al. 1991; Hurst and Schofield 1995; Livingston et al. 1997). Histological studies of western stock pre-spawners have shown that the proportion of fish spawning is variable. In 1992, 67% of female hoki aged 7 years and older were developing to spawn (Livingston et al. 1997) but in 1998, this was just 40% (Livingston and Bull 2000). The source of variability in proportion spawning is unknown, and the stock–recruitment relationship is unknown. The stock assessment model currently assumes a steep stock–recruitment relationship of 0.9 and stochastic environmental variables are assumed to drive survival and year-class strength (Francis 2005). The validity of these assumptions are unknown.

Growth and mortality

Growth of hoki is fairly rapid with juvenile hoki reaching about 27–30 cm TL at the end of the first year. There is some variability in growth rates, but hoki reach about 40–45, 50–55 and 60–65 cm TL respectively in the following three years. Most males appear to mature at 60–65 cm TL at 4–5 years, while females mature at 65–70 cm TL at 5–6 years. The growth rate of hoki in all stocks has increased by about 10% since 1983 (Bull and Livingston 2000; Francis 2005). Female growth rates are higher than males, reaching a maximum of 130 cm TL compared with 115 cm TL for males. Horn and Sullivan (1996) estimated growth parameters for eastern and western stocks separately (Table 12.1). Fish from the eastern stock

Table 12.1 Estimates of biological parameters used for stock assessment (M, natural mortality; weight = a (length) b*; K, growth coefficient; $L\infty$, maximal length; T_0, hypothetical age at zero length).

Stock	M	a	b	K	T_0	$L\infty$	Reference
Males	0.30						Sullivan and Coombs (1989)
Females	0.25						"
Combined		0.00479	2.89				Francis (2003)
Western males				0.261	−0.50	92.6	Von Bertalanffy (Horn and Sullivan 1996)
Eastern males				0.232	−1.23	89.5	"
Western females				0.213	−0.60	104.0	"
Eastern females				0.161	−2.18	101.8	"

*a and b are the parameters used in the length–weight relationship for hoki.

sampled in Cook Strait are smaller at all ages than fish from the west coast of South Island (WCSI). Maximum age is from 20 to 25 years, and the instantaneous rate of natural mortality in adults is estimated to be about 0.25 to 0.30 per year.

Stock structure

New Zealand hoki are currently assessed as two stocks centred on two main spawning grounds (the western stock spawning off the west coast of South Island, and the eastern stock spawning in Cook Strait). The current hypothesis is that regardless of where they are spawned, most young hoki move inshore to coastal waters and then to the Chatham Rise nursery grounds (shallow waters of 200–400 m) when they are one to two years of age. As they reach maturity (mostly four to six years of age), they move to deeper waters of the Chatham Rise (to form the eastern stock) or south to the Campbell Plateau in the sub-Antarctic (to form the western stock) where they continue to reside in their respective home grounds for the rest of their lives (Fig. 12.3; Livingston and Bull 2000).

Morphometric and growth rate differences between adult hoki taken from the two stocks clearly demonstrate that there are two subpopulations of hoki (Horn and Sullivan 1996; Livingston and Schofield 1996). Whether or not they reflect genetic differences or they are just the result of environmental differences between the Chatham Rise and sub-Antarctic is not known. No genetic differences have been detected with selectively neutral markers (Smith et al. 1981, 1996), but a low exchange rate between stocks could reduce genetic differentiation.

Currently, the stock assessment model is based on the assumption that hoki spawned in a western stock spawning ground will become part of the western stock as they reach maturity (Bull and Livingston 2000; Livingston et al. 2002).

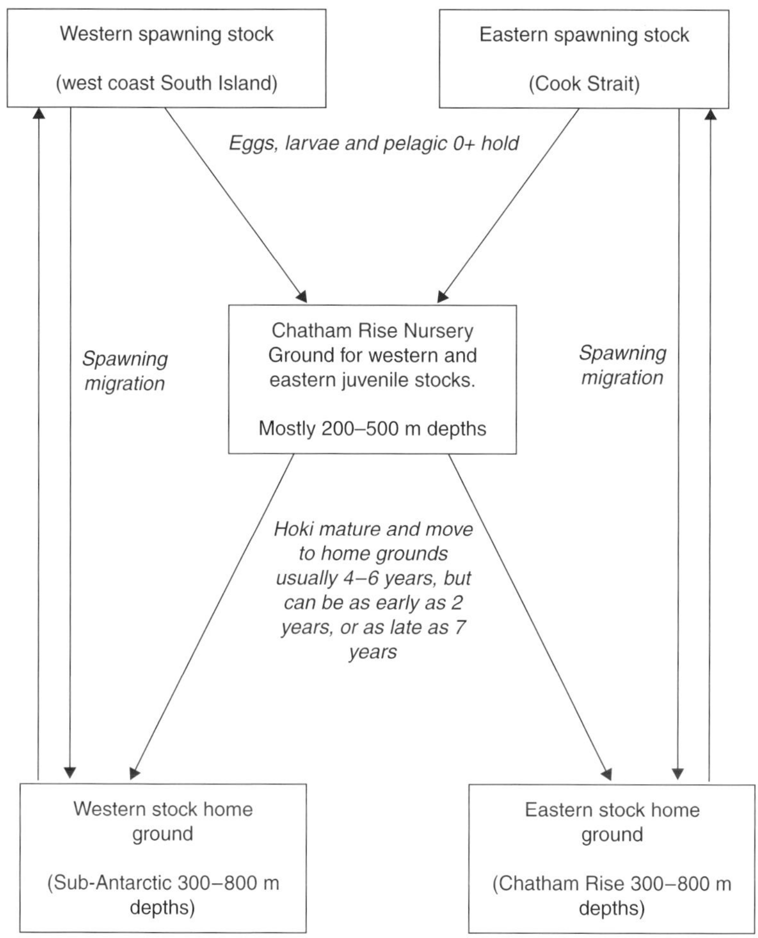

Figure 12.3 The conceptual life cycle of hoki (after Livingston et al. 2002).

History of the hoki fishery

In the early years of the fishery, before the declaration of the EEZ, and subsequently when the QMS was introduced in 1986, most of the catch was taken from spawning fish aggregated off the west coast of the South Island during winter (Fig. 12.4). Spawning aggregations begin to concentrate in depths of 300–700 m over the continental slope around the Hokitika Canyon in late June, spreading further north later in the season. Catches were close to 200 000 MT off the west coast in the late 1980s, with a large-scale *surimi* fleet operating in the area. *Surimi* is widely used in Japan to produce foods such as imitation crabsticks and scallops. The process involves mincing and leaching the fish and then mixing it with sugar and other additives before freezing it as blocks on board *surimi* factory vessels. *Surimi* vessels require very high catch rates, of more than 100 MT per tow, to be cost-effective

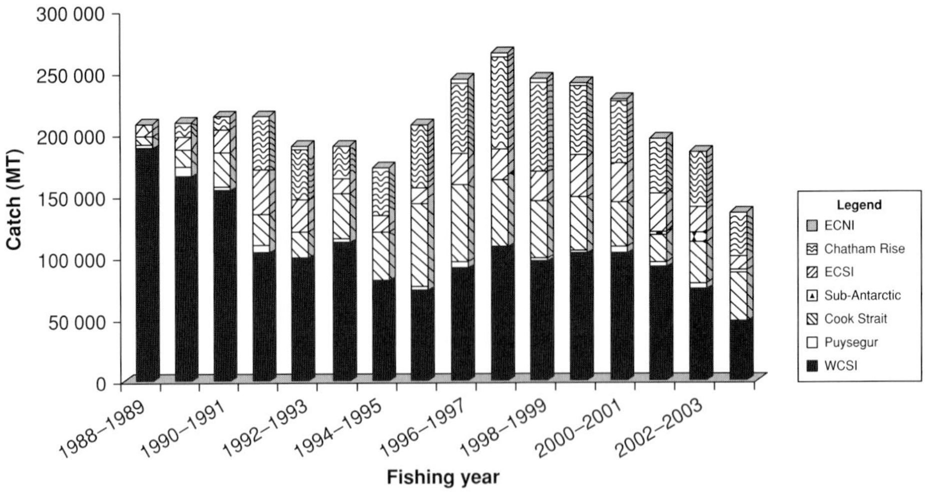

Figure 12.4 Estimated catch (metric tons) of hoki by area, 1988–89 to 2002–03. Fishing year runs from 1 October to 30 September.

and their ability to function in the west coast hoki fishery is indicative of the high catch rates possible at that time.

As accumulated biomass was fished down, catch rates dropped and *surimi* vessels were replaced by filleting vessels. Year-round fisheries developed in the feeding grounds across the Chatham Rise and sub-Antarctic in the early nineties (Fig. 12.4). At this time, the catch became increasingly recruitment-driven. The increased effort on the Chatham Rise resulted in more immature hoki being caught by the 100 mm commercial mesh size permitted on the Chatham Rise.

Catches from the west coast of the South Island spawning fishery declined to less than 50 000 MT in 2003–04 (Fig. 12.4). The head of Hokitika Canyon and the area south to Cook Canyon lie inside a 25 mile restricted fishing zone where only vessels less than 46 m overall length are allowed to fish. Until recently, this gave fish spawning inside the 25 mile zone some protection from disruption by the commercial fleet. In recent years, however, increasing numbers of smaller vessels have been working inside the 25 mile zone, and in the 2004 spawning season, 43% of the west coast catch came from inside the 25 mile zone (O'Driscoll et al. 2005). This almost certainly reflects the lack of fish outside the 25 mile zone, and the shrinkage in size of the main spawning schools.

After the discovery of spawning hoki in Cook Strait in 1986 (Livingston 1990; Murdoch et al. 1990) a second spawning fishery developed. Like the west coast, the spawning season in Cook Strait also runs during winter. This fishery lies entirely within a 25 mile restricted fishing zone and is only carried out by small vessels. Catches in the spawning Cook Strait fishery peaked around 67 000 MT in 1995/96 but have since steadily declined to about 30–40 000 MT (Fig. 12.4). Catches in Cook Strait continue to form about 30% of the total catch (Fig. 12.5). Small catches of

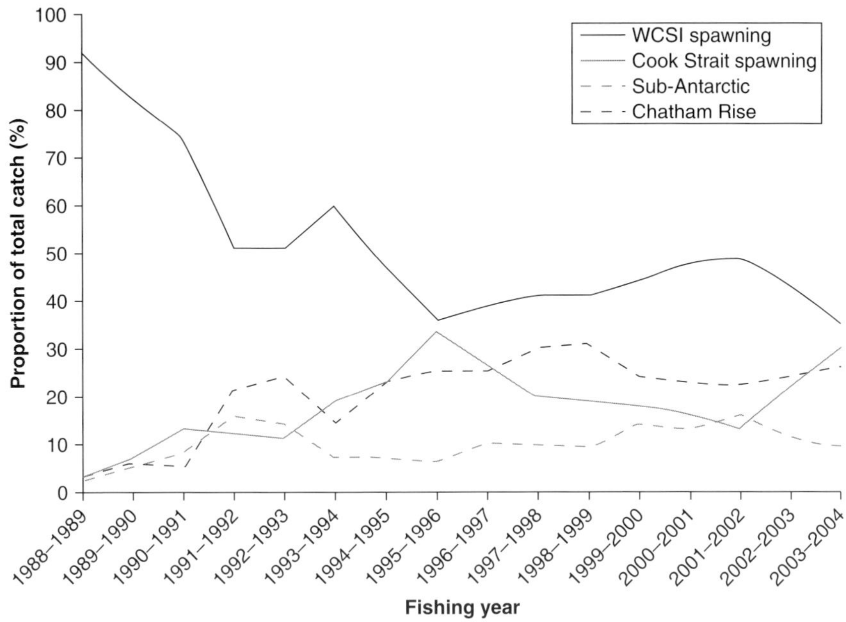

Figure 12.5 Proportions of annual catch in spawning and non-spawning fisheries.

spawning hoki are taken from other spawning grounds off the east coast of the South Island (ECSI) and late in the season at Puysegur Bank.

The changes in patterns of fishing for hoki are reflected in the proportions of the annual catch taken in each area (Fig. 12.5). In the early years, over 90% of the total catch was taken in the Westland spawning fishery. In 2003/04 about 35% of the annual catch came from Westland and 30% came from Cook Strait. The Chatham Rise and sub-Antarctic fishing ground catches have fluctuated over the years, but in 2003/04 they represented 26% and 9% of the annual catch respectively. Puysegur and ECNI fisheries remain at relatively low and stable levels, but east coast South Island catches more than doubled in 2002/03 (Fig. 12.4). This was largely due to increased fishing effort in Pegasus Canyon, a small spawning ground located to the west of Mernoo Bank on the Chatham Rise (Fig. 12.2) that is considered to be part of the eastern spawning stock (Ministry of Fisheries 2006).

Other changes to commercial operations include the type of trawl gear used and the type of processing carried out at sea. Following the decline in the *surimi* component of the catch, smaller vessels have formed an increasing portion of the fleet enabling flexibility in fishing location and reducing fuel and operational costs. The hoki fishery now operates throughout the year, producing high quality fillet product from both spawning and non-spawning fisheries (Dunn and Livingston 2004). More recently, the use of twin-trawl rigs (essentially two nets deployed from the same vessel) has been increasing in the non-spawning hoki fisheries to improve catch rates both of hoki and of the associated by-catch.

In spawning fisheries, vessels generally use mid-water trawl nets to catch the fish that are mostly schooled in mid-water to spawn. In some parts of the fishing grounds, this gear is used along the seafloor, similar to a bottom trawl. In recent years, the Hoki Fisheries Management Company has developed a code of practice to minimize exploitation of young hoki and also to reduce effort on spawning stocks in recent years (Ministry of Fisheries 2006).

Research and stock assessment

Research to assess the stock status of hoki has concentrated on the biology of the species (as outlined in the life history and productivity section) and monitoring the abundance of hoki. Methods to monitor abundance have included trawl surveys, acoustic surveys and the analysis of catch and effort (CPUE) statistics in the commercial fishery.

Trawl survey time series

Bottom trawl surveys of the Chatham Rise and sub-Antarctic have been undertaken annually since 1992 during summer, when hoki are assumed to have completed the return migration from spawning and to be at their most dispersed. The Chatham Rise surveys monitor juvenile biomass (both stocks combined – see Fig. 12.3) and eastern stock adult biomass (>7 years). The eastern stock abundance shows a steady decline to about 25% of initial trawl survey abundance levels in 1993 (Livingston and Stevens 2005; Fig. 12.6). The sub-Antarctic surveys monitor western stock adult biomass, and after being re-instated in 2000 following a seven year gap, (O'Driscoll and Bagley 2001), trends there are also consistently downward (Fig. 12.7). The time series are used in a relative sense and the catchability of surveys are calculated at 0.15–0.65 (Chatham Rise) and 0.17–0.61 (sub-Antarctic) (Francis 2005). This is relatively low, and for many years the trawl surveys were considered to be unreliable indicators of hoki abundance. As catch rates declined, however, the trends in both areas proved to be the most reliable indicators of stock abundance, and have been given a high relative weighting in the stock assessment model (Francis 2005).

Acoustic time series

Acoustic surveys have been carried out periodically on winter spawning grounds off the west coast of the South Island, and more regularly in Cook Strait. Acoustic methods were considered more useful for spawning hoki than dispersed hoki because, in theory, only a small proportion of the acoustic signals are from other species when the hoki are densely aggregated to spawn. Target strengths have not been refined sufficiently to allow absolute estimation of hoki abundance, however, by assuming that mean residence time on the spawning grounds and season length are constant, the acoustic indices were treated as a time series of relative abundance.

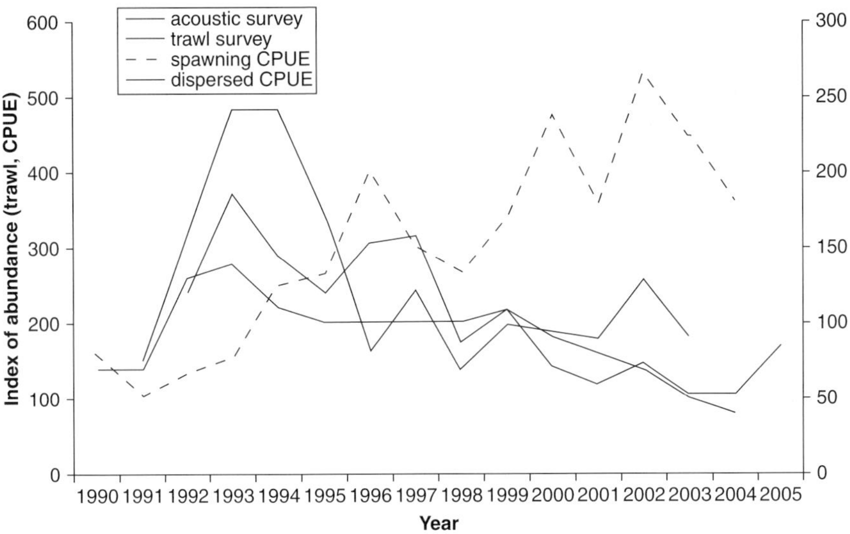

Figure 12.6 Abundance indices of the eastern hoki stock. (CPUE = unstandardized catch per unit effort).

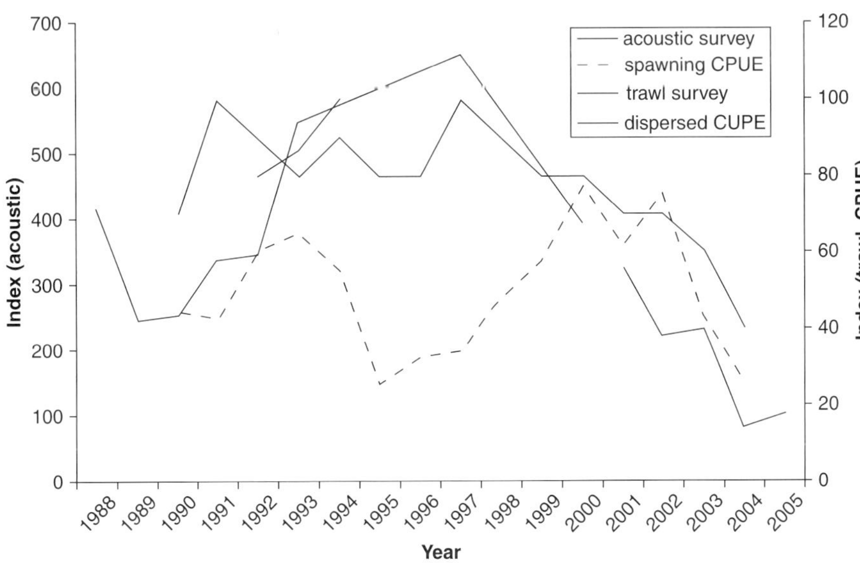

Figure 12.7 Abundance indices of the western hoki stock. Note break in line of sub-Antarctic trawl survey index from 1994 to 2001 due to absence of trawl survey data.

The Cook Strait acoustic surveys of spawning hoki have provided indices of abundance (O'Driscoll and McMillan 2004) that reflect the trends seen in the Chatham Rise trawl survey series, giving reasonable confidence in acoustic methods as a means of tracking abundance in the eastern hoki stock (Fig. 12.6). Acoustic abundance indices of hoki off the west coast of the South Island showed, however, an upward trend at a time when indices from other parts of the fishery were trending downward in 1997 (Fig. 12.7). Difficulties with estimating species mix in large areas of the west coast particularly to the north had a major effect on estimating western spawning abundance and were revised several times until the trawl survey time series of the western stock on their home grounds in the sub-Antarctic was reinstated in 2000 and the low level of the western stock was confirmed. The west coast acoustic surveys were then discontinued (Annala et al. 2003). Although acoustic estimates of relative abundance are used in the stock assessment, they are down-weighted compared with the trawl survey inputs.

Catch per unit effort

A range of statistical modeling processes for estimating standardized CPUE indices for each hoki fishery have been developed for both western and eastern spawning stocks and non-spawning stocks (Vignaux 1994; Langley et al. 2001; Dunn and Livingston 2004). However, in 2002, CPUE indices were finally rejected as reliable indices of abundance for input to the hoki stock assessment because high catch rates were maintained on spawning grounds even though biomass indices were declining elsewhere (Annala et al. 2003). Inclusion of model runs that used positive CPUE trends in the western stock in particular served to confuse the assessment results and give a more positive interpretation of stock status than was the case. Although high catch rates were maintained until 2001 (Fig. 12.7) they did eventually come down as the western stock became more severely depleted.

Unstandardized CPUE statistics are updated annually as part of the descriptive update on characterization of the hoki fishery, and are reported as metric tons per hour for each fishery (O'Driscoll et al. 2005). Although not used as an index of abundance in the stock assessment, the unstandardized indices on the Chatham Rise and sub-Antarctic do follow the same trends as the biomass indices (Figs 12.6, 12.7).

Reliability of abundance estimates

Three methods to estimate hoki stock abundance have been used within the history of formal stock assessments for hoki: trawl surveys on the home grounds, acoustic surveys of the spawning grounds and CPUE in each of the four fisheries. In the eastern stock, acoustic and trawl survey indices of abundance have followed similar trends. By continuing both time series, a cross-check on the link between spawning stock biomass and dispersed biomass was maintained (Fig. 12.6). In the western stock, however, trawl surveys for hoki during the dispersed phase were stopped as at that time acoustic surveys and CPUE indices of the western spawning stock were thought to provide adequate monitoring of the stock. Over time, it became apparent that there were difficulties associated with estimating the species mix in the northern part of the western spawning grounds, and that as the fishery changed from a *surimi* fleet to a filleting fleet with different fishing gear and different selectivities, CPUE indices also became difficult to interpret (Fig. 12.7). CPUE indices of the western spawning stock tracked downward between 1992–1995 but then steadily rose until 2001.

When trawl surveys of the sub-Antarctic were re-instated in 2000, abundance indices confirmed that biomass in the western stock had indeed declined considerably since 1994. As recruitment failed and trawl survey abundances continued to trend downward, CPUE indices off the west coast remained high until 2003 when the effect of the low abundance and collapse of the western stock fishery finally affected catch rates as well. In recent stock assessments, more weight has been given to the trawl surveys, which despite their catchability coefficients, appear to be reflecting relative changes in stock abundance more reliably than other indices (Francis 2005).

The confusion created by conflicting abundance trends in a very short time series by world standards, made it difficult for fisheries scientists and managers to reliably determine the status of the hoki stocks. Over time, it has become clear that even though the catchability of the trawl surveys is low, the surveys are actually the most reliable indices of hoki stock abundance.

Sustainability of the hoki fishery

Change to status of the hoki stocks

The general-purpose stock assessment program, named CASAL (Bull et al. 2003), was used to update the 2004 hoki stock assessment in 2005 (Francis 2005). The model is age-structured and partitions the population into four areas (Chatham Rise – CR, West Coast South Island – WC, sub-Antarctic – SA, Cook Strait – CS), and two stocks (east – E, west – W). It is based on the conceptual model shown in Fig. 12.3, and is structurally complex (Francis 2005). There is still some uncertainty over stock structure and maturity estimates for the two stocks, and migration estimates between the nursery grounds and the adult stocks. This all adds to uncertainty in determining stock status (Ministry of Fisheries 2006).

A major decline was seen in catches from western stock areas during 2003/04. Although the west coast spawning fishery remained the largest hoki fishery, the catch taken from this area in 2003/04 at 47 000 MT was the lowest since 1995/96 (Fig. 12.4). The difference between the outlook following 1995/96 and the current situation is that relatively strong juvenile year classes on the Chatham Rise from 1991–1994 were predicted to enter into the fishery in the years following 1995/96. Recruitment since then has, however, been relatively weak (Fig. 12.8), although there is cautious optimism about stronger year classes of one and two year olds on the Chatham Rise spawned in 2002 and 2003 (Livingston and Stevens 2005).

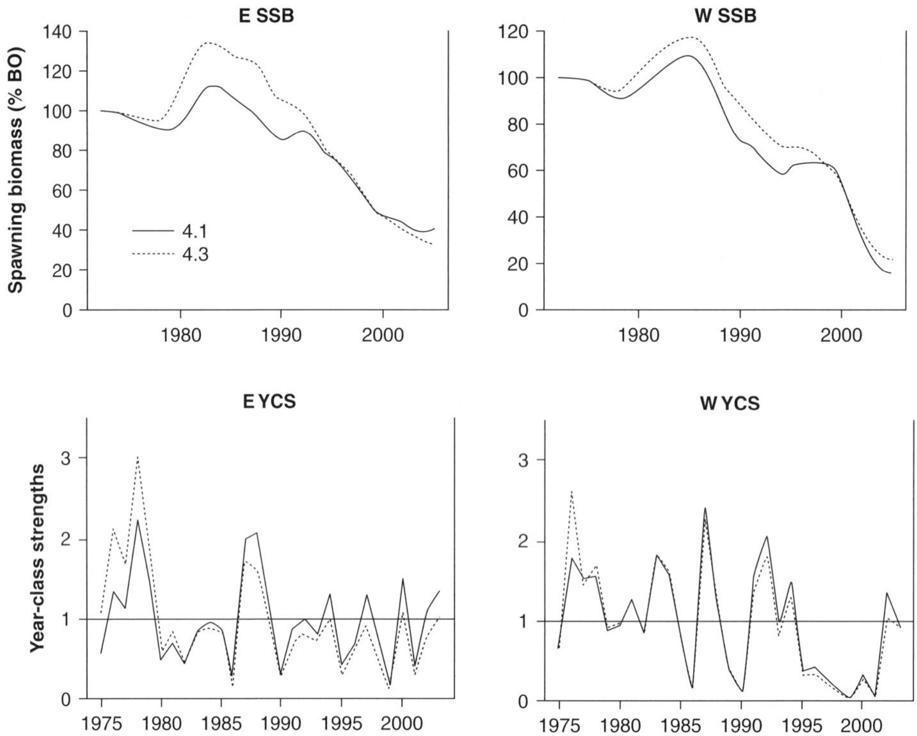

Figure 12.8 Year-class strengths and biomass trajectories from the 2005 hoki stock assessment (after Francis 2005). E = eastern stock, W = western stock, SSB = spawning stock biomass, YCS = year-class strength. Solid line indicates main base case run of model, dotted line shows an alternative run (Francis 2005).

However, there is a three to four year time lag before these fish recruit to their home grounds, and the proportion that will recruit to the western stock is unknown.

The most recent status of the western stock, as given in the Ministry of Fisheries Plenary Document resulting from the stock assessment of hoki in 2006 confirmed that the western stock is just above 20% of its virgin biomass (20% of what it would be if there was no fishing), while the eastern stock is estimated to be 33–49% of virgin biomass (Fig. 12.8). Following the period of poor recruitment from 1995 to 2001, there are signs of improved recruitment in 2002 and 2003 in the pre-recruit surveys of the Chatham Rise (Fig. 12.8), but model projections of biomass suggest only a slow rebuilding rate (Francis 2005; Ministry of Fisheries 2006), and the TAC will need to remain at 100 000 MT in the meantime to maintain sustainability and facilitate rebuilding.

Lessons learnt

It has become clear that to maintain the hoki fishery at a sustainable level, it is essential to monitor stock abundance using reliable, fishery-independent time series of abundance. Because the reliability of different abundance indices was not clear in earlier years, there was uncertainty about the steepness of the decline in abundance, and therefore a time lag in the management response. Once the effects of poor recruitment were clearly reflected in the two trawl survey abundance series and it became clear that CPUE indices were unreliable for monitoring spawning stock biomass, decisive and large cuts in quota were made. Such sudden and large cuts in quota, however, have had serious social and economic effects on the hoki fishery. Vessels have had to be sold, fishermen have been laid off, and there have been major adjustments in the fish management companies to rationalize overhead costs.

The reasons underlying the poor recruitment to the western stock that led to such a steep decline in abundance are not clear and remain a challenge for fishers, scientists, and fishery managers. It is likely that there is a climatic component to the decline in recruitment although this may not be the full picture. Early studies on climatic variation and year-class strength of hoki found a strong correlation between year-class survival and variables such as the Southern Oscillation Index (SOI), sea surface temperatures (SSTs), and westerly wind flow, suggesting that favorable conditions for hoki larvae occurred under cooler periods of greater westerly flow and El Niño conditions (Livingston 2000; Bull and Livingston 2001). Hoki larvae are adapted to low density food availability when they first hatch and move inshore toward more productive feeding grounds in the first few weeks of life (Murdoch 1992). Livingston (2000) hypothesized that the climate conditions that correlated with strong year classes promoted upwelling from deeper offshore waters towards the coast, and also resulted in a more rapid deepening of the mixed layer, leading to improved food availability for newly hatched larvae.

A recent and statistically more rigorous study that repeated and updated the model used by Bull and Livingston (2001), found that these climate variables and new inputs, such as nitrate concentration on the spawning grounds, were no longer valid predictors of model estimations of hoki year-class strength (Francis et al. 2005). Some of the discrepancies between this recent study and the earlier studies arose from increasing the statistical rigor, and from changes in the way that year-class strengths are estimated by the stock assessment model, and simply, the short period over which the original relationship was observed. It was noted, however, that some of the original predictors such as the SOI worked quite effectively for the juvenile one and two year old trawl survey abundance indices on the Chatham Rise. However, because the two stocks are mixed at this stage of the life cycle, it is difficult to interpret the results in terms of two separate spawning stocks (Francis et al. 2005).

It seems to us that environmental conditions must play a role in larval survival, even if the underlying mechanisms are unknown and predictive models cannot be derived from correlation studies. Long-term cycles such as the Interdecadal Pacific Oscillation (IPO) that can lead to regime shifts (Salinger et al. 2001) and shifts in the prevailing ecology of the marine environment may yet prove critical to the abundance of dominant species such as hoki, and the harvest strategy required to manage the fishery. Bull and Livingston (2001) summarized critical processes at different stages in the life history of hoki where vulnerability to changing environment conditions could increase mortality or contribute to poor recruitment, and although larval survival was identified as the most critical process, others such as food availability prior to spawning migrations may also be important.

There are a number of indicators, other than biomass and catch levels, that may also be useful in alerting scientists and managers to profound biological changes in the hoki population, and could have been taken into account when trying to interpret the status of the stocks. For example, growth rates of juvenile hoki in both stocks have risen by almost 10% since 1992 (Livingston et al. 2002; Francis 2005), and the size at maturity of the western stock may have changed to smaller size hoki in recent years (Livingston and Bull 2000). The age structure of the western stock has changed dramatically since the early fishery years (Francis 2005) and sex ratios on spawning grounds have become skewed towards females (Livingston et al. 2000). The combined effect of these changes on the spawning success of hoki, and therefore on recruitment success is essentially unknown. However, significant biological changes may have rendered the smaller stock more vulnerable to oceanic fluctuations and climate variability, and perhaps more vulnerable to physical disruption during the spawning season by the fleet. These and other population responses to fishing pressure such as changes in the duration and the timing of spawning season, changes in fecundity, the effect of size on egg viability, and a range of indices of fish condition have been observed in cod fisheries elsewhere, and are undergoing critical examination to identify key factors affecting fishery collapse (ICES 2005).

There is also a general concern among industry and the science community that the mortality on juvenile hoki may be much higher than the reported catch, and that this may also have contributed to a steeper decline in stock abundance than predicted by stock assessments. The Hoki Management Company has a voluntary code of practice in place that aims to not only reduce fishing effort on spawning grounds by partial closures (west coast of the South Island and Cook Strait) and full closures (Puysegur and Pegasus Canyon) during the spawning season, but also to reduce fishing effort on juvenile hoki on the Chatham Rise. A large area of the Chatham Rise where juvenile hoki occur in depths shallower than 450 m is now closed to target fishing. This precaution is to enable young fish to grow through to spawning maturity and to improve the yield from the fish (http://www.seafood.co.nz).

Future challenges for the hoki fishery

The hoki fishery has not reached the severely depleted levels seen in the northern cod. But there is little room for complacency. On the positive side, there is acceptance among industry and managers that long-term monitoring of fish stocks, and ongoing research to improve and investigate methodologies to assess stocks are an essential part of sustainable fisheries management. There is also broader acceptance in New Zealand that some of the effects of fishing may be deleterious both to habitats and to other organisms that are part of the ecosystem. Methods to monitor and evaluate broader changes are currently being explored worldwide (Blanchard et al. 2005; Fulton et al. 2005; Jennings and Dulvy 2005; Link 2005; Mueter and Megrey 2005; Yemane et al. 2005), and in New Zealand, fisheries managers and scientists are looking at methodologies that may be applicable to hoki and other fisheries. Recently, a comprehensive 'Strategy for Managing the Effects of Fishing' has been released by the Ministry of Fisheries, providing a framework for meeting the legislative requirements to address adverse effects of fishing on the environment (http://www.fish.govt.nz). The strategy outlines policies that are consistent with maximizing the value that New Zealanders can obtain through the sustainable use of fisheries resources and protecting the environment. The strategy allows for the development of environmental standards by assessing the effects of fishing on a fishery-by-fishery basis, and assessing the aggregate effects of all relevant fisheries.

Several research programs are now underway to determine the extent and effects of hoki and other fisheries on bottom habitat (Baird et al. 2002; Booth et al. 2002) and by-catch, including commercial and non-commercial species, (Livingston et al. 2003; Anderson and Smith 2005) and any threatened species such as the basking shark (Francis and Duffy 2002). The extent of bottom trawling has been determined (Baird et al. 2002) and scientists are developing metrics that provide relative indices for the level of impact (Clark and O'Driscoll 2003; O'Driscoll and Clark 2005). Information about bottom habitat and the associated communities is, however, patchy, and managers are currently working from first ecological principles to determine the likely immediate and aggregative effects of bottom trawling not only on the benthic habitat, but on ecosystem processes such as bentho-pelagic coupling, before determining what environmental standards can usefully be developed, and methods to achieve those standards (Cryer et al. 2005).

Hoki form such a dominant component of the mid-slope fish community (Bull et al. 2001) and there is concern that reduction in such a major component may have trophic implications to predator and prey relations. A new study to determine key trophic interrelationships among hoki and other mid-slope fishes on the Chatham Rise has recently been initiated (http://www.fish.govt.nz).

There has been a major drive to develop fishing practices in the hoki fleet that avoid or mitigate the incidental catch of seabirds by using tori lines, Brady bafflers and controlling offal wastes from the processed catch (http://www.hokinz.com). Sea lion escape devices (SLEDs) have also been trialed for use in trawl nets to minimize the incidental catch of mammals in trawl fisheries (Baird 2005a, b). These efforts have partly been in response to Marine Stewardship Council certification requirements, but also to meet New Zealand legislative requirements. The New Zealand Government has also developed a national plan of action to address seabird capture in fisheries.

Summary

Hoki forms New Zealand's largest fishery with annual TACs that were formerly around 200 000 to 250 000 MT. Recently, an extended period of poor recruitment has led to a rapid decline in stock biomass and fisheries managers have responded with large quota cuts down to 100 000 MT. In addition, the Hoki Fisheries Management Company has modified its voluntary code of practice to minimize capture of immature juvenile hoki, and reduce effort during the spawning season (http://www.seafood.co.nz). Fishing companies have had to diversify and reduce capacity within the fishery. The importance of reliable fishery-independent time series of relative abundance has been recognized and supported, and the hoki fishery, with its history of Marine Stewardship Council certification is now exploring ways that broader issues such as the effects of fishing on by-catch and on the environment can be assessed and minimized. The ongoing importance of the hoki fishery to New Zealand's fishing industry will set the pace and direction of this challenge over the next few years while policies and strategies to incorporate broader environmental concerns into fisheries management are developed. The Ministry of Fisheries has also demonstrated a commitment to adjusting catch levels as required, and to ensuring that adequate information to monitor the status of hoki stocks is maintained.

References

Anderson, O.F. and Smith, M.H. (2005) *Fish Discards and Non-target Fish Catch in the New Zealand Hoki Trawl Fishery, 1999–2000 to 2002–03*. New Zealand Fisheries Assessment Report 2005/3.

Annala, J.A. (Comp.) (1990) *Report from the Fishery Assessment Plenary, April–May 1990: Stock Assessments and Yield Estimates*. New Zealand Ministry of Fisheries Publication.

Annala, J.H., Sullivan, K.J. and Hore, A.J. (1991) Management of multispecies fisheries in New Zealand by individual transferable quotas. *ICES Marine Science Symposium*, **193**: 321–9.

Annala, J.H., Sullivan, K.J., O'Brien, C.J., Smith, N.W. and Grayling, S.M. (2003) *Report from the Fishery Assessment Plenary, April–May 2003: Stock Assessments and Yield Estimates*. New Zealand Ministry of Fisheries Publication.

Baird, S.J. (2005a) *Incidental Capture of Seabird Species in Commercial Fisheries in New Zealand Waters, 2002–03*. New Zealand Fisheries Assessment Report 2005/2.

Baird, S.J. (2005b) *Incidental Capture of New Zealand Fur Seals* (Arctocephalus forsteri) *in Commercial Fisheries in New Zealand Waters, 2002–03*. New Zealand Fisheries Assessment Report 2005/12.

Baird, S.J., Bagley, N.W., Wood, B.A., Dunn, A. and Beentjes, M. (2002) *The Spatial Extent and Nature of Mobile Bottom Fishing Methods within the New Zealand EEZ 1989–90 to 1998–99*. Final Research Report for MFish Project ENV2000/05.

Blanchard, J.L., Dulvy, N.K., Jennings, S., Ellis, J.R., Pinnegar, J.K., Tidd, T. and Kell, L.T. (2005) Do climate and fishing influence size-based indicators of Celtic Sea fish community structure? *ICES Journal of Marine Science*, **62**: 405–11.

Booth, J.D., Baird, S.J., Stevenson, M.L., Bagley, N.W. and Wood, B.A. (2002) *Review of Technologies and Practices to Reduce Bottom Trawl Bycatch and Seafloor Disturbance in New Zealand*. Final Research Report for MFish Project ENV2000/06.

Bull, B. and Livingston, M.E. (2000) *Hoki Migration Patterns: an Analysis of Commercial Catches in New Zealand Waters 1985–99*. NIWA Client Report 2000/63.

Bull, B. and Livingston, M.E. (2001) Links between climate variation and the year class strength of New Zealand hoki (*Macruronus novaezelandiae*): an update. *New Zealand Journal of Marine and Freshwater Research*, **35**: 871–80.

Bull, B., Livingston, M.E., Hurst, R.J. and Bagley, N. (2001) Upper-slope fish communities on the Chatham Rise, New Zealand, 1992–99. *New Zealand Journal of Marine and Freshwater Research*, **35**: 795–815.

Bull, B., Francis, R.I.C.C., Dunn, A., McKenzie, A., Gilbert, D.J. and Smith, M.H. (2003) *CASAL (C++ Algorithmic Stock Assessment Laboratory): CASAL User Manual v2.01-2003/8/01*. NIWA Technical Report 124.

Clark, M.R. and O'Driscoll, R.L. (2003) Deepwater fisheries and their impacts on seamount habitat in New Zealand. *Journal of Northwest Atlantic Fishery Science*, **31**: 441–58.

Cryer, M., Nodder, S., Thrush, S., Lohrer, D., Gorman, R., Vopel, V. and Baird, S. (2005) *Effects of Trawling and Dredging on the Benthopelagic Coupling Processes in the New Zealand EEZ*. New Zealand Aquatic Environment and Biodiversity Report ENV2001/09.

Dunn, A. and Livingston, M.E. (2004) *Updated Catch-Per-Unit-Effort Indices for Hoki* (Macruronus novaezelandiae) *on the West Coast South Island, Cook Strait, Chatham Rise, and Sub-Antarctic for the Years 1990 to 2002*. Draft New Zealand Fisheries Assessment Report 35.

Francis, M.P. and Duffy, C. (2002) Distribution, seasonal abundance and bycatch of basking sharks (*Cetorhinus maximus*) in New Zealand, with observations on their winter habitat. *Marine Biology*, **140**: 831–42.

Francis, M.P., Hurst, R.J., McArdle, B., Bagley, N.W. and Anderson, O.F. (2002) New Zealand demersal fish assemblages. *Environmental Biology of Fishes*, **62**: 215–34.

Francis, R.I.C.C. (2003) *Analysis Supporting the 2002 Stock Assessment of Hoki*. New Zealand Fisheries Assessment Report 2003/5.

Francis, R.I.C.C. (2005) *Assessment of Hoki* (Macruronus novaezelandiae) *in 2004*. New Zealand Fisheries Assessment Report 2005/35.

Francis, R.I.C.C., Grieve, J.M., Hadfield, M.G., Renwick, J.A. and Sutton, P.J.H. (2005) *Predicting Hoki Year-class Strengths*. Report WG-HOK-2005/14, Ministry of Fisheries, Wellington.

Fulton, E.A., Smith, A.D. and Punt, A.E. (2005) Which ecological indicators can robustly detect effects of fishing? *ICES Journal of Marine Science*, **62** (3): 540–51.

Horn, P.L. and Sullivan, K.J. (1996) Validated aging methodology using otoliths, and growth parameters for hoki (*Macruronus novaezelandiae*) in New Zealand waters. *New Zealand Journal of Marine and Freshwater Research*, **30**: 161–74.

Hurst, R.J. and Schofield, K.A. (1995) *Winter and Summer Trawl Surveys of Hoki South of New Zealand, 1990*. New Zealand Fisheries Technical Report 43.

ICES (2005) *Spawning and Life History Information for North Atlantic Cod Stocks*. ICES Cooperative Research Report 274.

Jennings, S. and Dulvy, N.K. (2005) Reference points and reference directions for size-based indicators of community structure? *ICES Journal of Marine Science*, **62**: 397–404.

Langley, A.D. (1993) *Spawning Dynamics of Hoki in the Hokitika Canyon*. New Zealand Fisheries Technical Report 34.

Langley, A.D., Dunn, A. and Ballara, S.L. (2001) *Determination of CPUE Indices from Catch Effort for Hoki* (Macruronus novaezelandiae) *for the 1986–87 to 1988–99 Fishing Years*. New Zealand Fisheries Assessment Report 2001/49.

Link, J.S. (2005) Translating ecosystem indicators into decision criteria? *ICES Journal of Marine Science*, **62**: 569–76.

Livingston, M.E. (1990) Spawning hoki (*Macruronus novazealandiae* Hector) concentrations in Cook Strait and off the east coast of the South Island, New Zealand, August–September 1987. *New Zealand Journal of Marine and Freshwater Research*, **34**: 55–69.

Livingston, M.E. (2000) Links between climate variation and the year class strength of New Zealand hoki (*Macruronus novazealandiae*) Hector. *New Zealand Journal of Marine and Freshwater Research*, **34**: 55–69.

Livingston, M.E. and Bull, B. (2000) *The Proportion of Female Hoki Developing to Spawn on the Southern Plateau, April 1998*. New Zealand Fisheries Assessment Report 2000/13.

Livingston, M.E. and Schofield, K.A. (1996) Stock discrimination of hoki (*Macruronus novaezelandiae* Merlucciidae) in New Zealand waters, using morphometrics. *New Zealand Journal of Marine and Freshwater Research*, **30**: 197–208.

Livingston, M.E. and Stevens, D.W. (2005) *Trawl Survey of Hoki and Middle Depth Species on the Chatham Rise, January 2004 (TAN0401)*. New Zealand Fisheries Assessment Report 2005/21.

Livingston, M.E., Uozumi, Y. and Berben, P.H. (1991) *The Abundance, Distribution and Spawning Condition of Hoki and Other Mid-slope Species on the Chatham Rise, July 1986*. New Zealand Fisheries Technical Report 25.

Livingston, M.E., Vignaux, M. and Schofield, K.A. (1997) Estimating the annual proportion of non-spawning adults in the New Zealand hoki, *Macruronus novaezelandiae* Hector. *Fishery Bulletin*, **95**: 99–113.

Livingston, M.E., Bull, B. and Gilbert, D.J. (2000) *The Observed Sex Ratios of Hoki* (Macruronus novaezelandiae) *in New Zealand, 1983–99*. New Zealand Fisheries Assessment Report 2000/24.

Livingston, M.E., Bull, B. and Stevens, D.W. (2002) *Migration Patterns During the Life-cycle of Hoki (Macruronus novaezelandiae): an Analysis of Trawl Survey Data in New Zealand Waters 1991–2002*. Final Research Report for Ministry of Fisheries Research Project HOK2000/01 Objective 6.

Livingston, M.E., Clark, M. and Baird, S.J. (2003) *Trends in Incidental Catch of Major Fisheries on the Chatham Rise for Fishing Years 1989–90 to 1998–99*. New Zealand Fisheries Assessment Report 2003/52.

Livingston, M.E., Stevens, D.W., O'Driscoll, R.L. and Francis, R.I.C.C. (2004) *Trawl Survey of Hoki and Middle Depth Species on the Chatham Rise, January 2003 (TAN0301)*. New Zealand Fisheries Assessment Report 2004/16.

Ministry of Fisheries (2006) *Report from the Fishery Assessment Plenary, May 2006*. New Zealand Ministry of Fisheries Publications.

Mueter, F.J. and Megrey, B.A. (2005) Distribution of population-based indicators across multiple taxa to assess the status of Gulf of Alaska and Bering Sea groundfish communities. *ICES Journal of Marine Science*, **62**: 344–52.

Murdoch, R.C. (1992) A review of the ecology of hoki *Macruronus novaezelandiae* (Hector), larvae in New Zealand waters. *Bureau of Rural Resources Proceedings*, **15**: 3–16.

Murdoch, R.C., Guo, R. and McCrone, A. (1990) Distribution of hoki (*Macruronus novaezelandiae*) eggs and larvae in relation to hydrography in eastern Cook Strait, September 1987. *New Zealand Journal of Marine and Freshwater Research*, **24**: 533–43.

O'Driscoll, R.L. and Bagley, N.W. (2001) *Review of Summer and Autumn Trawl Survey Time Series from the Southland and Sub-Antarctic Area 1991–98*. NIWA Technical Report 102.

O'Driscoll, R.L and Clark, M.R. (2005) Quantifying the relative intensity of fishing on New Zealand seamounts. *New Zealand Journal of Marine and Freshwater Research*, **39**: 839–50.

O'Driscoll, R.L. and McMillan, P.J. (2004) *Acoustic Survey of Spawning Hoki in Cook Strait During Winter 2003*. New Zealand Fisheries Assessment Report 2004/20.

O'Driscoll, R.L., Phillips, N.L., Ballara, S.L., Stevens, D.W. and Smith, M.H. (2005) *Catches, Size, and Age Structure of the 2003–04 Hoki Fishery, and a Summary of Input Data used for the 2005 Stock Assessment*. New Zealand Fisheries Assessment Report 2005/37.

Patchell, G.J. (1982) The New Zealand hoki fisheries 1972–82. *Fisheries Research Division Occasional Publication* **38**.

Patchell, G.J. and Coombs, R.F. (1986) Hoki. In: *Background Papers for the Total Allowable Catch Recommendations for the 1986–87 New Zealand Fishing Year*. Former New Zealand Ministry of Agriculture and Fisheries Publication.

Salinger, M.J., Renwick, J.A. and Mullan, A.B. (2001) Interdecadal Pacific Oscillation and South Pacific climate. *International Journal of Climatology*, **21**: 1705–21.

Schofield, K.A. and Livingston, M.E. (1998) Ovarian development and the potential annual fecundity of western stock hoki (*Macruronus novaezelandiae*). *New Zealand Journal of Marine and Freshwater Research*, **32**: 147–59.

Smith, P.J., Patchell, G. and Benson, P.G. (1981) Genetic tags in the New Zealand hoki *Macruronus novaezelandiae*. *Animal Blood Groups and Biochemical Genetics*, **12**: 37–45.

Smith, P.J., McVeagh, S.M. and Ede, A. (1996) Genetically isolated stocks of orange roughy (*Hoplostethus atlanticus*), but not of hoki (*Macruronus novaezelandiae*), in the Tasman Sea and southwest Pacific Ocean around New Zealand. *Marine Biology*, **125**: 783–93.

Sullivan, K.J. and Coombs, R.F. (1989) *Hoki Stock Assessment 1989*. New Zealand Fisheries Assessment Research Document 89/4.

Vignaux, M. (1994) *Catch Per Unit Effort (CPUE) Analysis of West Coast South Island and Cook Strait Spawning Hoki Fisheries 1987–93*. New Zealand Fisheries Assessment Research Document 94/11.

Yates, M. and Russell, D. (Eds) (1987) *Proceedings of the* Surimi *Symposium '86*. New Zealand Fishing Industry Board.

Yemane, D., Field, J.G. and Leslie, R.W. (2005) Exploring the effects of fishing on fish assemblages using Abundance Biomass Comparison (ABC). *ICES Journal of Marine Science*, **62**: 374–9.

Chapter 13
Community-based Fisheries in the Canadian Maritimes

Anthony Charles, Arthur Bull, John Kearney, and Chris Milley

Introduction

This penultimate chapter, while dealing with relatively developed fisheries within the Maritime Provinces of Canada's Atlantic coast, focuses, in particular, on the more small-scale and community-based components of those fisheries. Thus the chapter reflects something of a hybrid of the themes in the first section of the book, where the emphasis was on small-scale fisheries and local management, and those in later chapters dealing with developed fisheries. Indeed, the chapter demonstrates how moves toward greater participation of fishers and their communities in fishery management, common in developing regions of the world, are also emerging in a nation such as Canada.

This approach of community-based fishery management (Wiber et al. 2004) is gaining popularity worldwide as a mechanism with the potential, where appropriate, to support marine conservation and sustainable development in fisheries and to enhance the well-being of coastal communities. In particular, community-based management (CBM) in fisheries can potentially provide two major classes of benefits (Kearney 1999; Pinkerton 1999; Charles 2005):

(1) CBM can serve as a mechanism to orient fishery operations around community needs, improving the sustainability and resilience of both, by: (a) enhancing geographical, social, and cultural linkages between fisheries and coastal communities; (b) increasing efficiency and equity in the fishery and the community, in part through more effective mechanisms for resolving conflicting resource uses; and (c) countering trends toward concentration of fishery control outside the community, which naturally have negative effects on those communities.

(2) CBM can support marine conservation and sustainable development in fisheries, by: (a) better utilizing traditional ecological knowledge or local knowledge; and (b) empowering resource users, fisher associations, and

communities, thereby leading to better acceptance of conservation measures and thus improved compliance with management rules.

To what extent has community-based management been adopted in the fisheries of Canada's Maritime Provinces, and to what extent have the potential benefits described above actually been realized? This chapter explores these questions. We begin with an overview of fisheries in the region, paying particular attention to the prominent groundfish and lobster fisheries. Following this is the presentation of a set of three community-based fishery examples: community groundfish management boards; the Fundy Fixed Gear Council; and the aboriginal Mi'kmaq fisheries of Prince Edward Island. These examples are followed in turn by descriptions of a set of three fishery-related linkage and support mechanisms: marine resource centers; the Centre for Community-Based Resource Management; and the 'Turning the Tide' initiative. Finally, in the concluding section of the chapter, we provide a synthesis of these examples and a number of resulting conclusions on the state of and potential for community-based fishery management.

Fisheries in the Canadian Maritimes

The history of fishing in the Canadian Maritimes – the provinces of Nova Scotia, New Brunswick, and Prince Edward Island, on Canada's Atlantic coast (Fig. 13.1) – dates back to that of the aboriginal Mi'kmaq people who have inhabited the coastal Maritimes since long before the arrival of Europeans in the seventeenth century, and who depended greatly on the sea for food, trade, and travel. Today, fishing in the region continues to be widespread and important, taking place commercially, recreationally, and for subsistence. The contribution of current fisheries to employment, the role they play as engines of regional economic development, and their social, cultural, and historical place in the life of the region all contribute to their overall political importance.

Yet these fisheries have experienced large changes in the last few decades of the twentieth century: changes in the fish stocks, as well as in fishing methods, fishing communities, and fishery policies. The overview in this section first describes the major marine resources, the fishers and fishing communities, before examining a set of policy and management issues relevant to the subsequent discussion of community-based fisheries in the region. This analysis is based on, and updates, the coverage of these fisheries in Charles (1997).

The resources

The major species harvested in the Canadian Maritimes can be placed in three groups (Charles 1997). First, groundfish species – particularly cod, haddock, redfish, pollock, halibut, and various flatfishes – have historically provided about two-thirds of the total harvest by weight, and 40% or more of the landed value. This

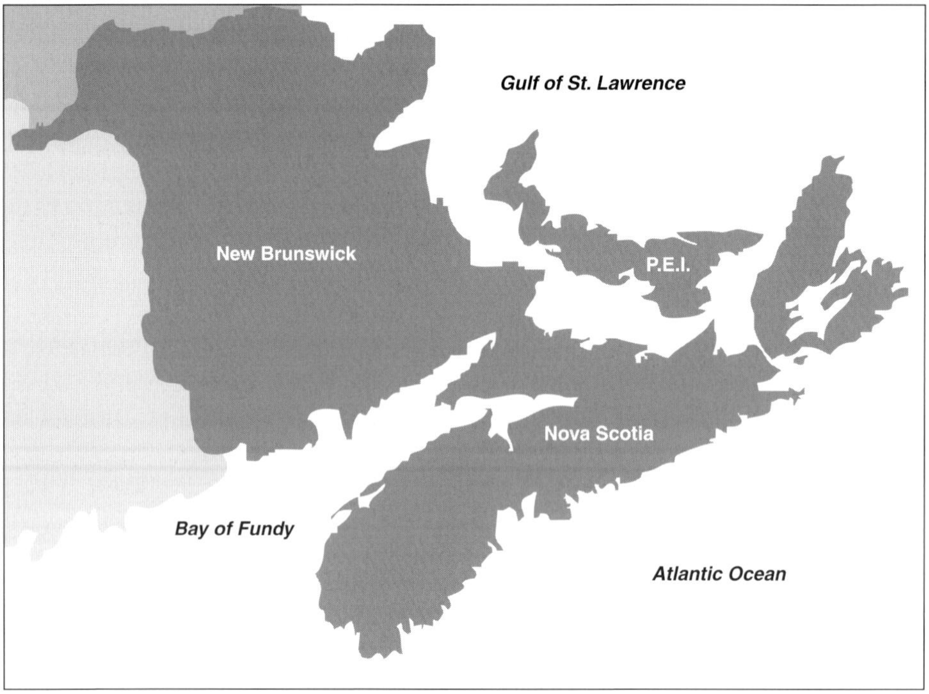

Figure 13.1 Canada's Maritimes, showing the three Maritime provinces and major ocean bodies. The fishery examples discussed in the chapter are located (1) in the Bay of Fundy and along the Atlantic coast of Nova Scotia, and (2) in the province of Prince Edward Island (designated as P.E.I. on the map).

contribution has, however, decreased greatly since the collapse or serious decline in many groundfish stocks in the early 1990s, described below. Second, shellfish species including lobster, crab, shrimp, and scallops, have collectively grown in importance over recent years, in relative terms compared to groundfish, and in absolute terms, with remarkable increases in both abundance and market price of lobster and crab. Third, pelagic species, including herring, mackerel, capelin, eels, skate, salmon, tuna, and swordfish, tend to be utilized in specific locales and by specific fishery participants.

Of particular importance on the Atlantic coast of Canada are the fisheries for groundfish and for lobster. The commercial groundfish fishery dates back roughly 500 years and, over much of that time, groundfish resources of the region were among the most plentiful in the world. However, the current reality in this fishery has been driven by two major collapses suffered in the second half of the twentieth century (Charles 1997, 2001). The first, in the early and mid-1970s, was largely due to heavy fishing pressure by foreign vessels. With extended fisheries' jurisdiction and strong year classes recruiting to the fishery, groundfish stocks rebuilt considerably in the late 1970s and early 1980s. Nonetheless, stock assessment and management efforts proved unable to properly monitor and control fishing pressure. A second groundfish collapse occurred in the late 1980s and early 1990s, caused fundamentally by excessive levels of domestic and foreign fishing mortality, both legal and illegal, leading to a fragile situation where the stocks were unable to overcome the onset of unfavourable environmental conditions. The first groundfish fishery to be closed was that for northern cod, off the coast of Newfoundland and Labrador, north of the Maritimes. In 1993 and the next few years, Maritimes fisheries were closed on most cod stocks and a considerable number of other groundfish species. Only the most southerly part of the region, near the US border, avoided most closures – this area will be discussed below. Today, most closed fisheries have been reopened, but none can sustain past levels of harvesting. Indeed, for some stocks, such as cod on the Scotian Shelf, recent biomasses remain at only 10% of those in the mid-nineteenth century (Rosenberg et al. 2005).

Lobsters, a more sedentary resource than groundfish, are present throughout most of the Maritimes, but there is great variation in catches between areas, with the greatest production in southwest Nova Scotia and the southern Gulf of St Lawrence. While considerable information is available about some aspects of lobster biology, the amount of research on lobster is much less than that on groundfish (Pringle and Burke 1993). The commercial lobster fishery developed in the mid-1980s, to serve a market for canned lobster. As the fishery developed in the late 1800s, catch levels declined steadily over several decades, but then were relatively stable over the period from mid-1910 to mid-1970. Subsequently, in the 1980s and 1990s, catches began rising, perhaps in response to environmental conditions or the decline of their groundfish predators, reaching a peak in the early 1990s, then falling somewhat. Today, the Maritimes lobster fisheries – while always at risk, as with any other fishery – remain in a far healthier state than their groundfish counterparts, and continue to serve as the 'backbone' of many Maritimes fishing communities.

Fishers and fishing communities

There are tens of thousands of fishers located across the Canadian Maritimes. The lobster fishery in particular provides seasonal employment to a large percentage of these fishers, and is considered to represent a relatively stable core of the overall fishery in the Maritimes (Pringle and Burke1993). Within the coastal commercial fishery (as opposed to the more industrially-oriented 'offshore' fishery), fishers often fish for multiple resources, to reduce reliance on any one fish stock, and to provide more fishing time. However, the methods and gears vary between fisheries. Some fleets, particularly for shellfish such as lobster and crab, tend to be structurally quite uniform, involving the same general configuration of vessels and gear across the region. Other fisheries involve a variety of gear types and vessel sizes – for example, the groundfish fleet is very heterogeneous, involving vessels ranging from under 12 to well over 35 m in length, and involving several different gear types.

The coastal fisheries support over 550 coastal communities across the region (Moore et al. 1993). Many other economic activities in coastal communities are dependent, at least in part, on the fishery – these range from closely-connected activities such as boat-building to sectors such as tourism and aquaculture, which can benefit from fishery infrastructure and scientific expertise. Increasingly, the communities of indigenous peoples in the region – the Mi'kmaq, as well as Maliseet and Passamaquoddy – are adding to traditional subsistence fishing by playing an expanding role in the commercial fisheries. This evolution is described below and is the focus of a case study later in the chapter.

Fishery management

Canada's national government – in particular, its Department of Fisheries and Oceans (DFO) – has responsibility for the management of Canada's ocean fisheries and the conservation of marine resources. This takes place within specific administrative zones established through some combination of political factors, natural boundaries such as between major ecological zones, as in the Gulf of St Lawrence and the Scotian Shelf, and areas used to compile statistics, such as for the groundfish fishery of the international Northwest Atlantic Fisheries Organization (NAFO). The management of all commercial fisheries in Canada has certain common features – such as the use of 'limited entry licensing' to limit participation in the fishery. However, many differences exist between specific fisheries – indeed, the examples explored in this chapter largely revolve around groundfish and lobster fisheries, for which management approaches differ considerably, reflecting differences in the nature of the resource, the gear used, and the management style.

With respect to the groundfish fishery, in areas where fishing has continued, or has reopened after closure, the focus of management remains on the same approach in place at the time of the collapse, namely total allowable catches (TACs) set for each stock in an attempt to control harvests (Charles 1995, 2001). Once the federal Minister of Fisheries and Oceans sets the TAC for each stock, it is subdivided into allocations for each of the fishery's various 'sectors', where groupings are defined by location, gear type, and vessel size. For example, a certain fraction of the TAC for cod in NAFO area 4X would be assigned to a sector defined as those vessels using otter trawls and of 14–19 m in length. Given these quota allocations, each sector of the fishery engages in a co-management process by developing its own conservation harvesting plan, a set of management measures agreed upon by the corresponding fishers in that sector, and subsequently approved by DFO fishery managers, who retain final authority over fishing plans. In those plans, single-species quotas are typically supplemented by gear restrictions, such as mesh and hook sizes, to enhance the selectivity of fishing, and other measures such as closed areas and closed seasons, to protect spawning and nursery grounds.

In contrast to groundfish, the lobster fishery operates on a more localized geographic scale, with one's lobster fishing taking place near one's home community. Hence, management issues tend to be discussed locally. There is a long history of spatial management, which was institutionalized in the early 1980s by dividing the region into lobster fishing areas (LFAs), each with its own management plan adjusted to local conditions. Unlike 'quota management' for groundfish, lobster fishing is regulated largely through input controls. The key to such controls are: limited entry, where a license is required to participate; limits on the number of traps per fisher; and limits on the fishing season. In addition, biologically-oriented controls – including closed areas, a prohibition on capture of egg-bearing females, and size limits – are designed to keep the size and maturity of individuals in the harvest to desired levels. The size limits are regulated through a slot size where lobsters below and above a certain size must be returned to the water – the minimum size protects the juveniles while the maximum protects the most mature and fertile female lobsters. Recent conservation concerns have led to efforts to increase future lobster recruitment by increasing the overall egg production, adjusting management measures to decrease harvesting of egg-bearing lobsters.

Fishery policy

The manner by which fishery management was implemented in Canada and in many locations during the second half of the twentieth century created a polarized view of the world, pitting government regulators (as 'protectors of the resource') against fishers (seen as 'selfish exploiters'). This view led to the exclusion of fishers from management decisions, which in turn led to a poor acceptance of regulations among fishers, creating a situation where enforcement efforts were unable to prevent illegal fishing and overharvesting (Charles 1997). When this dysfunctional situation was eventually recognized, DFO moved to consultative co-management, where government discussed management measures with the industry prior to implementation. However, attending consultations did not equate to decision-making power and, as a result, the government-imposed regulations did not receive broad acceptance. This was clear in the groundfishery, where in the 1980s, extensive anti-conservationist behavior such as dumping and high-grading, as well as under- and misreporting of catches, was reported (Angel et al. 1994).

As fishery management has evolved globally, the weaknesses of such approaches have been recognized, leading to more participatory management. Thus 'co-management' becomes more than consultation – it implies the creation of suitable institutions where stakeholders and governments work together to develop and enforce regulations (Berkes et al. 1989; Jentoft 1989; Pinkerton 1989; Charles 2001). The Canadian government has some form of co-management in many fisheries, but in most cases, this is organized by fishery sectors, reinforcing the established divisions of the fishery by gear type and vessel size. This 'sector-based co-management' was pursued at the expense of attention to more broad-based 'community-based co-management' approaches (Charles 2001), incorporating community institutions alongside governments and fishery sectors. In many parts of the world, the latter approaches have effectively increased the efficiency of management by more fully incorporating local knowledge, promoting local enforcement of regulations, and maximizing community pressure in support of conservation measures. Despite the lack of attention in Canada to community involvement in fisheries, there are some positive initiatives on this front – some of those in the Maritimes are the focus of the examples presented below.

Closely related to this policy issue was the key event of September 1999, when the Supreme Court of Canada issued the so-called 'Marshall decision' that recognized the treaty-based right of the Mi'kmaq, Maliseet, and Passamaquoddy peoples in the Maritimes to fish commercially. This court decision has led to a much increased involvement of these 'First Nations' in the fisheries of the Canadian Maritimes, as the Mi'kmaq in particular move to finally build their livelihoods through the use of natural resources that they have traditionally harvested, but that they lacked access to in recent decades. The Mi'kmaq also view the Marshall decision as an opportunity to advance their role in management of local fishery resources, in accordance with their self-governance aspirations and building on their long tradition of community management. Unfortunately, this goal has been inhibited by a lack of attention to the potential of community management systems as effective means to promote conservation and sustainable fisheries (Milley and Charles 2001). However, progress in this direction is indicated in one of the case studies below.

Examples of community-based fishery management

The above discussion of fisheries in the Canadian Maritimes highlighted the evolution of the region's groundfish and lobster fisheries, and of corresponding fishery management and policy in the region. In particular, it was noted that government policy has yet to embrace local-level fishery management initiatives – a minority of fisheries are more industrial and large-scale, and thus not conducive to such management, but for most, the lack of management on a local or community basis seems to reflect an implicit preference within government for the conventional 'sector-based' approach referred to earlier.

Despite this situation, there have been advances in the direction of community fisheries across Canada's Maritimes region, including a range of local initiatives. There is, for example, a history of local-level management in lobster fisheries, both in Canada and the United States; these fisheries, as noted earlier, tend to operate on a more local level than others, with regulation often occurring outside the framework of formal governmental management (Acheson 1975; Brownstein and Tremblay 1994). More recent examples of community fisheries in the Maritimes are also to be found. In this section, we present three such examples: (1) community groundfish management boards, a community-based approach in the groundfish fishery; (2) an example of such a board, the Fundy Fixed Gear Council; and (3) aboriginal (Mi'kmaq) fisheries in the province of Prince Edward Island. These examples are complemented later in the paper by discussion of a set of linkage and support mechanisms for such community fisheries.

Community management in the Scotia–Fundy groundfishery

As noted earlier, despite the now-famous groundfish collapse on the Atlantic coast of Canada, noted above, there is one region of the Maritimes – the more southerly part of what is known as the Scotia–Fundy management region (the Scotian Shelf and Bay of Fundy) – in which the fisheries for most species of groundfish remained open, albeit at a low level. The following describes how the small-boat fixed gear (hook and line and gill net) fishery for groundfish in that region was transformed in the direction of community-based management. Later, we examine the development and operation of one particular player in this community management system, the Fundy Fixed Gear Council, on the Nova Scotian side of the Bay of Fundy.

The story begins following the groundfish collapse in the early 1990s, when debates ensued over the appropriate management of fisheries that remained open. The debate originated around the choice between managing by input controls such as fishing effort, or controls on outputs such as quotas. The fixed gear groundfish fishery was originally managed by a combination of effort and biological controls, similar to the lobster fishery management described above. Indeed, a high proportion of inshore, fixed gear groundfish fishers are also lobster fishers who are aware of the benefits of such management. Fishers see effort controls as keeping with fixed gear technology, for which catch rates are more closely related to abundance of the stocks and are further limited by available fishing time, given the migratory behavior of groundfish species. Furthermore, as the lobster fishery protects juveniles through minimum size limits, this is achieved in the groundfishery by the hook and mesh sizes employed. In addition, just as the lobster fishery protects spawning females, it appears that some groundfish species do not feed during reproduction and will thus not bite a baited hook.

There is, however, another important function of effort and biological controls besides resource conservation, namely social equity. The former suite of management measures in the inshore, fixed gear groundfishery provided a level playing field that permitted what was seen as an equitable distribution of fishery wealth in the community; with differences in income levels considered to arise reasonably, through the degree of hard work, knowledge of the resource, and fishing skill. When the DFO began to impose output controls on this fishery in the late 1980s and early 1990s, the fishers insisted that these controls should also limit the output of individual fishing enterprises per fishing trip so that more highly capitalized fishers could not take an inordinately high share of the total catch allocation. But then, in the fall of 1994, the DFO announced that they would no longer enforce these trip limits. Fishers feared that the larger boats that could fish during rough winter conditions would catch the entire quota before the vast majority of license-holders in the region could begin their fishing.

After a long period of negotiations, in 1995, the fishers devised their own system of distributing the fish quotas allocated to them through trip limits within 18 management units, based on the 18 quota groups and varying by geography, vessel, and gear characteristics of the license-holders. One of these 18 plans included a 'community quota' for the village of Sambro near Halifax, Nova Scotia (Loucks 1998). The fishers' organizations enforced their fishing plans through contractual agreements with individual fishers, arrangements for independent monitoring of catch levels, and an industry sanctions committee that determined the appropriate punishment for any fisher found to have violated the rules.

In September 1995, the fixed gear groundfish organizations held a workshop to develop a vision of their fishery. They agreed to operate the fishery based on the goals of:

(1) Sustaining groundfish resources through use of non-destructive gear, a code of conduct, inclusion of ecological factors, reduction of waste, and improvement in yield from the available harvest.
(2) Sound science incorporating an ecosystem approach and greater involvement by fish harvesters.
(3) Sustaining coastal people through economic viability, stability of incomes, equitable sharing of the benefits of resource exploitation, and maintaining values of independence, inclusiveness, commitment, and flexibility.
(4) Improving communication and educational programs for fishers and the general public.

These goals were to be achieved through community-based management – understood as a system of management authorities where each would manage the fleets rather than the fish in a specific geographical area. Although many fishers would have preferred a community-based system based on effort and biological controls, they recognized that quotas now lay at the heart of the government's management of groundfish. Thus, using the Sambro experiment during the previous season as a base model, a request was made to the government to sub-allocate the total Scotia–Fundy fixed gear allocation of cod, haddock, and pollock into a number of discrete community quotas. The sub-allocation was to be guided by the historical catch performance of the fleet based in a particular geographical area, usually a county or pair of adjacent counties.

The development of community-based management was soon threatened by the real or perceived threat that the federal government, in collaboration with a group of license-holders comprising the largest boats in the fleet, would scuttle the efforts by imposing a system of individual transferable quotas (ITQs). Having seen the high cost of access and the concentration of ownership in a regional trawl fishery that moved to ITQs, most fixed gear fishers strongly opposed such a system. It appeared to them as the antithesis of their fundamental orientation toward an independently owned and operated, family-based, and socially equitable fishery.

This concern led to weeks of large-scale protests, resulting in a promise by the government that there would be a one year moratorium on ITQs for fixed gear groundfish, an agreement to allocate fixed gear groundfish quotas by area rather than by fishing sectors, and support for establishment of five community management boards to determine the allocation. By 1997, there were seven such boards, and ten by 1999. The fundamental shift to community management did not, in the end, involve a change from quota to effort controls, but rather from quota allocations based on fishery sectors, defined by vessel size and gear type, to allocations by area, reflecting the real ties of fishers to their home communities and regions.

Each of the groundfish management boards operates differently, reflecting one of the goals of community-based management, to allow enough flexibility to manage fish according to local ecological, economic, and social conditions. However, all or most management boards share a number of characteristics:

(1) The boards were established and are run by fish harvesting organizations, and strive for inclusive decision-making processes.
(2) The boards sub-allocate the community quota among different gear types and devise rules for all license-holders in the form of a community management plan.
(3) The management plans are enforced through contractual arrangement between the board, the license-holders, and the catch monitoring companies.
(4) Management plans are consistent with basic conservation requirements set out by DFO and each license-holder must follow the conditions of license as determined by the government.
(5) Management boards have infractions committees to judge alleged violations of management plans and impose penalties.
(6) Seasonal adjustments are made to management plans and in a number of cases these adjustments include the sale or trade of unused quota between different management boards.
(7) Individual license-holders can still choose to fish under a generic management plan devised by DFO for the whole Scotia–Fundy region, instead of under a community management plan devised for local conditions. However, more than 98% of license-holders choose to fish under the community management boards each year.

This approach can enhance community sustainability by allowing each community to decide for itself how to utilize its quota. Loucks' (1998) description of the goals of Sambro's community quota arrangement could equally apply to all of the boards: 'The objective of the Sambro community quota fishing plan was to protect community access rights to the fishery, to ensure it lasted throughout the year, and to maximize the value of a scarce resource through flexible decision making . . . The community shared the burden of the reduced quota allocations, and of enforcing regulations, and tried to adapt as a collective by participating in the fisheries management planning process and maintaining a sustained commitment to follow it'.

This community management board system was undoubtedly a compromise. As noted, the fishers would have preferred effort-controlled rather than quota-controlled management. Another limitation of the system is its orientation towards managing the quota of a particular local fleet rather than managing the geographic area where that fleet operates. While in a sense this was the breakthrough approach that allowed community management to take place for a fishery on a mobile resource, this has limited the capability to take an ecosystem approach to management. Mobile gear fleets have moved around the region opportunistically, depleting fish substocks in shallow water inshore grounds first (Graham et al. 2002). This phenomenon is at least partially responsible for the more severe effect of groundfish declines on the hand line fishers who fish closer to shore and use a less intensive fishing strategy. Without geographic control of the overall harvesting practices on local fishing grounds and more general powers to exercise ecological stewardship, community-based management is handicapped in the benefits it can provide.

Nevertheless, community-based management in the Scotia–Fundy fixed gear groundfish fishery is a positive development. It has served to bring management more to a local level, thus improving the acceptance of management, and drawing on inherent moral persuasion within the communities to improve compliance. A fuller description of this tendency is illustrated in the following case study of one management board.

The Fundy Fixed Gear Council

The Fundy Fixed Gear Council (FFGC) is an example of how the process described above has produced community-level fishery management. The FFGC is the management board for the approximately 160 inshore fishing enterprises comprising the fixed gear fleet of vessels under 15 m in length – some of which are shown in Fig. 13.2 – fishing on the Nova Scotian side of the Bay of Fundy, for groundfish, notably cod, haddock, and pollock, but also species such as halibut, hake, and dogfish. The FFGC was established in 1996 (Bull 1998), following the large-scale protest noted above, which included fishers in the Digby–Annapolis area who saw this as an opportunity to move ahead with community-based management. Although the FFGC manages a relatively small fishery with limited access to the resource, its work has involved many aspects of community-based fishery management, such as policy, science, research, compliance, access, conservation and harvesting.

The FFGC governance structure is designed to give equal representation for the two fishers associations in the area, the Bay of Fundy Inshore Fishermen's Association, and the Maritimes Fishermen's Union, Local 9. Representatives of these bodies sit on a decision-making council that sets the overall direction for the local fixed gear fishery, subject to quota limitations and management constraints imposed by the DFO. Under the Council, three gear-specific committees, for long line, hand line and gill net fishers, elected by the membership of each, design and implement fishing plans for each gear type, and carry out the hands-on management of the plans, including setting catch limits, to reach the goals set by the membership. In addition to these committees, there are elected port representatives who act as the contacts and communication between the council and the local fishers. This whole

Figure 13.2 Harbor on Brier Island, a community along the Nova Scotian side of the Bay of Fundy, within the boundaries of the Fundy Fixed Gear Council. The photograph shows fishing boats, coast-guard patrol boats (foreground) and fish plant (background).

community management system works thanks to the institutionalized requirement that, in order to take part in the fishery, each fisher must sign a civil contract with the FFGC and, to do that, a fisher must belong to one of the two associations and agree to be bound by the FFGC fishing plan.

The following analysis identifies several major features of the FFGC, in terms of accomplishments and failures, and lessons learned over the past 10 years.

(1) The FFGC has managed each year to (a) negotiate a set of allocations of the available fish quotas among its gear sectors and (b) write and ratify annual fishing plans. This continuing operation can be seen as an accomplishment in itself, one that can be traced in part to the development of a democratic governance system with transparent and participatory decision-making. However, that system continues to be prone to major periodic conflicts over quota allocations between the major gear sectors involved (hook and line and gill net fishers).

(2) The FFGC, as well as other community management boards, has shown that community-based fisher organizations are able to deal internally with those who break agreed-upon fishing plans. This has been accomplished through a major innovation, an infractions committee, reporting to the FFGC council. This committee is responsible for ensuring that members comply with the fishing plans, and has the authority to give out sanctions. Each fisher must

sign a contract agreeing to be subject to the infractions committee if he or she contravenes any part of the plan, for example, by exceeding a weekly quota. The committee is made up of fishers, but to ensure fairness, membership changes regularly and is unknown to members external to the committee. In addition, the files of fishers under review by the committee are unlabeled so that committee members do not know the individual who is being sanctioned. While this system has not eliminated rule breaking, it is producing what are generally considered fair and appropriate sanctions.

(3) The FFGC has played a role in the basic survival of the local fishery. Without an effective management board, many more fishers would have been driven out of the fishery than was the case. In particular, the gill net and long line fleets have been maintained, with both able to continue to fish, keeping within allowable catch limits, despite the low level of those limits. On the negative side, the FFGC has not been able to prevent a reduction of the hand line fleet from about 159 to about 25 boats, as a result of declining quotas and fishing on local stocks by outsiders, which has prevented hand line fishers from finding sufficient fish locally.

(4) The FFGC instituted innovative mechanisms for developing community-based management, such as an outward-looking research and advisory committee, and establishment of three seats on the FFGC's council for non-fishery community members. However, most of the innovations did not continue beyond the FFGC's first few years – this would appear to be due to a lack of staff and a continual need to deal with pressing management issues, such as fishing plan conflicts and new DFO measures. (To some degree this is speculation, because the FFGC did not build in a process of evaluation and reflection.) Nevertheless, a community spirit continues in the FFGC, and there continue to be efforts to build its community focus.

(5) The FFGC has had success with 'outreach' on two fronts. First, it recognized the need for research to meet its information needs, and has successfully initiated research projects from studies of marine protected areas, to fishing fleet profiles, to mapping of local cod spawning areas, to the assessment of local dogfish stocks. Second, the FFGC served as a catalyst for several other initiatives supporting local management and conservation, such as the Bay of Fundy Marine Resource Centre and Saltwater Network, which are described below.

While the FFGC has a number of accomplishments to its credit, it has also identified some important goals that it has yet to achieve:

- Marketing. The hook-and-line fishers have yet to capitalize on the fact that they catch a premium product that is environmentally friendly, and could receive a higher price if sold into niche markets.

- Conflict resolution. The conflict over allocation continues and, without a built-in conflict resolution mechanism, this is a constant internal threat to the FFGC's integrity.
- Policy change. The overall policy of DFO is not supportive of community-based management and the FFGC has not successfully mobilized a policy change strategy.
- Evaluation. Clear goals and objectives were never set out, nor were the mechanisms for the FFGC to recognize and document its achievements.

One tentative conclusion we might reach from this sketch of the FFGC experience is that there was a need for an investment early on in developing the social capital and research that might have helped advance the FFGC's visions for community-based management. Because the ongoing demands of managing are intense, and the financial resources that the fishers can provide are limited, the FFGC has yet to fully capitalize on its early successes. On the other hand, the FFGC has managed to develop mechanisms to encourage conservation and compliance among fishers, and indeed its continued existence and operation may be seen in itself as a success story.

PEI Mi'kmaq fisheries

As noted earlier, the indigenous peoples of Canada's Maritimes, the Mi'kmaq, Maliseet, and Passamaquoddy 'first nations', play an important role in the region's history and culture, and are now playing an increasing role in the commercial fisheries. This section describes how the two Mi'kmaq communities in the province of Prince Edward Island – Abegweit and Lennox Island – have been developing their commercial fisheries in recent years. As shall become apparent, this recent situation is linked inextricably to the historic relationship between the Mi'kmaq and the colonial (British) and post-colonial (Canadian) governments, a relationship first defined through treaties between the Mi'kmaq and the British Crown, and subsequently by Canada's federal Indian Act.

Historically, the fishery was a central part of Mi'kmaq society, a major source of protein and a subject of trade and commerce. Fishing for subsistence has continued into the present and, in the early 1990s, a ruling by Canada's Supreme Court (the so-called Sparrow case) recognized the rights of Canada's aboriginal peoples to fish for food, social, and ceremonial purposes. Nonetheless, this did not halt the dwindling of Mi'kmaq participation in the commercial fishery, something that can be attributed to implementation of government policies, such as fleet licensing. The first nations' desire to implement a commercial fishery was realized in 1999, with the Supreme Court of Canada's Marshall decision recognizing the Mi'kmaq and Maliseet right to fish commercially (based on treaty rights) in a communal 'livelihood fishery'.

As noted earlier, this has led to an increased Mi'kmaq involvement in the region's fishery (Wiber and Kennedy 2001), facilitated by a government initiative to provide financial support for first nations' transition into the fishery, and interim fisheries agreements between DFO and individual Mi'kmaq and Maliseet communities. The focus of these agreements was primarily on harvesting activities, with little attention paid to other aspects of the fishery, in particular, management activities. However, Mi'kmaq leadership recognizes the nation's responsibility for management of the resources, and traditional management systems were integral within the Mi'kmaq culture. These are based on the principle of *Netukilmk* – a traditional world view where all resources are seen as part of a web of connected living and non-living things, providing benefits to the community as a whole. Today, Mi'kmaq resource management efforts remain founded in the principles of *Netukilmk*, and Mi'kmaq leaders still recognize the management responsibilities associated with the right to harvest natural resources.

After the Marshall decision, the PEI first nations began a process of planning their participation in the communal livelihood fishery. It is important to note that in each community, fishery management (as with other governance matters) is ultimately the responsibility of the band council. Thus, each band set up internal consultations to determine the desired vision for their fishery and to develop band-level management systems. The Mi'kmaq Confederacy of PEI (MCPEI), the common forum for the two bands, coordinated this vision development. The PEI first nations established long-term priorities for their fisheries: (1) protection of Mi'kmaq treaty and aboriginal rights to access natural resources for the benefit of their communities and their members; (2) establishment of secure systems for food production to meet the needs of local band members; (3) development of commercial fishing ventures that will provide band members with stable and effective employment, such as providing incomes sufficient to meet the moderate needs of fishers and their families; (4) establishment of local governmental and administrative structures and mechanisms regarding fishery decision-making; and (5) development and maintenance of harmonious relations with adjacent fishing communities.

An important tenet that the PEI first nations have built into their plans for fisheries development is that of pursuing a goal of employment and not profit generation. In other words, the objective is to establish and maintain a fishery where the number of band members employed as harvesters, fish handlers, and marketers, and the duration of work during any calendar year, are maximized subject to the band's catch quotas. This development strategy has greatly increased the level of employment and accordingly reduced the level of dependence in the community on social programs. The bands have negotiated with the Canadian government to acquire access to multiple resources so that as the season for one species ends, fishers will change gear to target an alternative species and extend the duration of employment, all within conservation limits. This strategy requires that the species targeted be viable enough to provide returns sufficiently greater than the cost of fishing, since the fishery must not be an economic burden on the community but must generate enough revenue to meet the costs of fishing, provide enough revenue to meet the personal needs of the fishers, and provide sufficient surplus revenue to cover the management costs incurred by the community.

In developing the fishery management system, an arms-length approach to management decision-making was established, using community organizational structures to develop band-level management plans based on community needs and priorities, resource abundance and health, and market opportunities. These plans include policies and regulations that govern all aspects of the fishery including entry requirements, harvesting methods and seasons, distribution and sharing of benefits, and penalties for non-compliance.

An organizational structure was established in each community to ensure community control over the fishery – so individual band members, or non-native interests outside the community, could not compromise agreed-upon management objectives. In each community, the band council, through its fisheries sub-committee, appoints community committee members, reviews policy documents and fishery plans, and approves operational procedures and work plans. Band council approval of fishing plans occurs only after discussion and endorsement by the wider community, through open community meetings. This reflects the reality that all aspects of the Mi'kmaq fishery involve the wider community, beyond just the fishers, in protection, planning, allocation, research, and monitoring of the fishery resources.

The actual preparation of annual fishery management plans, as well as sets of rules by which commercial, food, social and ceremonial fisheries are conducted, is the responsibility of a community fisheries management planning committee. Finally, a harvesters' committee, comprised of all fishers, meets specifically to review and discuss proposed plans and regulations, sharing arrangements, markets and other day-to-day matters. This committee ensures that those most affected by management rules and regulations have a significant input into the decision-making process, but not full control of that process.

There have been three ongoing challenges in this process. First, despite best efforts to maintain a politically-independent management regime, the potential for political intervention has remained a reality in these small communities. The transition to an independent planning and management system has been greatly facilitated by the advisory and technical support from the MCPEI, but diligence on the part of band council and committee members has been necessary to prevent local political intervention in the decision-making process. Second, the PEI Mi'kmaq experience in establishing local management capacity has not been without resistance from the federal government. Due to the first nations' dependence on federal funding programs, efforts to make local management decisions that may be in the best interest of the communities have been impeded by the need to accommodate policies and plans of the federal fisheries department that have a broader geographical focus. The DFO has been slow to recognize and adopt the first nations' plans and to assist in enforcing band-level rules and regulations. Third, the rapid increase in first nations' involvement in the commercial fishery since the Marshall decision has placed a considerable demand on the human resource capacity within the PEI. As a result, some of the development efforts have been slow in progressing, as band members acquire the knowledge and skills necessary to be effective participants in the fishery. In some instances, bands have engaged external, non-native staff or entered into business arrangements with non-native companies that have created new dependencies for the community that have not always been well received by some sectors of the communities.

On the positive side, a significant development in building local fishery management capacity has been a community justice program (Customary Justice Group) that was established to ensure infractions to band and DFO management plans and regulations are dealt with in a manner that will promote compliance, be respectful to the communities as a whole, and ensure effective penalties are imposed where required. This is an example, like that of the community groundfish management boards described earlier, of how community-based fishery management can support conservation through improved mechanisms for enforcement and compliance.

In addition to developing fishery harvesting capacity and annual harvest planning, PEI bands have initiated business relationships in fishery marketing and marine supply activities. This is an important component of the community management system designed to promote conservation, treaty rights, employment and social wellbeing of the communities. In addition, band participation in marine research is promoted to strengthen their involvement in resource management. Finally, bands have undertaken coordinated management of fisheries, aquaculture, wharves, and ecotourism, recognizing the broader context in which the fishery fits into integrated marine and coastal management. In this regard, there has also been increased first nations participation in management activities with adjacent non-native communities and industry.

Linkage and support mechanisms for community fisheries

In this section, we examine three linkage and support mechanisms for community-based fisheries and fishery management in Canada's Maritimes region: marine resource centers; the Centre for Community-Based Resource Management; and the Turning the Tide initiative.

Marine resource centers

Discussions of local-level and community-based fisheries management naturally focus to a considerable extent on fishers and their organizations and, in cases such as the Mi'kmaq fisheries described above; it is also crucial to explore the role of the broader community. There is often a third component in community fishery management that is less obvious but still crucial – the local community-based institutions that provide the capacity-building support for the fishers and their communities. This case study describes a set of marine resource centers established in the Bay of Fundy and Gulf of Maine area as mechanisms supporting local-level fishery and coastal management and conservation.

The first such initiative, in 1999, was the Bay of Fundy Marine Resource Centre (MRC), established by the Fundy Fixed Gear Council, a local community economic development agency, and a number of marine-related community groups, harvesters' organizations, first nations, and others, all seeing the need for a local organization to support the marine economy and ecosystem in Nova Scotia's Digby–Annapolis region. The MRC helps local marine-oriented groups develop the skills needed for defining problems, establishing goals and working cooperatively, and provides support services for these groups in conflict resolution, organizational development, facilitation, technical support, geographic information systems (GIS), and research. The MRC's work focuses on six broad areas: (1) community-based resource management; (2) aquaculture; (3) marine tourism promotion and marketing; (4) digital data storage and retrieval; (5) training for aquaculture and fisheries; and (6) ecological, market, legal, technological, and social research relating to marine resources.

The MRC has provided support to a range of fisher organizations, fish processors, aquaculturalists, marine ecotourism operators, first nations, environmental groups, researchers, and governments. A major focus has been on fishery themes, to organizations such as the Fundy Fixed Gear Council, the Digby–Annapolis Clam Management Board, and the new Management Board in Lobster Fishing Area 34 of southwest Nova Scotia, as well as to Bear River First Nation, providing assistance in planning its fishery. The MRC has also worked at the scale of the entire Bay of Fundy in: (a) developing a Community Learning Network to provide technical, research, educational, and informational resources, and (b) supporting the Writing the Rules project to develop ecosystem-based management principles for the Bay of Fundy fisheries, based on input from fishers and their organizations.

The MRC experience points to some lessons on how to support fishery management at the community level. As a community-run institution, it essentially provides a 'tool box' for local groups interested in local management. Emerging community fishery groups can draw on institutional capacity-building support (organizational, technical, or research-based) from an organization that does not have interests or an agenda of its own – it is purely an enabling agency. Without a local capacity-builder of this kind, it is very hard for a fishery to move forward. Indeed, the MRC has helped link local fishers to universities, networks, first nations groups, and other fisheries organizations as needed. It has also supported the development of organizations that address wider regional and bio-regional issues, such as the Saltwater Network, described below, and the Bay of Fundy Fisheries Council, which involved more than a dozen fisheries groups, and was able to create a plan for ecosystem-based fisheries in the Bay of Fundy. All these efforts seem likely to have increased the resiliency of community-based management in the Digby and Annapolis area and beyond.

There remains an ongoing challenge relating to funding. While the MRC provides valued services to local organizations, those organizations do not always have the financial capability to pay for those services. The MRC has relied on a range of funding sources, including rental income from offices rented to community groups and government agencies, 'fee-for-service' income, for providing GIS mapping, technical support, facilitation and other support, funding from Canadian and US foundations, and funding from government grants provided for specific project activities. Nonetheless, there will always remain a concern about financial sustainability, and thus a need to explore new funding strategies. Some possible strategies include: (a) creating 'one stop shopping' capabilities to reduce costs and increase cooperation among groups; (b) working with university students, interns, or summer students to carry out research for and with communities; or (c) looking for additional fee-for-service opportunities. As with many other community-based organizations, risk in this continual search for funding lies in its potential to limit the options and/or drive the working of the MRC.

Funding limitations lie in part behind the failure of the MRC to achieve some of the goals it had identified at the outset – such as setting up a research facility with a wet lab, and a fully-staffed marine information and referral center. On the other hand, the lack of these facilities may be because local groups were not demanding this form of support. Fundamentally, the question is: has the MRC's work had a significant effect on the health of economies and ecosystems of the region? Beyond the obvious benefits to local organizations, broader benefits are much harder to identify, and to measure. This may point to the need for a greater focus on evaluation of MRC performance in the long term.

The MRC is part of an expanding network of citizen-run place-based marine resource centers in the Bay of Fundy and Gulf of Maine region. Efforts are underway to establish three new marine resource centers in the Canadian Maritimes: in St Andrews, New Brunswick; the Upper Bay of Fundy region of Nova Scotia; and the Acadian region of southwest Nova Scotia. There are also marine resource centers formed or forming in the US, such as the Cobscook Bay Resource Centre in Eastport, Maine, and another in Stonington, Maine. This provides the potential for a coordinated local-level approach to marine conservation and management in the Gulf of Maine.

Indeed, an important outcome of the collaborative work between resource centers in the Gulf of Maine region has been the creation of Saltwater Network. This organization was established to support the development and growth of community-based ocean and coastal management in the Gulf of Maine, to increase peer learning and cross-fertilization among organizations involved in community-based management, and to fill the gap of a lack of significant government support for this work on either side of the Canada–US border. Saltwater Network provides small grants to resource centers, fisher groups and others, through funds from Canadian and US foundations, to support community-based management in the Gulf of Maine, and to support convening, capacity-building, and access to learning opportunities, including using bursaries to attend learning events. Established to support local activities within a broad ecosystem, it is still in its early days, but has so far been able to successfully channel financial support to a range of local-level conservation and management initiatives.

Centre for Community-Based Resource Management

In 1998, in large part in response to the initiation of community-based management in the Scotia–Fundy fixed gear groundfishery described earlier, the Centre for Community-Based Resource Management (CCBRM) was established at St Francis Xavier University. The CCBRM drew on the university's strong history (since the 1920s) of supporting the development of adult education, cooperatives, and credit unions in the Maritimes, through the 'Antigonish Movement'. Indeed, university outreach staff saw the present-day implementation of community-based management in the Scotia–Fundy groundfish fishery as a natural evolution of the cooperative movement in the face of globalization and increasing privatization of natural resources.

One of the first initiatives of the CCBRM in the 1990s was a needs assessment with the groundfish management boards in order to evaluate ways to advance their capacity for community-based management (Kearney and MacIntosh 1998). The Centre also supported in various ways the formation and operations of the Bay of Fundy Marine Resource Centre, and is currently leading the establishment of the new marine resource center in St Andrews, New Brunswick, referred to above. In addition to its work in the Bay of Fundy, the CCBRM also supported community-based fisheries management in Guysborough County, Nova Scotia. This involved working with the Guysborough County Women Fisheries Enhancement Association, an organization of women fishers and fisher spouses that provided support for the groundfish community management board in that area, among other initiatives (GCWFEA 2001).

As with marine resource centers, the CCBRM plays a linking and support role for community groups, as an independent third-party organization that is able to facilitate multi-stakeholder meetings and act as a mediator in conflict resolutions. It also provides training and capacity-building resources, a training program in community-based resource management, and support for community-based research – such as a cod tagging study in the Bay of Fundy, and a compilation of fisher knowledge on spawning areas and local stocks (Graham et al. 2002).

A rather unique contribution of the CCBRM in the Maritimes arises out of its dual role in working with local organizations and communities and, at the same time, working with their counterparts in developing countries. This enables the CCBRM to provide north–south linkages between fishers in the Maritimes and those in developing nations. Such linkages to fishers and countries where community-based management is the dominant form of management and is supported by government policy have proven extremely important for Canadian fishers – given their situation within an environment of little government support for community-based management and a largely privatized Canadian fishing industry. This networking not only offers new ideas and tools for community management but also helps to situate Canadian fishers in a wider global setting of conflicting economic, social, and ecological values.

One of the potential advantages of a university-based resource center is its larger funding base and long-term presence in a region. However, cutbacks in governmental funds available to universities in Nova Scotia placed the CCBRM in the position of having to externally raise much of both its core and program funding. It thus developed many of the same issues around funding described above for the Bay of Fundy Marine Resource Centre. To make matters worse, the community-based and university-based resource centers were essentially competing for a limited pool of funds from similar sources. Even when there was clear differentiation in the roles played by each, there was not enough Maritimes-oriented funding to support both types of centers in local work. In response, the CCBRM turned more and more to international work, where funding was more readily available, which alleviated competing for funding with its community partners.

Another disadvantage of a university-based resource center is its lack of accountability to the broader community. At least in this particular case, decision-making is entirely internal to the university administration. Even when there is community input, it is of an advisory nature. Thus, if the CCBRM is to provide a significant and stable contribution to community-based fisheries management, the university will need to find a way to provide a secure funding base and develop a partnership with the community such that it can have a real influence over decision-making.

Turning the Tide

The 1999 Marshall decision created real opportunities among Mi'kmaq communities, as described earlier, with significant social and economic development through increased fishery access. It also created opportunities for participants in the Maritimes inshore small-boat fishery, many of whom are involved in the community management boards described earlier. These fishers had undertaken major efforts to develop community-based management, but with little official support. They are finding common ground with some Mi'kmaq communities in the region, ones with an expanding role in commercial fishing and a shared desire for more locally-based fishery management and more emphasis on place-based and ecosystem-based management.

This has led to new linkages and cooperative endeavours between Mi'kmaq first nations and community-based non-native fishing organizations. One such linkage is 'Turning the Tide: Communities Managing Fisheries Together', which has sought to support capacity-building for community-based fishery management, and at the same time to build linkages between native and non-native participants in the fishery.

Turning the Tide involves several participating organizations described earlier in this paper – the Fundy Fixed Gear Council, the Mi'kmaq Confederacy of Prince Edward Island, the Bay of Fundy Marine Resource Centre and the Centre for Community-Based Resource Management – as well as several other organizations and communities: Acadia First Nation and Bear River First Nation (the two Mi'kmaq bands in southwest Nova Scotia), the Bay of Fundy Inshore Fishermen's

Association and the Maritime Fishermen's Union, Local 9 (the two fisher associations making up the Fundy Fixed Gear Council) and Saint Mary's University, in Halifax, Nova Scotia. Turning the Tide provides these participants with a forum to discuss issues, principles, and values relating to community-based management, as well as to identify common issues, priorities, and strategies. It also supported educational and capacity-building efforts, such as workshops, study tours, training courses, and resource materials, the establishing of regional, national and international links, and concrete activities ranging from the development of fishery management plans to the production of a community fishery management handbook.

Turning the Tide participants – coming from native and non-native communities that may well have had little interaction historically with one another – see the initiative as a helpful 'safe place' for participants to meet, with no strings attached. In particular, this was apparent in a set of study tours carried out within the initiative, where Maritimes participants travelled together to Canada's Pacific coast and around the Gulf of Maine. These trips helped to develop bonds among participants, to develop linkages regionally and nationally, and to produce a set of new community fishery approaches, through discussions in mobile workshops held along the route of the study tours (Fig. 13.3). The team-building phenomenon also led to local work on specific initiatives, as in the support provided to Bear River First Nation by other participants – from facilitation of a process to formulate their fishery development and management plan, to providing a fishing vessel for their first post-Marshall involvement in the commercial lobster fishery.

Turning the Tide differs from the support mechanisms described earlier in that it has been funded entirely through one source – a Pew Fellowship project of

Figure 13.3 A workshop held on approaches to community-based fishery management, facilitated by the Bay of Fundy Marine Resource Centre as part of a Turning the Tide tour of the Gulf of Maine region.

the US-based foundation, Pew Charitable Trusts. This has had both positive and negative implications. On the one hand, it enabled activities to take place independent of governmental constraints and, it could be argued, these activities might have been simply impossible if government were involved. On the other hand, because the federal government in Canada has nearly absolute power over fishery management, it is essentially impossible for an activity such as this to take on actual management efforts in the absence of governmental involvement – although it certainly was possible to engage in useful capacity-building and knowledge-oriented work. Furthermore, the lack of interaction with government may have caused the initiative to miss out on larger-scale effects, such as on government policies and related government work. Thus the type of independent initiative reflected in Turning the Tide cannot be seen as a model for local self-management by itself, but rather as an effective tool for relationship-building and grass roots capacity-building.

Conclusions

At the start of this chapter, two major benefits of community-based management in fisheries were noted:

(1) CBM can serve as a mechanism to orient fishery operations around community needs, one that helps improve efficiency and equity in the fishery and the community, resolve conflicting resource uses, and counter trends toward the centralization of control over the fishery that can have a negative effect on coastal communities.
(2) CBM can support marine conservation and sustainable development in fisheries, by better utilizing traditional and local ecological knowledge, and empowering resource users, fisher associations and communities, leading to better acceptance of conservation measures and thus improved compliance.

This closing section examines the extent to which the examples provided in the preceding section have demonstrated these benefits.

Orienting fisheries around community needs

There are indications that both the Mi'kmaq fisheries and the community management boards have improved efficiency and equity in their fisheries – better avoiding the 'rush for the fish' and better distributing fishing opportunities, to meet a locally-determined sense of equity. On the other hand, there is an increased burden on the organizations and individuals involved, in terms of time and resources needed to maintain the new institutional arrangements. Communities have also been supported through limits on the centralization of fishery control that have come from (a) maintaining the small-boat fishery, in the case of the boards, or (b) building the economic base of rural aboriginal communities, in the case of Mi'kmaq fisheries. It would also appear that in responding to the needs of community fisheries, the support mechanisms described earlier played a constructive role.

The question of whether the case studies reflect success in 'orienting fishery operations around community needs' requires additional exploration. Recall that development of commercial fishery activities in aboriginal Mi'kmaq communities arose as a result of a Supreme Court of Canada decision recognizing the right of Mi'kmaq to fish for a 'moderate livelihood'. The court decision in fact recognized the authority of Mi'kmaq communities to administer this right, thus entrenching the community-based nature of Mi'kmaq fisheries. This forced the federal government, which is much more accustomed to working with individual fishers and their associations, to deal directly with communities. This reality contrasts with the community management boards that operate within geographically-defined areas, as in any community-based approach, but do not necessarily involve those in the local communities apart from the groundfish fishers themselves. In some cases, as with the Fundy Fixed Gear Council, there is an allowance for involvement of non-fishers in the boards, but in general it is reasonable to conclude that Mi'kmaq fisheries have an inherently higher level of community involvement. Indeed these fisheries, as in the case of the two Prince Edward Island first nations, are managed through community governance bodies, rather than through a fisher organization. This strong community governance is reflected in the close linkage between fishery operations and community objectives, so that in such cases it is clear that fishery operations are indeed explicitly oriented to meeting 'community needs'.

Supporting marine conservation

While this chapter has focused primarily on the human dimensions of community-based fishery management, underlying these aspects is the fundamental importance of CBM to conservation of ocean resources, and the resulting sustainability of resource use over time. As noted above, CBM can support marine conservation through a combination of ingredients.

First, community-based management can fully incorporate the local understanding of the fish stocks and ocean environment. The Fundy Fixed Gear Council represents a strong example of this: its member fishers, operating in the unique ecosystem of the Bay of Fundy, have the familiarity with that system on which wise management depends. The FFGC is able to directly build on the local 'traditional ecological knowledge' of the fishers in designing its management plans. Both the FFGC and the PEI Mi'kmaq communities are also involved in research activities, in some cases in conjunction with marine resource centers and university support mechanisms, demonstrating a further involvement in the management system and a link between the local knowledge and scientific research.

Second, the fishery case studies described in this chapter are strong examples of how empowerment of fishers and their associations and communities can improve the acceptance of conservation measures, and thus the compliance with management rules. In the case of the FFGC and other community management boards, the development of infractions committees involving board members themselves, which penalize the breaking of the fishing rules they set, has been a highly significant and successful innovation. In the case of the PEI Mi'kmaq fisheries, similar home-grown processes to deal with infractions are incorporating aboriginal values, thus respecting the cultural aspects of the communities.

Third, the progress toward community fisheries described here also provides an indirect conservation benefit in supporting a shift toward a more diversified livelihood base for fishers, households and communities. The idea is that diversifying income sources reduces pressure on individual fish stocks by spreading out the fishing effort, and reduces pressure on the fishery as a whole by diversifying to other sources of livelihood (Charles 2004). This is particularly apparent in the use by Mi'kmaq communities of fishery development funds to develop a diverse economy through (a) acquiring access to a range of fish stocks, and (b) initiating a diverse set of coastal economic activities.

Synthesis

The three fishery case studies of this chapter can be usefully compared and contrasted. Major similarities are that: (1) all represent successes, in their own way, in the implementation of community-based fishery management; (2) all developed initially within a context of turmoil (for the community management boards, following a period of widespread protest, and for the Mi'kmaq fisheries, following the court's Marshall decision); and (3) all involved a management framework that allowed for a strong diversity of approaches (between boards or between first nations), reflecting the desired capability for grass-roots adjustments to reflect local conditions.

Two key differences are important to note between the cases examined here. First, while the community management boards have been widely accepted and institutionalized, their existence has not been enshrined in legislation, and therefore remains somewhat at the whim of the government. In contrast, the Mi'kmaq fisheries reflect a formal right, and a Supreme Court decision. This may well have provided the Mi'kmaq bands with a greater level of confidence in their negotiations with the federal government than might have been the case with the boards.

Second, a major difference exists with respect to financial support provided by the government. On the one hand, the boards have received little such support, and indeed, it is often argued that their formation served the government's goal of downsizing its management services and downloading management costs onto fishery participants. The situation was very different in the case of the Mi'kmaq fisheries, with the federal government recognizing the need to redress a situation where Mi'kmaq communities had been excluded from the fishery for many years, and thus lacked the capital for fishing once their right to do so was recognized. Therefore, government negotiated with each band an agreement by which, in return for the band agreeing to abide by the government's fishery management scheme, a certain number of boats and fishing licenses (purchased from non-native fishers) were assigned to the band, and further funds were provided for the purchase of equipment. In this way, the communities generated the capacity not only to enter various fisheries, but – as described earlier – to become involved also in a range of related activities, such as fish processing, tourism, and other economic development. The success of these efforts suggests that providing such support for broad-based developmental initiatives within other fishing communities would be a useful addition to government policy.

Overall, the examples in this chapter reflect progress toward community-based fishery management in Canada's Maritimes. The groundfish management boards, while limited in the scope of their management responsibilities and in their level of community involvement, have proven to be resilient organizations for local or geographically-defined management, ones maintained without outside financial support. The Mi'kmaq fisheries, while initially reliant on federal government financing, are becoming self-reliant, are strongly community-based, and are successful in their key goal of providing sustainable livelihoods for their community members. Significantly, both the boards and the Mi'kmaq communities operate their fisheries within the context of co-management and within the constraints of conservation. This has been supported by development of marine resource centers and a broad support and communications network reflected in the cases of this chapter.

As noted earlier, the cases described here do not reflect the typical experience in the Maritimes region, where the majority of fisheries are not managed on a community basis. Indeed, the federal government's fishery management regime has tended to resist movement in that direction. To the extent that the cases here reflect successful community-based management, it is important to recognize two key determinants. One major factor has been pressure on the government, arising through a large-scale protest by fishers on the one hand, and a fundamental court decision forcing the government to take a community approach in dealing with aboriginal fisheries on the other hand. A second key factor has been the support mechanisms available in the Maritimes, as described in this chapter, which are important vehicles for facilitation, conflict resolution and knowledge gathering needed to make community approaches work, particularly in a context of limited governmental backing. The availability of such mechanisms, combined with the knowledge and determination of the region's fishing peoples, may well continue to produce successful conservation-minded community approaches to fishery management into the future.

References

Acheson, J.M. (1975) The lobster fiefs: economic and ecological effects of territoriality in the Maine lobster fishery. *Human Ecology*, **3**: 183–207.

Angel, J.R., Burke, D.L., O'Boyle, R.N., Peacock, F.G., Sinclair, M. and Zwanenburg, K.C.T. (1994) *Report of the Workshop on Scotia–Fundy Groundfish Management from 1977 to 1993*. Canadian Technical Report of Fisheries and Aquatic Sciences 1979.

Berkes, F., Feeny, D., McCay, B.J. and Acheson, J.M. (1989) The benefits of the commons. *Nature*, **340**: 91–3.

Brownstein, J. and Tremblay, J. (1994) Traditional property rights and cooperative management in the Canadian lobster fishery. *The Lobster Newsletter*, **7**: 5.

Bull, A. (1998) The Fundy Fixed Gear Council: implementing a community quota. In: L. Loucks, A.T. Charles and M. Butler (Eds) *Managing Our Fisheries, Managing Ourselves*. Gorsebrook Research Institute, Saint Mary's University, Halifax.

Charles, A.T. (1995) The Atlantic Canadian groundfishery: roots of a collapse. *Dalhousie Law Journal*, **18**: 65–83.

Charles, A.T. (1997) Fisheries management in Atlantic Canada. *Ocean and Coastal Management*, **35**: 101–19.

Charles, A.T. (2001) *Sustainable Fishery Systems*. Blackwell Science, Oxford.

Charles, A.T. (2004) Sustainability and resilience in natural resource systems: policy directions and management institutions. *Encyclopaedia of Life Support Systems (EOLSS)*. Developed under the auspices of the UNESCO, Eolss Publishers, Oxford. http://www.eolss.net

Charles, A.T. (2005) Turning the Tide: toward community-based fishery management in Canada's Maritimes. *Proceedings of the 4th World Fisheries Congress. American Fisheries Society and University of British Columbia Fisheries Centre. Vancouver.*

GCWFEA (Guysborough County Women Fisheries Enhancement Association) (2001) *Women Working Together and Making a Difference: a Story of Women's Participation and Leadership in Enhancing Local Fisheries*. Centre for Community-Based Resource Management, Antigonish, Canada.

Graham, J., Engle, S. and Recchia, M. (2002) *Local Knowledge and Local Stocks: An Atlas of Groundfish Spawning in the Bay of Fundy*. Centre for Community-Based Resource Management, Antigonish, Canada.

Jentoft, S. (1989) Fisheries co-management. Delegating government responsibility to fishermen's organizations. *Marine Policy*, **13**: 137–54.

Kearney, J. (1999) *Community-Based Management as an Alternative to Privatization; from Maritime Fishing Communities to a New Globalization*. The 1999 W.C. Desmond Pacey Memorial Lecture, University of New Brunswick.

Kearney, J. and MacIntosh, P. (1998) *A Needs Assessment of Scotia–Fundy Fixed Gear Groundfish Management Boards*. Report to the Management Boards. Extension Department, St Francis Xavier University, Antigonish, Canada.

Loucks, L. (1998) Sambro community quota fisheries management: a case of innovative community based decision-making. In: L. Loucks, A.T. Charles and M. Butler (Eds) *Managing Our Fisheries, Managing Ourselves*. Gorsebrook Research Institute, Saint Mary's University, Halifax.

Milley, C. and Charles, A.T. (2001) *Mi'kmaq Fisheries in Atlantic Canada: Traditions, Legal Decisions and Community Management*. People and the Sea: Maritime Research in the Social Sciences: An Agenda for the 21st Century. University of Amsterdam and SISWO (Netherlands Institute for the Social Sciences), Amsterdam.

Moore, W.P., Walsh, D., Worden, I. and MacDonald, J.D. (1993) *The Fish Processing Sector in Atlantic Canada: Industry Trends and Dynamics*. Prepared for the Task Force on Incomes and Adjustment in the Atlantic Fishery. Department of Fisheries and Oceans, Ottawa.

Pinkerton, E.W. (Ed.) (1989) *Co-operative Management of Local Fisheries*. University of British Columbia Press, Vancouver.

Pinkerton, E.W. (1999) Directions, principles, and practices in the shared governance of Canadian marine fisheries. In: D. Newell and R. Ommer (Eds) *Fishing Places, Fishing People: Traditions and Issues in Small-scale Fisheries*. University of Toronto Press, Toronto.

Pringle, J.D. and Burke, D.L. (1993) The Canadian lobster fishery and its management, with emphasis on the Scotian Shelf and the Gulf of Maine. In: L.S. Parsons and W.H. Lear (Eds) *Perspectives on Canadian Marine Fisheries Management. Canadian Bulletin of Fisheries and Aquatic Sciences*, **226**.

Rosenberg, A.A., Bolster, W.J., Alexander, K.E., Leavenworth, W.B., Cooper, A.B. and MacKenzie, M.G. (2005) The history of ocean resources: modeling cod biomass using historical records. *Frontiers in Ecology and Environment*, **3**: 78–84.

Wiber, M. and Kennedy, J. (2001) Impossible dreams: reforming fisheries management in the Canadian Maritimes in the wake of the Marshall decision. *Law and Anthropology*, **11**: 282–97.

Wiber, W., Berkes, F., Charles, A.T. and Kearney, J. (2004) Participatory research supporting community-based fishery management. *Marine Policy*, **28**: 459–68.

Part IV
Conclusion

Chapter 14
Healing Fisheries

Tim R. McClanahan and Juan Carlos Castilla

Introduction

This book and general summaries of the status of the world's fisheries indicate that there is a global fisheries crisis where demand and yields are outpacing the natural productive potential of fish (Pauly et al. 2002; Chapter 1). This has provoked a wide variety of solutions and alternatives, and the chapters in this book outline some of the current efforts to improve management in order to stave off this impending crisis. Various approaches are presented that are unique to the ecological, socio-political, and economic conditions in each fishery. The basic tools of wild-caught fisheries management are always the same: limits on effort, time, space, species, size, and gear (Fig. 1.1), but the way that these tools are adopted and blended in various institutional arrangements is part of the art of fisheries management that should consider the local and the particular social, economic, and ecosystem context. In this chapter we have organized the discussion along four major principles of problem solving, namely the complexity of management, differential solutions for different diagnoses, the need for holistic solutions, and the need for increasing the professional approach to fishing and management. We hope these guidelines and the discussion below create flexible guidelines for managers who wish to artfully cobble together their own progressive programs.

Ecosystems and cultural systems are complex

Wild caught resources are part of complex ecosystems that can exhibit surprising responses to environmental change and human interactions (Myers 1996; Lessard et al. 2005). The full complexity of these ecosystems will probably never be fully understood and may support the possibility that they are essentially unmanageable (Thompson and Trisoglio 1997). At the level of highly specific control this is likely to be true and managing is more the ability to cope with the moment without jeopardizing the future and creating alternate future visions, rather than the ability to precisely control the above tools and predict the outcomes of management decisions (Walters 1997; Waltner-Toews and Kay 2005). Nearly all the chapters in this book describe a historical trial and error process where solutions to problems are addressed as they become dire. This is unlikely to change, and there is, therefore, a need to develop a set of simple principles for coping with crises and creating future vision that will guide the management out of or away from these crisis situations (Waltner-Toews and Kay 2005).

Ecosystems are hierarchically organized and humans are most often harvesting at the top of this hierarchy (Pauly et al. 1998). Similarly, human organization and value is hierarchical and decisions are often made at the top of this hierarchy (Odum 1988). The consequences of both are, however, not predictable cascading effects because the bottom supports the top and there are a diversity of pathways that create interdependency and surprise (Polis and Strong 1996). This results in complex organization that point to some of the following principles for management.

Need to save ecosystem parts, interactions, and processes

The first rule of resource-use tinkering is to keep all the ecosystem parts. Surprises may often lie ahead and the loss of parts or processes is likely to undermine the organization and ability to create alternatives and adaptability. We saw one good example of this in Chapter 4 where the removal of shells while dredging for scallops reduced suitable settlement substratum, which undermined the long-term sustainability of the scallop harvest. Saving the parts is a fundamental principle of ecosystem management (Browman and Stergiou 2004; Pikitch et al. 2004). Although most ecologists agree on this need, there is less of a consensus on how it will be achieved (Browman and Stergiou 2005; Castilla and Defeo 2005). Ecosystem management has, however, become largely synonymous with the call for the creation of marine protected areas that exclude resource extraction, as this is seen as a key way to save the parts (Jackson et al. 2001; Balmford et al. 2004; Pandolfi et al. 2005). Provided it is socially and economically feasible to exclude extraction from areas large enough to assure viability this is certainly one of the simplest ways to ensure that the ecological parts and processes are preserved.

Support for marine reserves is cultural and variable among resource users and managers (Mascia 2003; Christie 2004; Martin-Smith et al. 2004; Cinner et al. 2005; McClanahan et al. 2005a, b; Richardson et al. 2005). For example, in Chapter 8 we learned that the perceived benefits of closed areas were higher among resource managers than users, but that resource users' perceptions of benefits increased with the age of the closed area (McClanahan et al. 2005a). In Chapter 7 we saw that fishers had more positive views of area management when they believed they were involved in management decisions, even when there was little evidence for an increase in their catches. This may, however, differ with gear use as recent studies indicate that permanent trap fishers have more positive perceptions of area-based conservation than mobile fishers (Chapter 13; McClanahan et al. 2005b; Richardson et al. 2005). Additionally, small closed areas in the Philippines appear to be more accepted in the Philippines than Kenya (Martin-Smith et al. 2004; McClanahan et al. 2005a; Chapters 5, 7). There are not enough comparative studies to make sweeping conclusions, but issues such as the management structure, education, involvement, size, permanence, location, and spacing will influence perceptions towards closed areas and probably be influenced by culture, politics, professions, and the poverty of the stakeholders.

Halpern and Warner (2002) have argued that some of the friction between resource managers and users can be reduced if 20–50% of the area is closed and covered by variable-sized closed areas at intervals that allow a mix of conservation and resource use. Small reserves where rotating closures are used may be popular and easily managed by resource users (Cinner et al. 2005), but they only develop under certain socio-economic conditions (Cinner 2005), are usually much smaller than 20% of the fishing area, and may have limited ability to preserve the full diversity, biomass, and processes of the ecosystem (McClanahan and Graham 2005; Russ et al. 2005; McClanahan et al. 2006).

An example of the problems of these different closed area designs is exemplified by a comparison of the collaborative management system described in Chapter 7 with the larger and permanent closed area system in Chapter 8 (Fig. 14.1). Studies in Kenya's marine protected areas have found that the maximum fish biomass of

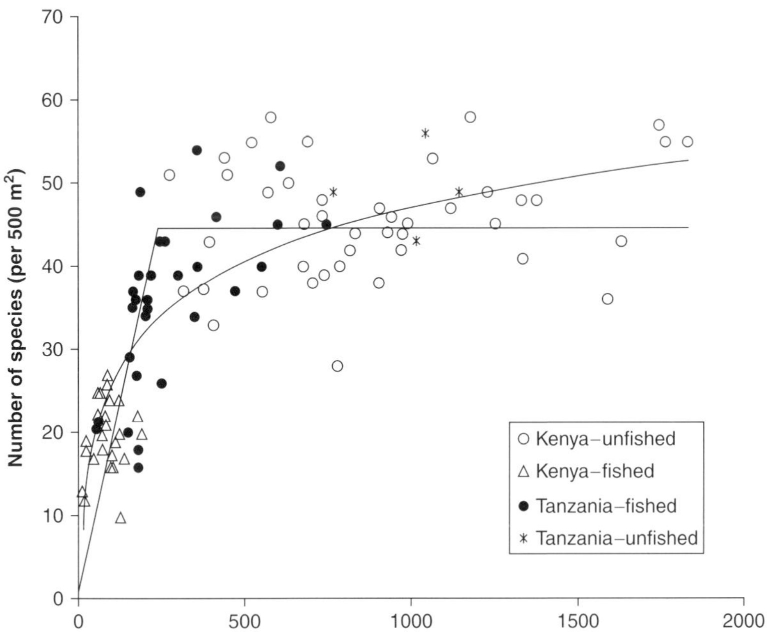

Figure 14.1 Scatter plots of numbers of fish species as a function of their fishable biomass (>10cm lengths) in Kenya and Tanzania based on the average of four 500 m^2 transects per study site. The results contrast the Kenyan situation where there are large differences between unfished marine parks and fished reefs and Tanzania where there are some collaborative fisheries management programs with intermediate biomass.

120 tons/km^2 was achieved after 20–25 years of closure and that other ecological processes such as calcification may take even longer (McClanahan and Graham 2005). An East African fishery that is managed for maximum sustained yield would, therefore, have a target fish biomass of around 45–60 tons/km^2, which is similar to values found in the Tanga collaborative program. Comparative studies indicate that the Tanga system had lower diversity of fish and corals, particularly rare and sensitive species, and had lower levels of herbivory and predation (McClanahan et al. 2006). This is despite having a near-optimal biomass and the use of some small but temporary closures that are supported by resource users. The solution to management would appear to lie somewhere between the polarized Kenyan situation of permanent closures and overexploited fisheries, and collaborative systems that may optimize yields but do not support the full diversity of species and processes of full closures. This is as much a social organization as a technical design problem that will require achieving a balance between the desires of users and managers of biodiversity and their perception of the value of nature. The balance is likely to change in the light of experience and appreciation of the benefits versus the cost of protected areas.

Diverse opinions

Most of the case studies in the book exemplified the importance of the forum as a key means towards problem solving. It is likely that the complexity of ideas and desires find their expression in these meetings and this increases the chances for their reconciliation. In the absence of forums, there is a tendency for resource users to largely consider their selfish interests and management authorities to become isolated and inflexible (Ostrom et al. 1999; Holling 2001). Selfish interests of resource users is expected to result in the optimal foraging model where expenditures are minimized and resource capture is maximized (Aswani 1998), which is expected to lead to an equilibrium between opportunity costs and expenditures. Where opportunities are few, as in many poor countries, this is expected to result in the reduction of resources below an optimal yield.

Where the creation of a forum with symmetric relationships, backed by guided and integrated processes are present or encouraged, the expectation is that successful rules will develop based on their experience and culture, that are easy to learn, follow, and monitor, that reduce the complexity of management, and are perceived to have joint benefits (Ostrom et al. 1999). Forums are the basis for co-management as they socially rationalize what may otherwise become self-interested activities. Achieving symmetric relationships can be a problem as there will be a tendency for those with the greatest wealth and incentives to be better organized and influence decisions (Scheffer et al. 2001). An evaluation of co-management institutions in the small-scale South African line and invertebrate fishery described in Chapter 6 found that the most strongly-correlated factors for success were that benefits must exceed costs, the existence of training and empowerment, and the existence of people with long-term commitment to the process (Napier et al. 2005). The Philippines example in Chapter 5 indicates that forums alone will not necessarily lead to improvements unless they are guided and integrated into a planning process that is instituted by local governments. Consequently, the forums that build on these principles are more likely to produce effective co-management and reduce the chances that common-property resources will be unmanaged. In the Chilean example described in Chapter 1, the co-management system has been institutionalized and successfully applied to small-scale fishery activities, mainly for those based on diving activities.

Assume a humble and precautionary approach

Precaution is difficult to achieve, as it requires resolving the conflicting short-term interests of individuals with the long-term interests of society. This is exemplified in the tension between ecological and economic-based management and their different view of discounting, or the future value of resources (Chichilnisky 1997). Economic theory sees the future to be lower than the present value because of factors such as inflation and competition. This assumption may be true where resources are replaceable, but ecologists will see the limits to replacement, particularly non-economic or market values such as species and ecological processes that recycle waste and produce clean water and air. Where these services were free or cheap they become increasingly expensive and become part of the economic costs (Costanza et al. 1997; Balmford et al. 2002). An economist will see this as creating new markets where as an ecologist is more likely to see this as an increased ecological overhead that will reduce the net production of a process (Odum 1996).

Ecologists will see future values as greater than present, as the values of species, their services, and ecological processes needs to be considered over their long-term use by people over generations. The loss of species and ecological processes will result in long-term losses of ecosystem services. There are limits to the replacement of species, particularly when they are lost or reduced to such levels that they are no longer valuable or when non-renewable resources are diminished to the point where they cannot replace the service of renewable resources (Odum and Odum 2001). Some ecosystems and fisheries are irreversible or slow to reverse and this can create significant lags in production and lost utility (Carpenter et al. 1999; Hutchings 2000). Beyond this utilitarian view, there is the ethical issue of depleting species and processes to levels where their future survival is jeopardized.

The precautionary approach is appealing and socially rational but, because it is in conflict with the selfish rational position, it requires active management and agreement among stakeholders. Optimal solutions or maximum sustainable yields are often near the edge of catastrophes (McClanahan 1992; Scheffer et al. 2001). Slight overestimates of yield potential or underestimates of effort can lead to a runaway decline in resources. The political process, where influential resource users can argue for raising yields or quotas, can lead to overestimates. We saw how this tendency to overestimate yields in meso-scale fisheries resulted in readjustments and declines in allowable yields. Less obvious are more passive processes such as the increase or creep in the efficiency of effort due to improved technology, as was described in the Australian invertebrate fisheries (Chapters 10, 11). Political compromise is seen as fair in decision-making, but it has the potential to flirt with ecological catastrophes. Integrating needs within the limits of ecosystems is more likely to provide optimal and long-term solutions, but requires a more extensive forum and management framework (Scheffer et al. 2001).

Where change is ecologically irreversible or very slow, the consequences of incremental but irreversible change need to be avoided. Strong discounting is particularly risky when ecosystems are irreversible, as utility could be permanently lost (Carpenter et al. 1999; Hutchings 2000). When irreversibility is suspected, lowering discount rates and supporting investigations into the irreversibility and time dynamics is a possible management solution. Consequently, if fisheries yields are calculated to be optimal at some level, the precautionary approach would be to assume some error in the estimates and to use the more conservative estimates, say two standard deviations to the lower end of the yield estimate. There is also the need for simple thresholds or cut offs where resource use is curtailed. Correction of a political bias, fast response to problems, closed areas, moratoriums, conservative estimates, and monitoring resources and recovery rates are important tools to implement the precautionary approach.

Complexity requires differential diagnoses

Artisanal versus industrial fishing

Most of the fishery examples included in this book represent cases of small to medium-scale fisheries. No universal definitions of small-scale and large-scale or industrial fisheries exist, but authors have synonymized the former with 'artisanal' or 'traditional' (Berkes et al. 2001; Castilla and Defeo 2001; Defeo and Castilla 2005). There are numerous characteristics differentiating them. Most artisanal fisheries are low to very low investment, supporting important subsistence, recreational, and commercial activities in terms of number of people directly or indirectly engaged (Berkes et al. (2001) estimates over 500 million worldwide), while industrial fisheries are long-term and large investments. Improvements in fishery technology have been documented for artisanal fisheries (diving gear, outboard engine, air compressor); nevertheless, the amount of improvement and investment in technology in the industrial fishery is substantially higher and the fleet capacity cannot be easily adjusted in years when resources are declining. Artisanal activities occur on coastal grounds, therefore the feasibility of implementing management options drastically differs between them. Territorial users rights for fisheries, the allocation of community fishery quotas and spatial use of exclusive rights for communities to manage and exploit resources, and co-management strategies, are fishery management tools that are more easily implemented in small-scale than in industrial or meso-scale fisheries (Chapters 2, 3, 6, 7).

National versus community organization

Within a country, the national versus community organization dilemma rests on the so-called 'tragedy of the commons' (Hardin 1968). Because common property and individual rationality would lead to a tragedy, Hardin's proposition suggested basically two solutions to the management of common resources: (1) top-down management by the central government; or (2) privatization of free enterprises. This pessimistic view, where fishers are trapped into acting independently and inflexibly is based on a top-down management strategy dictated by an external authority. In fisheries, central management or top-down strategies are mostly used for the administration of industrial, meso-, or large-scale fisheries (Chapters 10, 11, 12). The opposite view, particularly valid for small-scale fisheries, is that resource management is in fact 'people management', and that fishers are indeed one of the key actors in the management process (Chapters 5, 6, 7, 13). This management has been framed as 'co-management', and represents a spectrum of resource managerial arrangements, ranging from the so-called 'consultative co-management', where the government creates mechanisms of dialogue with resource users and informs those users of management decisions; to the other extreme, called 'delegated co-management'. Here, the government transfers all the managerial decisions to user groups (community, communal group, tribe), which then informs the government of decisions made at local level (White et al. 1994; Johannes 2002; McConney et al. 2003; Castilla in press).

Several cases in the book include the implementation of bottom-up fishery managerial actions for the rational use of coastal resources, where the users have been directly incorporated into the planning process such as in the Philippines (Chapter 5). We suggest that in small-scale fishery activities and, in some cases also in medium-scale fisheries (Chapter 13), an important part of the success regarding long-term sustainability of resources lies in the implementation of co-management actions, where the users are incorporated at an early stage in the processes. For these fisheries, the avoidance of the 'tragedy of the commons' must include the implementation of bottom-up management strategies and, whenever possible, of communal actors (Berkes et al. 2001; Castilla and Defeo 2001).

Pristine versus mature fisheries

Fisheries management and the associated perceptions will depend greatly on the position along the resource use continuum, where limits are not perceived early on but where perception and agreement on limits are critical to successful management at later or mature stages of resource use (Smith 1988). The failure to make this transition is the basis for the catastrophes and crises that we are currently experiencing and trying to rectify. In the early stages, the principles of freedom of the sea, also lack of individual, community, or national control, may be justified as the resource has the potential to tolerate and respond to additional sources of mortality. In the early stages of management it is, however, important to establish a baseline for the ecosystem and determine key life-history characteristics of the resources, as this will allow a projection of the potential and ultimately limits for harvesting. The lack of a baseline and its shift over time plagues fisheries management and it is largely through various indirect methods that estimated baselines are established (Pauly et al. 1998; Jackson et al. 2001; Saenz-Arroyo et al. 2005).

During the early stages of fisheries development, profits can be high and the capital is reinvested. The inability to project or admit that there are limits to production is what causes the overcapitalization that leads to so many fisheries problems (Smith 1988). Overcapitalized fisheries can led to severe depletion of resources because even when there are no net profits, the lost depreciation of capital is slower for low than no fishing. This causes fishing to continue well beyond optimal levels and has frequently caused governments to intervene and often subsidize fishing, which further removes the incentives for conservation and sustainable use. Consequently, it has been suggested that incentive-based systems where fishers feel the pain of overexploitation and gains of conservation are preferable to subsidies and associated disincentives (Hilborn et al. 2005). Similarly, resource managers should have similar incentive systems where their decisions and actions have clear consequences (Walters and Martell 2004).

This carrot and stick approach may be an improvement on subsidies as it will limit the perverse effects on the fisheries, but it is less preferable to an early recognition of limits where overcapitalization and effort creep are slowed prior to natural disincentives and catastrophes. Based on the many examples of capital and effort overshoot, it may be naive to believe that fishers and managers can foresee limits and act in a way that prevents these natural disincentives. However, we have seen examples of foresight from Australian invertebrate fisheries where monitoring, feedbacks, and incentives can improve the chances to stay within limits, and even account for the passive upward creep in effort (Chapters 11, 12). Unfortunately, there are not many pristine fisheries at present but some, such as deep seamounts, need to undertake the necessary research and management steps to project sustainable yields and avoid what is currently unsustainable mining of fish resources (Branch 2001). There is, however, considerable compiled information on fish life histories and for ecosystem modeling that can help estimate levels of exploitation and ecological consequences (http://www.fishbase.org).

Mature fisheries need to be managed with significant caution as they are likely to lie on a precipice where there is a mix of under-, over-, and maximally-exploited species (Hilborn et al. 2004). Here, precaution and the recognition of firm thresholds, limits to effort, gear, time, and the need for periodic closures are required. There is also the need to avoid runaway competition for dwindling resources that can lead to the use of gear that is overly competitive or destructive to fish habitats, as described in Chapter 8 (Pauly et al. 1989; McClanahan and Mangi 2004). Small mesh and drag nets are among the most destructive gear that will dominate fisheries where there is competition for smaller and smaller fish. The chance of achieving consensus and elimination of destructive gear is improved if resource users and managers can agree, but still requires forums and active management to achieve compliance (Chapters 4, 7, 8). Because fishing gear is rooted in tradition, and easy to see and enforce, their management may be one of the most acceptable regulations for fisheries in poor countries where restricting effort or catch is difficult or politically unacceptable (Wilson et al. 1994; McClanahan and Mangi 2004).

All permanent solutions are holistic solutions

Fishery crises of the world are not a singular crisis, but a series of diverse, complex, multifaceted and multiscale crises, embedded in different societies. Fish overexploitation, as due to world fleet overcapitalization and the race for fish, and solutions via the implementation of ecosystem-based fishery management (EBFM) considerations, marine protected areas (Castilla 2000), and use of precautionary approaches have been widely advocated among the main solutions. We believe that new perspectives for solutions to world fishery crises for small- as well as large-scale fisheries, require more holism and shifts in management focus, notably including effective governance and sharing of management roles and responsibilities between governments and local organizations (Castilla and Defeo 2005). Operational fishery prescriptive drivers, including catch and fleet limits, global and individual quotas, and size limits need to be set under precautionary approach policies, including unfished MPAs and EBFM considerations.

In view of the much larger number of artisanal fishers, associated generation of employment, low annual fuel consumption and annual food catch, it appears equally important to implement societal adaptations, including territorial user rights in fisheries (TURFs), self-governance and co-management strategies that acknowledge cultural and political backgrounds (Chapters 2, 3, 4, 6, 7). Furthermore, fishery management measures for fixing developed and rich consumer societies will not necessarily match those required for developing societies. Even within societies, measures to fix fisheries may vary substantially. Solutions for fish-oriented societies will differ from those applied to non fish-oriented cultures and consideration of cultural, economic, and geographical contexts are essential.

In developing countries where there are often multiple pressing priorities for action, of which fisheries is only one, an integrated management approach for coastal resources that is embedded in local and national government mandates may offer hope for improved long-term management. Such efforts in the Philippines (Chapter 5) are proving quite successful in addressing multiple issues that add up to improved planning for management, more rigorous coastal law enforcement and other actions for the benefit of coastal resource protection and sustainable use of resources (White et al. 2005).

Long-term fluctuations in artisanal and industrial fisheries are partly related to large-scale environmental variations and climatic oscillations, so fixing them may require pragmatic measures, such as incentives and disincentives. To this aim, the crafting of new regional and inter-regional institutional arrangements is urgently required, a key point usually forgotten (Botsford et al. 1997).

The household context of resource users

Even in the most traditional fishing societies, it is unusual for fishing to be the only form of household income or resource acquisition (Cinner et al. 2005). Agriculture, husbandry, and various types of trade or formal or informal employment usually accompanies fishing. Division of labor also occurs in the household based on age, gender, and seasons, and the sum of all activities will gauge the wealth status of the household and the decisions they make and undertake. While some fishers are permanently occupied by fishing, many fishers will move between the alternative occupations based on the success or profits of the various alternatives.

Change and the existence of options are part of the resilience of the household and efforts to undermine this resilience without alternatives will be met with resistance. Conversely, development efforts to increase specialization and reliance on fishing through improved gear and training will increase dependency on fishing and reduce the occupations and resilience in the absence of good fishing. Consequently, successful poverty alleviation may be made more successful by increasing the pluralism of occupations rather than specialization and increased capitalization around a dwindling resource. An example from this book is Gilutongan Island, Philippines, where many poor and previously illegal fishers are now growing seaweeds and managing a highly popular MPA for scuba divers (Chapter 5). Finding and reducing restrictions to occupational pluralism may be one of the better alternatives to livelihood and therefore fisheries management. Development projects that increase the non-fishing capital and alternatives of the household may be more effective than the usual attempts at bolstering the fishing occupation (Allison and Ellis 2001).

Strong interactions in ecosystems

Not all interactions in ecosystems are the same, and in fact, an analysis of one of the better-studied food webs, the Caribbean coral reef, found that they varied over six orders of magnitude (Bascompte et al. 2005). Although the complexity of ecosystems such as coral reefs can be daunting, the recognition that there are considerably fewer strong interactions can considerably reduce the complexity of both understanding and managing ecosystems. Identifying the strong interactions in ecosystems can be one of the major challenges of ecologists that can assist management. Where important species or functional groups that are critical to maintaining function and structure of ecosystems are identified and managed, it is expected that many of the dependent species would be supported by the maintenance of these key species or functional groups (Walker 1995). MPAs are one of the ways to discover strong interactions, as they act as controls for areas where resource use has altered species interactions. Active control of species is one way to simplify management considerably but it must be appreciated that simplistic predator–prey or trophic relationships can often lead to overly simple predictions as these species are embedded in more complex systems (McClanahan 1995). For example, it has been suggested that the termination of harbor seal hunting in British Columbia has increased seal populations such that they are reducing survival of commercial salmon. Reinstating a harvest on seals would be expected to increase commercial salmon catches, except that seals feed on other prey including the hake that feed on juvenile salmon. A simulation model created to test the strengths of these interactions found that seal culling was predicted to reduce commercial catches of salmon due to their indirect effect through the hake populations (Lessard et al. 2005). Examples such as these point towards the need to find the strong interactions in ecosystems and manage them to maintain processes of production and associated diversity. The combination of field studies, models, and experimentation will help to uncover these strong interactions and will be important for identifying species and functional groups that require special status in fisheries.

Change is part of the system

Resource users and managers are well aware that nature and ecosystems are dynamic, changing at many scales of space and time and that strict control is seldom possible. Accepting change is, therefore, both philosophically and practically valuable but it should be appreciated that there are different types of change and some, such as irreversible change, need to be avoided when there is lost ecological utility (Carpenter et al. 1999). There is a fine line between managing to avoid irreversible change and for tolerating or adapting to the inevitable environmental change. In fact, as above, the stability of key species and functional groups is critical to ecosystem functioning, and their loss is expected to result in shifts in ecosystem structure, function, and utility. In some cases the loss of key species can be replaced or compensated for by other species of lower dominance but potentially functionally redundant taxa; but in other cases, this is unlikely. Consequently, maintaining the full diversity of taxa and functional groups is a key aspect of maintaining ecological functions (Naeem and Wright 2003). In principle, it may be inevitable that the species change, but it is important that processes and utility are maintained. Therefore, this distinction between players and processes is needed when evaluating, managing, and accepting change. Management should maintain all species in the system while recognizing that their dominance will change and give special conservation status to important functional taxa that have roles and utility that is not easily replaced. Similarly, ecological structure and processes are often associated with the size spectra of organisms and there is concern that fishing is changing the size spectra of fish, largely by removing the larger individuals (Jennings et al. 2002; Graham et al. 2005); this is expected to have evolutionary consequences for species and their interactions (Palumbi 2001).

Monitoring and evaluation are essential

Holistic monitoring of nature, people, and their compliance

Monitoring national or trans-boundary industrial fisheries in the open ocean requires national and multinational arrangements, including global quotas and technical and societal consensus on acceptable fishing effects and management measures. In artisanal fisheries, global quotas, limits to fishery entry, and fleet limits are difficult to enforce, due to easy access to stocks, even by unauthorized and recreational users, and the high enforcement costs at landing sites. Most developing countries cannot afford effective monitoring, but this is essential at the local level to develop effective localized fishery management systems (Chapters 3, 5, 6, 7, 8). Artisanal fisheries represent data-poor situations, precluding the quick implementation of sophisticated management measures (Johannes 2002). There is a lack of reliable data regarding resource spatial structure and environmental processes and high enforcement costs may preclude management effectiveness. On the other hand, societal incentives such as TURFs, temporary closures, co-management, and community catch quotas, under the framework of traditional or institutionalized arrangements have been shown to have some success (Chapter 2, 3, 6, 7). At these reduced spatial scales, technical as well as societal consensus are needed but, due to the importance of traditions, more than economic incentives are needed to achieve compliance (Castilla and Defeo 2001, 2005). In addition, integrated coastal management regimes at the local scale can achieve improved fisheries management if accomplished within the existing legal and institutional context of the country (White et al. 2005).

Need for simplicity

Costs of management are often unrealistically high for the benefits of management and this creates the need to simplify but also learn from management. Because of the costs of large-scale field experiments and managing many people, experimental management has played only a minor role in developing fisheries management theory. Adaptive management is the closet approach to field fishery experimentation and if practiced within some simple guidelines it can reduce management costs (Holling 1978; Walters 1986). Adaptive refers to a dynamic management strategy by which a sequential reassessment is done as more accurate information about the fishery is obtained. Walters and Holling (1990) defined three adaptive management approaches: 'deferred action', for a fishery that is managed by trial and error decisions until sufficient information on the stock and the fishery of the resource is achieved; 'passive adaptive', where the previous information about the fishery is used to build a deterministic fishery model and to perform management actions; and 'active adaptive', where the short- and long-term fishery performance is continuously improved by experiments consisting of different and adjusted management policies. Mainly passive and active adaptive management requires the coordination of actions, modeling, and policies and are therefore likely to have the most direct management expenses as they actively engage fishers. Deferred action has opportunity costs of potentially lost fishing income when decisions lead to losses in either productivity or fishing time. Because decisions involve the agreement and coordination of resource users, the adaptive management processes are thought of as co-management and not top-down control (Castilla and Defeo 2001).

Need for targets and threshold measures

Simple management heuristics or thresholds that have been established from prior scientific studies or the experiences of resource users may be one of the better hopes for managing the fisheries that lack the resources for complex fisheries management. Monitoring is useful when testing for the long-term effects of specific management systems and to evaluate the effectiveness of management. In practice, it is difficult to implement the results of monitoring because there is often missing information and lags in time between recommended changes and compliance with management regulations. Managers need to find the mix of information, stakeholder involvement, and targets or thresholds where management action is needed to facilitate the adaptive process. In poor and locally-managed fisheries there is a need for simple information collection and management decisions. For example, if a particular size or catch per day of a key target species is below a certain level, then this target species would be protected from fishing for a period before testing for this threshold at some later date. Cinner and colleagues (2005) describe a system of periodic reef closures in Papua New Guinea where decisions to open closed areas are made based on the time since the last closure and need for food for feasts. They conclude that in the absence of more elaborate monitoring and management systems this type of village-level social heuristic can have the potential to maintain fish stocks.

Another simple way to manage a fishery is to reduce effort whenever the total catch for a specific area declines. This requires care in determining that the catch came from a specified area. Too often, catch per effort is the dominant statistic used for management decisions but this has problems because as many investigators have shown, effort can creep or be quite flexible and hard to fix and measure (Chapters 3, 4, 10, 11, 12). For example, catch per boat that is often the measure and management system for quotas, are bypassed by discarding low value catch or by increasing the travel distance or time for the boat as resources become less abundant. It is possible for the total catch at a landing site to be constant or to increase far beyond the maximum sustained yield of the local fishery because the fished area increases and fishers are expending more daily effort. A constant CPUE associated with increased movements can reflect serial depletion and is a worry for many fisheries as described for sea urchins and scallops in Chapters 3 and 4. Consequently, in a declining fishery, fishers will travel further and fish longer and this makes it necessary to change the effort and area parts of this calculation to get unbiased catch per area estimates. If these factors are considered and the conditions of the environment that influence production do not change greatly between years, then monitoring the catch per area can give good estimates of sustainable fish yields and indicate thresholds for closures or other management restrictions.

If managers have some indication of what the optimal or pristine biomass of the fishery is, then biomass targets can be useful for establishing thresholds as seen for the hoki in Chapter 12. One concern of biomass thresholds is the state of the rest of the ecosystem at that threshold. In the example given for East Africa, the 45–60 tons/km^2 threshold also appears to be associated with high biodiversity, although near the threshold, diversity declines precipitously (Chapter 8; Fig. 14.1). Consequently, rare and sensitive species are lost at that threshold and therefore protected areas that protect rare and vulnerable taxa are needed in conjunction with thresholds, to ensure the protection of all biodiversity (McClanahan et al. 2006).

Fishing and fisheries management is a profession

Many observers too often see fisheries as being the last chance effort of poor and desperate people and that fishing is a means to absorb this unskilled and surplus labor. There is some truth to this belief in some fisheries, but there are many fisheries, even among those considered very poor, where this is far from the truth and where fishers have considerable commitment to the fishing lifestyle, have considerable knowledge and skills, and the potential for profits are large and have produced considerable wealth that is better than many employment alternatives (Pollnac et al. 2001). Similarly, the fisheries management profession might be observed to be a highly circumscribed form of applied mathematics or field biology that is not as lucrative or as challenging as other fields in mathematics and biology. However, the fisheries manager is often dealing with complex problems that transcend any particular discipline and requires a holistic view of the problem and problem solving. The problems of fisheries will not be lessened until there is an awareness of the need to professionalize the institutions of resource use and management. In some developing countries, such as Philippines and Indonesia, most of the responsibility for fisheries and coastal resource management has been devolved to local governments and is largely separate from the national resource management agencies (Lowry et al. 2005). This presents both challenges and opportunities for improved management in small-scale fisheries.

One of the hallmarks of professionalism is the existence of professional standards and means to ensure these standards are maintained by the society of professionals through democratic processes. Where professional standards are high, maintained, and respected, the chances for success are greatly increased. This generally requires some form of formal or informal certification, meetings where knowledge is shared, and principles of organizations are learned and enforced in democratic ways, the maintenance of rights of the users and managers, and the means to enforce compliance with the profession and the fishery itself.

Incentives, disincentives, and certification of people and fisheries

The productive capacity of fisheries creates its own incentives and the limits to production create disincentives. These natural positive and negative feedbacks have the potential to regulate the fishery without the needs for complex management and social organization. In fact, it has been argued that politically-motivated management that artificially buttresses fishing have often created perverse subsidies that override nature's negative feedbacks with catastrophic consequences to their production (Hilborn et al. 2004, 2005).

This raises the question as to whether it would be best to have laissez-faire or simply a different type of management that does not encourage perverse subsidies and where limits to production are recognized as being the cornerstone of management. There are many cases, particularly in the developed fishery of wealthy countries, where no active management would be an improvement on subsidies, but this buttressing is less often the case in poor countries that lack funds to subsidize fishing. In the case studies from Australia we have seen that where managers have seen a limit to production and the incentive to maintain certification, management has set up restrictions rather than subsidies, with positive consequences. Conversely, in the seine net case study of Kenya, the position of the government fisheries department that increased effort would increase catches and reduce poverty led to an uncritical and laissez-faire approach towards gear. This allowed the proliferation of seine nets despite their unpopularity with the majority of fishers (McClanahan et al. 1997, 2005b). Here, once the permanent traditional fishers' view was heard and recognized as valid, they were able to override and get partial support from managers, which led to an increase in catch. Consequently, the issue is less a case of having or not having managers but more of one where stakeholders recognize limits and the need for some form of management, local or national, and incentives are created to enforce compliance with those limits.

A key aspect of professionalism is the development of understanding and recognition of cause and effect relationships within the profession's area of focus. The profession should be constantly vigilant in testing the assumptions behind this cause and effect but they should also have some confidence in the predictive ability of their profession and insist that management address data and causation rather than the politicizing of decisions. This professionalism requires that those people involved in the profession share some similar and tested ideas of causation, such as the limits to production, and that standards are maintained through professional bodies and certification methods. This also means that members stick to their professional principles and are not malleable or become corrupted by political processes that can dominate in the absence of professionalism. This requires that both managers and fishers are responsible for the consequences of their decisions and that there are decision-making processes that create incentives and disincentives for good and bad decision-making.

Experienced observers of fisheries management note that smaller and more permanent management structures allow for easier identification of decision-makers and consequences of decisions, and also permits easier allocation of responsibility (Walters and Martell 2004; Hilborn et al. 2005). Fishing and fisheries management may benefit from certified expertise in various resources and ecosystems, where training, internships, certification, and membership in professional specialties are seen as prerequisites to professional membership, similar to the way that medical specialties are maintained. The creation of professional bodies that operate with positive and negative feedbacks, similar to other successful professions, will help to improve the quality of the profession and management. Generally speaking, crystallizing these principles appears to be much more difficult in poor than in rich countries but there are examples where certification of local management can be made by the national government or NGOs, encouraging compliance with basic standards.

Meetings

The success stories in many of the chapters in this book started with a meeting followed by a series of meetings and the integration of the resolutions into the management system. Consequently, this must be seen as one of the key elements of success. Meetings where positions and opinions are expressed are the first step towards integrating the various needs of the stakeholder and the ecosystem. Some of the forums that naturally occur are households, landing sites, and fishery business and management offices. At this level there is a limited exchange and integration of needs that has the potential to lead to a myopic view of needs and trade-offs in decision-making. Most of the successes in the book occurred when one or most of these organic or cellular forums met with other key individuals or forums. Consequently, a critical step towards truly integrative solutions will be the creation of these intercellular meetings. These are likely to arise from conflict between cells, but with some foresight by key individuals they can be planned and maintained to avoid these conflicts. Attending meetings within and between these cells should be seen as one of the measures of professionalism described above.

In the absence of meetings, rigid control or management structures can develop and lead to strong adherence to institutional views that can become disengaged from field realities (Holling 2001). This occurs at the cost of lost time and associated resource use (Scheffer et al. 2003). Command-and-control management can lead to polarization of resource users and managers that leads to poor integration and feedback between the user and the resource (Holling and Meffe 1996). Hilborn and colleagues (2004, 2005) have argued that an incentive system is preferable to the command-and-control system. Nonetheless, central decision-making is expected to reduce the time to respond to problems, is an important element in change and can rectify problems if it has the ability to detect and respond (Scheffer et al. 2003). Therefore, there is a need to integrate, centralize, and enforce without falling into the administrative trap of command-and-control, and meetings are one important way to insure a direct link between the managers and users of resources.

Democratic principles

The creation of negative and positive feedbacks and associated self-organization of fisheries and its management is likely to be improved with the creation and maintenance of basic democratic principles. Professional societies and industry and management committees are more likely to change and adapt if there are forums for debate and the decisions and leaders are democratically elected. Like all democratic institutions they will need to develop systems to moderate the wealthy, well organized, and vested interests of particular stakeholders. This is most likely to succeed if the debate is open, includes maximum participation and is guided by professions with ethical standards that promote integration over special interests. In addition, because fisheries are part of the use of ecosystems that have many values, particularly non-use values, it is important that forums and debates include representatives of these non-use values. These are currently a mix of government personnel, also citizen and environmental groups. Finding the right mix of representation and maintenance of ethical standards will be critical to the integration and long-term success of the fisheries.

Rights and compliance

Without compliance with management initiatives there is little chance for achieving limits. Compliance is, however, most likely to be achieved if there is some sense of user and manager rights that create incentives for compliance as shown in a number of chapters. For example, TURFs or exclusive rights are one tool where area or resources are allocated permanently or for a given period (Castilla 2000; Castilla and Defeo 2001). The importance of fisher co-ownership and responsibility in management decisions of these TURFs is directly linked to the establishment of resource co-management strategies. This co-management is an effective institutional arrangement by which fishers and managers can interact to improve quality of the fishery regulatory process. Rights are given to fishers as long as they follow some compliance of management that is seen to maintain sustainability. The allocation of TURFs is a powerful tool to engage communities in the management process and to achieve accountability. In places where this management tool has been experimentally tried (Castilla and Defeo, 2001), it has been welcomed by fisher communities and has improved the status of small-scale fisher communities. The enforcement aspect of TURFs is attractive since the responsibility for the maintenance and adequate use of the TURF area is transferred from the government to fishers, with the possible loss of the resource as an incentive to promote the expected social organization.

Conclusion

Most of the examples included in this book refer to small-scale fisheries and problems related to micro- to meso-stock resources. They also refer to fishers' communities, associations, and communal organizations and to innovative management approaches adopted in different parts of the world under very different societies and political and cultural arrangements. Most describe a history of trial and error processes that have alleviated problems, but have seldom been entirely successful and, in some cases; success has created its own new set of problems. This historical organization has developed to solve long-term fishery sustainability problems via the incorporation of three basic policy drivers: (1) bottom-up or co-management approaches; (2) adaptive management; and (3) use of implicit or explicit incentives.

In conclusion, we suggest that, as cleverly illustrated by Prince (2003) in connection with the solution to the shortage of medical doctors in rural China and the implemented 'barefoot medical doctor solution', and Prince's fishery equivalent, the 'barefoot ecologist solution', countries, international organizations, and mostly fishery biologists need to incorporate the diagnosis that there is not a single fisheries crisis but a complex series of small-scale crises, all requiring different recipes and medicines. Local healers with widely-tested basic principles may be the best people to solve the local crises. We believe that fishery recipes and medicines as well as the basic policies and management drivers used in the examples presented in this book, are the way to approach the complex small-scale fishery crises in the world. Furthermore, appropriate management systems need to and can evolve in the context of local situations. Certainly, there is hope for the future as shown in these chapters. We remain optimistic that these crises are becoming the opportunity to develop permanent solutions.

References

Allison, E.H. and Ellis, F. (2001) The livelihoods approach and management of small-scale fisheries. *Marine Policy*, **25**: 377–88.

Aswani, S. (1998) Patterns of marine harvest efforts in south-western New Georgia, Solomon Islands: resource management or optimal foraging? *Ocean and Coastal Management*, **40**: 207–35.

Balmford, A., Bruner, A., Cooper, P., Costanza, R., Farber, S., Green, R.E., Jenkins, M., Jefferiss, P., Jessamy, V., Madden, J., Munro, K., Myers, N., Naeem, S., Paavola, J., Rayent, M., Rosendo, S., Roughgarden, J., Trumper, K. and Turner, R.K. (2002) Economic reasons for conserving wild nature. *Science*, **297**: 950–3.

Balmford, A., Gravestock, P., Hockley, P., McClean, C.J. and Roberts, C.M. (2004) The worldwide costs of marine protected areas. *Proceedings of the National Academy of Sciences USA*, **101**: 9694–7.

Bascompte, J., Melian, C. and Sala, E. (2005) Interaction strength combinations and the overfishing of a marine food web. *Proceedings of the National Academy of Sciences USA*, **102**: 5443–7.

Berkes, F., Mahon, R., McConney, P., Pollnac, R. and Pomeroy, R. (2001) *Managing Small-scale Fisheries: Alternative Directions and Methods*. IDRC publication. http://www.idrc.ca/en/ev-9328-201-1-DO_TOPIC.html

Botsford, L.W., Castilla, J.C. and Peterson, C.H. (1997) The management of fisheries and marine ecosystems. *Science*, **277**: 509–15.

Branch, T.A. (2001) A review of orange roughy (*Hoplostethus atlanticus*) fisheries, estimation methods, biology and stock structure. *South Africa Journal of Marine Science*, **23**: 181–203.

Browman, H.I. and Stergiou, K.I. (2004) Perspectives on ecosystem-based approaches to the management of marine resources. *Marine Ecology Progress Series*, **274**: 269–303.

Browman, H.I. and Stergiou, K.I (Eds) (2005) Politics and socio-economics of ecosystem-based management of marine resources. *Marine Ecology Progress Series*, **300**: 241–96.

Carpenter, S.R., Ludwig, D. and Brock, W.A. (1999) Management of eutrophication for lakes subject to potentially irreversible change. *Ecological Applications*, **9**: 751–71.

Castilla, J.C. (2000) Roles of experimental marine ecology in coastal management and conservation. *Journal of Experimental Marine Biology and Ecology*, **250**: 3–21.

Castilla, J.C. (expected 2007) Shoreline economics. In: M.W. Denny and S.D. Gaines (Eds) *Encyclopedia of the Rocky Intertidal*. University of California Press, Berkeley.

Castilla, J.C. and Defeo, O. (2001) Latin American benthic shellfisheries: emphasis on co-management and experimental practices. *Review in Fish Biology and Fisheries*, **11**: 1–30.

Castilla, J.C. and Defeo, O. (2005) Paradigm shifts needed for world fisheries. *Science*, **309**: 1324–5.

Chichilnisky, G. (1997) What is sustainable development? *Land Economics*, **73**: 467–91.

Christie, P. (2004) Marine protected areas as biological successes and social failures in southeast Asia. *American Fisheries Society Symposium*, **42**: 155–64.

Cinner, J. (2005) Socioeconomic factors influencing customary marine tenure in the Indo-Pacific. *Ecology and Society*, **10**: 36.

Cinner, J.E., Marnane, M.J. and McClanahan, T.R. (2005) Conservation and community benefits from traditional coral reef management at Ahus Island, Papua New Guinea. *Conservation Biology*, **19**: 1714–23.

Costanza, R., d'Arge, R., de Groot, R.M., Grasso, F.S., Hannon, B., Limburg, K., Naeem, S., O'Neill, V., Paruelo, J., Raskin, R.G. and Sutton, P. (1997) The value of the world's ecosystem services and natural capital. *Nature*, **387**: 253–60.

Defeo, O. and Castilla, J.C. (2005) More than one bag for the world fishery crisis and keys for co-management successes in selected artisanal Latin American shellfisheries. *Review in Fish Biology and Fisheries*, **15**: 265–83.

Graham, N.A.J., Dulvy, N.K., Jennings, S. and Polunin, N. (2005) Size-spectra as indicators of the effects of fishing on coral reef fish assemblages. *Coral Reefs*, **24**: 118–24.

Halpern, B.S. and Warner, R. (2002) Matching marine reserve design to reserve objectives. *Proceedings of Royal Society B*, **270**: 1871–8.

Hardin, G. (1968) The tragedy of the commons. *Science*, **162**: 1243–8.

Hilborn, R., Punt, A.E. and Orensanz, J. (2004) Beyond band-aids in fisheries management: fixing world fisheries. *Bulletin of Marine Science*, **74**: 493–507.

Hilborn, R., Orensanz, J.M. and Parma, A.M. (2005) Institutions, incentives and the future of fisheries. *Philosophical Transactions of the Royal Society B*, **360**: 47–57.

Holling, C.S. (1978) *Adaptive Environmental Assessment and Management*. John Wiley and Sons, New York.

Holling, C.S. (2001) Understanding the complexity of economic, ecological, and social systems. *Ecosystems*, **4**: 390–405.

Holling, C.S. and Meffe, G.K. (1996) On command-and-control, and the pathology of natural resource management. *Conservation Biology*, **10**: 328–37.

Hutchings, J.A. (2000) Collapse and recovery of marine fishes. *Nature*, **406**: 882–5.

Jackson, J.B.C., Kirby, M.X. et al. (2001) Historical overfishing and the recent collapse of coastal ecosystems. *Science*, **293**: 629–38.

Jennings, S., Pinnegar, J.K., Polunin, N.V.C. and Warr, K.J. (2002) Linking size-based and trophic analyses of benthic community structure. *Marine Ecology Progress Series*, **226**: 77–85.

Johannes, R.E. (2002) The renaissance of community-based resource management in Oceania. *Annual Review of Ecology and Systematics*, **33**: 317–40.

Lessard, R.B., Martell, S.J.D., Walters, C.J., Essington, T.E. and Kitchell, J.F. (2005) Should ecosystem management involve active control of species abundances? *Ecology and Society*. http://www.ecologyandsociety.org/vol10/iss2/art1/

Lowry, K., White, A.T. and Courtney, C. (2005) National and local agency roles in integrated coastal management in the Philippines. *Ocean and Coastal Management*, **48**: 314–35.

Martin-Smith, K.M., Samoilys, M.A., Meeuwing, J.J. and Vincent, C.J. (2004) Collaborative development of management options for an artisanal fishery for seahorses in the central Philippines. *Ocean and Coastal Management*, **47**: 165–93.

Mascia, M.B. (2003) The human dimension of coral reef marine protected areas: recent social science research and its policy implications. *Conservation Biology*, **17**: 630–2.

McClanahan, T.R. (1992) Resource utilization, competition and predation: a model and example from coral reef grazers. *Ecological Modelling*, **61**: 195–215.

McClanahan, T.R. (1995) A coral reef ecosystem-fisheries model: impacts of fishing intensity and catch selection on reef structure and processes. *Ecological Modelling*, **80**: 1–19.

McClanahan, T.R. and Graham, N.A.J. (2005) Recovery trajectories of coral reef fish assemblages within Kenyan marine protected areas. *Marine Ecology Progress Series*, **294**: 241–8.

McClanahan, T.R. and Mangi, S. (2004) Gear-based management of a tropical artisanal fishery based on species selectivity and capture size. *Fisheries Management and Ecology*, **11**: 51–60.

McClanahan, T.R., Glaesel, H., Rubens, J. and Kiambo, R. (1997) The effects of traditional fisheries management on fisheries yields and the coral-reef ecosystems of southern Kenya. *Environmental Conservation*, **24**: 105–20.

McClanahan, T.R., Davies, J. and Maina, J. (2005a) Factors influencing resource users and managers' perceptions towards marine protected area management in Kenya. *Environmental Conservation*, **32**: 42–9.

McClanahan, T.R., Maina, J. and Davies, J. (2005b) Perceptions of resource users and managers towards fisheries management options in Kenyan coral reefs. *Fisheries Management and Ecology*, **12**: 105–12.

McClanahan, T.R., Verheij, E. and Maina, J. (2006) Comparing management effectiveness of a marine park and a multiple-use collaborative fisheries management area in East Africa. *Aquatic Conservation: Marine and Freshwater Ecosystems*, **16**: 147–65.

McConney, P.A., Pomeroy, R. and Mahon, R. (2003) *Guidelines for Coastal Resource Co-management in the Caribbean: Communicating the Concepts and Conditions that Favor Success*. Caribbean Conservation Association and University of West Indies, Kingston.

Myers, N. (1996) Two key challenges for biodiversity: discontinuities and synergisms. *Biology and Conservation*, **5**: 1025–34.

Naeem, S. and Wright, J.P. (2003) Disentangling biodiversity effects on ecosystem functioning: deriving solutions to a seemingly insurmountable problem. *Ecology Letters*, **6**: 567–79.

Napier, V.R., Branch, G.M. and Harris, J.M. (2005) Evaluating conditions for successful co-management of subsistence fisheries in KwaZulu-Natal, South Africa. *Environmental Conservation*, **32**: 165–77.

Odum, H.T. (1988) Self-organization, transformity, and information. *Science*, **242**: 1132–9.

Odum, H.T. (1996) *Environmental Accounting, Emergy and Decision Making*. John Wiley, New York.

Odum, H.T. and Odum, E.C. (2001) *A Prosperous Way Down: Principles and Policies*. University of Colorado Press, Boulder.

Ostrom, E., Burger, C.B., Field, C.B., Norgaard, R.B. and Policansky, D. (1999) Sustainability – revisiting the commons: local lessons, global challenges. *Science*, **284**: 278–82.

Palumbi S.R. (2001) Humans as the world's greatest evolutionary force. *Science*, **293**: 1786–90.

Pandolfi, J.M., Jackson, J.B.C., Baron, N., Bradbury, R.H., Guzman, H.M., Hughes, T.P., Kappel, C.V., Micheli, F., Ogden, J.C., Possingham, H.P. and Sala, E. (2005) Are U.S. coral reefs on the slippery slope to slime? *Science*, **307**: 1725–6.

Pauly, D., Silvestre, G. and Smith, I. (1989) On development, fisheries and dynamite: a brief review of tropical fisheries management. *Natural Resource Modeling*, **3**: 307–29.

Pauly, D., Christensen, V., Dalsgaard, J., Froese, R. and Torres, F., Jr (1998) Fishing down marine food webs. *Science*, **279**: 861–3.

Pauly, D., Christensen, V., Guenette, S., Pitcher, T.J., Sumaila, U.R., Walters, C.J., Watson, R. and Zeller, D. (2002) Towards sustainability in world fisheries. *Nature*, **418**: 689–95.

Pikitch, E.K., Santora, C., Babcock, E.A., Bakun, A., Bonfil, R., Conover, D.O., Dayton, P., Doukakis, P., Fluharty, D., Heneman, B., Houde, E.D., Link, J., Livingston, P.A., Mangel, M., McAllister, M.K., Pope, J. and Sainsbury, K.J. (2004) Ecosystem-based fishery management. *Science*, **305**: 346–7.

Polis, G.A. and Strong, D.R. (1996) Food web complexity and community dynamics. *American Naturalist*, **147**: 813–46.

Pollnac, R.B., Pomeroy, R.S. and Harkes, I.H.T. (2001) Fishery policy and job satisfaction in three southeast Asian fisheries. *Ocean and Coastal Management*, **44**: 531–44.

Prince, J.D. (2003) The barefoot ecologist goes fishing. *Fish and Fisheries*, **4**: 359–71.

Richardson, E.A., Kaiser, M.J. and Edward-Jones, G. (2005) Variation in fishers' attitudes within an inshore fishery: implications for management. *Environmental Conservation*, **32**: 1–13.

Russ, G.R., Stockwell, B. and Alcala, A.C. (2005) Inferring versus measuring rates of recovery in no-take marine reserves. *Marine Ecology Progress Series*, **292**: 1–12.

Saenz-Arroyo, A., Roberts, C.M., Torre, J., Carino-Olvera, M. and Enriquez-Andrade, R.R. (2005) Rapidly shifting environmental baselines among fishers of the Gulf of California. *Proceedings of the Royal Society B*, **272**: 1957–62.

Scheffer, M., Brock, W. and Westley, F. (2001) Socioeconomic mechanisms preventing optimum use of ecosystem services: an interdisciplinary theoretical analysis. *Ecosystems*, **3**: 451–71.

Scheffer, M., Westley, F. and Brock, W. (2003) Slow response of societies to new problems: cause and costs. *Ecosystems*, **6**: 493–502.

Smith, T.D. (1988) Stock assessment methods: the first fifty years. In: J.A. Gulland (Ed.) *Fish Population Dynamics*. John Wiley and Sons, Chichester.

Thompson, M. and Trisoglio, A. (1997) Managing the unmanageable. In: L.A. Brooks and S.D. VanDeveer (Eds) *Saving the Seas: Values, Scientists, and International Governance*. Maryland Sea Grant, College Park, Maryland.

Walker, B. (1995) Conserving biological diversity through ecosystem resilience. *Conservation Biology*, **9**: 747–52.

Walters, C. (1986) *Adaptive Management of Renewable Resources*. Macmillan Publishing Co., New York.

Walters, C.J. (1997) Challenges in adaptive management of riparian and coastal ecosystems. *Conservation Ecology*, **1**. http://www.consecol.org/vol1/iss2/art1

Walters, C.J. and Holling C.S. (1990) Large-scale management experiments and learning by doing. *Ecology*, **71**: 2060–8.

Walters, C.J. and Martell, J.D. (2004) *Fisheries Ecology and Management*. Princeton University Press, Princeton, New Jersey.

Waltner-Toews, D. and Kay, J. (2005) The evolution of an ecosystem approach: the diamond schematic and an adaptive methodology for ecosystem sustainability and health. *Ecology and Society*, **10**: 38.

White, A.T., Hale, L.Z., Renard, Y. and Cortesi, L. (Eds) (1994) *Collaborative and Community-based Management of Coral Reefs: Lessons from Experience*. Kumarian Press, West Hartford, Connecticut.

White, A.T., Christie, P., D.'Agnes, H., Lowry, K. and Milne, N. (2005) Designing ICM projects for sustainability: lessons from the Philippines and Indonesia. *Ocean and Coastal Management*, **48**: 271–96.

Wilson, J.A., Acheson, J.M., Metcalfe, M. and Kleban, P. (1994) Chaos, complexity and community management of fisheries. *Marine Policy*, **18**: 291–305.

Index